P9-DVU-845

My
iPhone®
for Seniors

SEVENTH EDITION

Brad Miser

My iPhone® for Seniors, Seventh Edition

Copyright © 2021 by Pearson Education, Inc.

AARP is a registered trademark.

No part of this publication may be reproduced, stored in a retrieval system, or transmitted in any form or by any means, electronic, mechanical, photocopying, recording, scanning, or otherwise, except as permitted under Section 107 or 108 of the 1976 United States Copyright Act, without the prior written permission of the Publisher. No patent liability is assumed with respect to the use of the information contained herein. For information regarding permissions, request forms, and the appropriate contacts within the Pearson Education Global Rights & Permissions Department, please visit www.pearson.com/permissions.

Limit of Liability/Disclaimer of Warranty: While the publisher, AARP, and the author have used their best efforts in preparing this book, they make no representations or warranties with respect to the accuracy or completeness of the contents of this book and specifically disclaim any implied warranties of merchantability or fitness for a particular purpose. No warranty may be created or extended by sales representatives or written sales materials. The advice and strategies contained herein may not be suitable for your situation. You should consult with a professional where appropriate. The publisher, AARP, and the author shall not be liable for any loss of profit or any other commercial damages, including but not limited to special, incidental, consequential, or other damages. The fact that an organization or website is referred to in this work as a citation and/or a potential source of further information does not mean that the publisher, AARP, and the author endorse the information the organization or website may provide or recommendations it may make. Further, readers should be aware that Internet websites listed in this work may have changed or disappeared between when this work was written and when it is read.

ISBN-13: 978-0-13-682441-1

ISBN-10: 0-13-682441-2

Library of Congress Control Number: 2020917063

1 2020

Trademarks

All terms mentioned in this book that are known to be trademarks or service marks have been appropriately capitalized. Que Publishing cannot attest to the accuracy of this information. Use of a term in this book should not be regarded as affecting the validity of any trademark or service mark.

Unless otherwise indicated herein, any third party trademarks that may appear in this work are the property of their respective owners and any references to third party trademarks, logos, or other trade dress are for demonstrative or descriptive purposes only. Such references are not intended to imply any sponsorship, endorsement, authorization, or promotion of Que Publishing products by the owners of such marks, or any relationship between the owner and Que Publishing or its affiliates, authors, licensees, or distributors.

Screenshots reprinted with permission from Apple, Inc.

Warning and Disclaimer

Every effort has been made to make this book as complete and as accurate as possible, but no warranty or fitness is implied. The information provided is on an "as is" basis. The author and the publisher shall have neither liability nor responsibility to any person or entity with respect to any loss or damages arising from the information contained in this book.

Special Sales

For information about buying this title in bulk quantities, or for special sales opportunities (which may include electronic versions; custom cover designs; and content particular to your business, training goals, marketing focus, or branding interests), please contact our corporate sales department at corpsales@pearsoned.com or (800) 382-3419.

For government sales inquiries, please contact governmentsales@pearsoned.com.

For questions about sales outside the United States, please contact intlcs@pearson.com.

Editor-In-Chief
Mark Taub

Director, ITP Product Management
Brett Bartow

Executive Editor
Laura Norman

Sponsoring Editor
Chhavi Vig

Marketing
Stephane Nakib

Director, AARP Books
Jodi Lipson

Development Editor
Charlotte Kughen

Technical Editor
Karen Weinstein

Managing Editor
Sandra Schroeder

Project Editor
Mandie Frank

Copy Editor
Charlotte Kughen

Indexer
Ken Johnson

Proofreader
Debbie Williams

Editorial Assistant
Cindy Teeters

Designer
Chuti Prasertsith

Compositor
Tricia Bronkella

Contents at a Glance

Table of Contents

7   Customizing How Your iPhone Looks   **229**

8   Managing Contacts   **265**

About the Author

Brad Miser has written extensively about technology, with his favorite topics being the amazing "i" devices, especially the iPhone, that make it possible to take our lives with us while we are on the move. In addition to *My iPhone for Seniors*, Seventh Edition, Brad has written many other books, including *My iPhone*, Twelfth Edition. He has been an author, development editor, or technical editor for more than 60 other titles.

Brad is or has been the manager of a team of sales engineers and proposal writers, a principal sales engineer, the director of product and customer services, and the manager of education and support services for several software development companies. Previously, he was the lead proposal specialist for an aircraft engine manufacturer, a development editor for a computer book publisher, and a civilian aviation test officer/engineer for the U.S. Army. Brad holds a bachelor of science degree in mechanical engineering from California Polytechnic State University at San Luis Obispo and has received advanced education in maintainability engineering, business, and other topics.

Brad would love to hear about your experiences with this book (the good, the bad, and the ugly). You can write to him at bradmiser@icloud.com.

About AARP

AARP is a nonprofit, nonpartisan organization, with a membership of nearly 38 million, that helps people turn their goals and dreams into *real possibilities*™, strengthens communities, and fights for the issues that matter most to families such as healthcare, employment and income security, retirement planning, affordable utilities, and protection from financial abuse. Learn more at aarp.org.

Dedication

To those who have given the last full measure of devotion so that the rest of us can be free.

Acknowledgments

To the following people on the *My iPhone for Seniors* project team, my sincere appreciation for your hard work on this book:

Laura Norman, who is my current acquisitions editor and who was the development editor on many prior versions. We developed the original concept for *My iPhone* together (many years ago now!) and she works very difficult and long hours to ensure the success of each edition. Laura and I have worked on many books together, and I appreciate her professional and effective approach to these projects. Thanks for putting up with me yet one more time! Frankly, I have no idea how she does all the things she does and manages to be so great to work with given the incredible work and pressure books like this one involve!

Charlotte Kughen, my development and copy editor, who helped craft this book so that it provides useful information delivered in a comprehensible way. Charlotte is an extremely professional and skilled editor and somehow manages to be very pleasant to work with at the same time. That is indeed a rare combination. Thanks for your work on this book!

Karen Weinstein, the technical editor, who caught a number of mistakes I made and who made useful suggestions along the way to improve this book's content.

Mandie Frank, my project editor, who skillfully managed the hundreds of files and production process that it took to make this book. Imagine keeping dozens of plates spinning on top of poles and you get a glimpse into Mandie's daily life! (And no plates have been broken in the production of this book!)

Chhavi Vig, my sponsoring editor, who coordinated the many files among the different people who worked on this book. Chhavi also handled the administrative tasks for the book.

Chuti Prasertsith for the cover of the book.

Que Publishing's production and sales team for printing the book and getting it into your hands.

We Want to Hear from You!

As the reader of this book, *you* are our most important critic and commentator. We value your opinion and want to know what we're doing right, what we could do better, what areas you'd like to see us publish in, and any other words of wisdom you're willing to pass our way.

We welcome your comments. You can email or write to let us know what you did or didn't like about this book—as well as what we can do to make our books better.

Please note that we cannot help you with technical problems related to the topic of this book.

When you write, please be sure to include this book's title and author as well as your name and email address. We will carefully review your comments and share them with the author and editors who worked on the book.

Email: community@informit.com

Reader Services

Register your copy of *My iPhone for Seniors* at informit.com/que for convenient access to downloads, updates, and corrections as they become available. To start the registration process, go to informit.com/register and log in or create an account.* Enter the product ISBN, 9780136824411, and click Submit.

*Be sure to check the box that you would like to hear from us in order to receive exclusive discounts on future editions of this product.

Credits

Figure	Chapter	Page Number	Credit Attribution
Figure 3-13	3	77	Screenshot © 1992-2020 Cisco
Figure 3-14	3	77	Screenshot © 1992-2020 Cisco
Figure 4-31	4	134	Screenshot © Google LLC
Figure 4-32	4	134	Screenshot © Google LLC
Figure 4-42	4	138	Screenshot © 2020 Facebook
Figure 4-44	4	139	Screenshot © 2020 Facebook
Figure 4-45	4	140	Screenshot © 2020 Facebook
Figure 4-46	4	140	Screenshot © 2020 Facebook
Figure 4-47	4	140	Screenshot © 2020 Facebook
Figure 4-48	4	141	Screenshot © 2020 Facebook
Figure 4-49	4	141	Screenshot © 2020 Facebook
Figure 4-50	4	141	Screenshot © 2020 Facebook
Figure 8-28	8	278	Rolando Da Jose/123RF
Figure 8-29	8	279	Rolando Da Jose/123RF
Figure 14-6	14	500	Screenshot © TWC Product and Technology LLC 2014, 2020
Figure 14-8	14	501	Screenshot © TWC Product and Technology LLC 2014, 2020
Figure 14-11	14	503	Screenshot © United States Department of Commerce
Figure 14-13	14	504	Screenshot © United States Department of Commerce
Figure 14-21	14	512	Screenshot © 2020 Jocko Podcast
Figure 14-39	14	521	Screenshot © Netflix
Figure 17-35	17	644	Screenshot © 2020 Starbucks Coffee Company
Figure 17-36	17	644	Screenshot © 2020 Starbucks Coffee Company
Figure 17-37	17	645	Screenshot © 2020 Starbucks Coffee Company
Figure 17-38	17	646	Screenshot © 2020 Starbucks Coffee Company
Figure 17-39	17	646	Screenshot © 2020 Starbucks Coffee Company
Cover			Yuliia Petrenko/Shutterstock

Using This Book

This book has been designed to help you transform an iPhone into *your* iPhone by helping you learn to use it easily and quickly. As you can tell, the book relies heavily on pictures to show you how an iPhone works. It is also task-focused so that you can quickly learn the specific steps to follow to do lots of cool things with your iPhone.

Using an iPhone involves lots of touching its screen with your fingers. When you need to tap part of the screen, such as a button or keyboard, you see a callout with the step number pointing to where you need to tap. When you need to swipe your finger along the screen, such as to browse lists, you see the following icons:

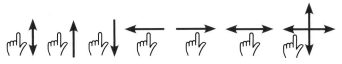

The directions in which you should slide your finger on the screen are indicated with arrows. When the arrow points both ways, you can move your finger in either direction. When the arrows point in all four directions, you can move your finger in any direction on the screen.

To zoom in or zoom out on screens, you unpinch or pinch, respectively, your fingers on the screen. These motions are indicated by the following icons:

You can touch and apply slight pressure on the screen to activate certain functions, such as to open the quick action menus for apps or to preview an email. The following icon indicates when you should touch and apply slight pressure to the screen:

When you should rotate your iPhone, you see this icon:

As you can see on its cover, this book provides information to help you use iPhone models that can run iOS 14. These models are: 12, 12 mini, 12 Pro, 12 Pro Max, 11, 11 Pro, 11 Pro Max, Xs, Xs Max, Xʀ, X, 8, 8 Plus, 7, 7 Plus, 6s, 6s Plus, and SE (1ˢᵗ and 2ⁿᵈ generations). Each of these models has specific features and capabilities that vary slightly (and sometimes more than slightly!) from the others. Additionally, they have different screen sizes with the SE being the smallest and the 12 Pro Max model being the largest.

Because of the variations between the models, the figures you see in this book might be slightly different than the screens you see on your iPhone. For example, the iPhone 12 Pro has settings that aren't on the SE. In most cases, you can follow the steps as they are written with any of these models even if there are minor differences between the figures and the screens on your iPhone.

When the model you're using doesn't support a feature being described, such as Face ID that is on the iPhone 12 but not on an 8 Plus, you can skip that information.

As you saw from the list above, there are many models of iPhone that run iOS 14. These models fall into two general groups: models without a Touch ID/Home button and models with that button.

The models of iPhone that run iOS 14 and don't have a Touch ID/Home button are referred to throughout the book as the "iPhones without a Home button;" these are the 12, 12 mini, 12 Pro, 12 Pro Max, iPhone 11, 11 Pro, 11 Pro Max, X, Xs, Xs Max, and Xʀ. The models in this group use Face ID when user authentication is required, such as when you unlock the phone. These are the models that are primarily used for the tasks throughout this book. Where there are variations on tasks with other models (such as using Touch ID instead of Face ID), you see those differences noted in the text.

Models that have a Touch ID/Home button are referred to (cleverly I must say) as "iPhones with a Home button." There is quite a bit of variety in this group.

For example, the iPhone 7 Plus and iPhone 8 Plus have dual cameras on the backside (which enable additional photographic capabilities, such as portrait mode) while other models in this group have only one backside camera.

Fortunately for this book's purposes, most of the tasks you need to do are the same or very similar among all the models. (When there is a difference, it is called out so you'll know.) So, no matter which iPhone model you use, this book helps you make the most of it.

Getting Started

Learning to use new technology can be intimidating. Don't worry; with this book as your guide, you'll be working with your iPhone like you've been using it all your life in no time at all.

There are several ways you can purchase an iPhone, such as from an Apple Store, from a provider's store (such as AT&T or Verizon), or from a website. You might be upgrading from a previous iPhone or other type of cell phone, in which case you're using the same phone number, or you might be starting with a completely new phone and phone number. However you received your new phone, you need to turn it on, perform the basic setup (the iPhone leads you through this step-by-step), and activate the phone.

If you purchased your phone in a physical store, you probably received help with these tasks, and you're ready to start learning how to use your iPhone. If you purchased your iPhone from an online store, it came with basic instructions that explain how you need to activate your phone; follow those instructions to get your iPhone ready for action.

For this book, I've assumed you have an iPhone in your hands, you have turned it on, followed the initial setup process it led you through, and activated it.

With your iPhone activated and initial setup complete, you're ready to learn how to use it. This book is designed for you to read and do at the same time. The tasks explained in this book contain step-by-step instructions that guide you; to get the most benefit from the information, perform the steps as you read them. This book helps you learn by doing!

As you can see, this book has quite a few chapters and lots of pages. However, there are only a few that you definitely should read as a group as you get started. You can read the rest of them as the topics are of interest to you. Most of the chapters are designed so that they can be read individually as you move into new areas of your iPhone.

After you've finished reading this introduction, I recommend you read and work through Chapter 1, "Getting to Know Your iPhone," Chapter 2, "Getting Started with Your iPhone," Chapter 3, "Using Your iPhone's Core Features," and Chapter 4, "Setting Up and Using an Apple ID, iCloud, and Other Online Accounts" in their entirety. These chapters give you a good overview of your iPhone and help you set up the basics you use throughout the rest of the book.

Next, read Chapter 5, "Customizing Your iPhone with Apps and Widgets," to learn how to add capabilities to your iPhone by downloading and installing apps from the App Store.

From there, read the parts of Chapter 6, "Making Your iPhone Work for You," and Chapter 7, "Customizing How Your iPhone Looks," that are of interest to you (for example, in Chapter 7, you find out how to change the wallpaper image that you see in the background of the Home and Lock screens).

After you've finished these core chapters, you're ready to explore the rest of the book in any order you'd like. For example, when you want to learn how to use your iPhone's camera and work with the photos you take, see Chapter 15, "Taking Photos and Video with Your iPhone," and Chapter 16, "Viewing and Editing Photos and Video with the Photos App."

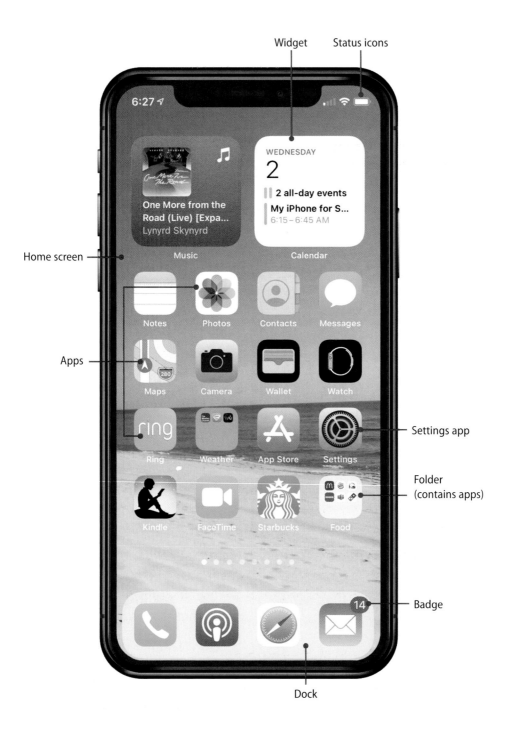

Widget

Status icons

6:27

WEDNESDAY
2

|| 2 all-day events
| My iPhone for S...
6:15 – 6:45 AM

Home screen

Music

Calendar

Notes

Photos

Contacts

Messages

Apps

Maps

Camera

Wallet

Watch

ring

Ring

Weather

App Store

Settings

Settings app

Folder
(contains apps)

Kindle

FaceTime

Starbucks

Food

Badge

Dock

Getting to Know Your iPhone

Your iPhone is one of the most amazing handheld devices ever because of how well it's designed. For most of the things you do, you just use your fingers on your iPhone's screen (which just seems natural), and the iPhone's consistent interface enables you to accomplish most tasks with similar steps.

Getting Started

In this chapter, you learn fundamental concepts that move you along your iPhone journey. So, let's get started with the following:

- **iOS**—This term refers to the software that runs the iPhone. More generally, this type of software is called an operating system because it controls the operations of the device. All computers, smart phones, and similar devices have operating system software. The major operating system software you've likely heard of includes macOS, Windows, Android, and, of course, iOS.

  iOS is used on all iPhones. (iPads run iPad OS, which is a bit different than the iOS). Like all computer software, iOS is updated to add features, take advantage of new hardware capabilities, make improvements, and fix problems. iOS receives a major update every year.

There are many minor updates between major updates; these usually add minor features, fix problems, and so on.

The current version of iOS is 14 (it's the fourteenth major version of the software), and that is the version covered in this book. If you use a different version of iOS, some of the information in this book won't match what you experience with your iPhone. I recommend you upgrade to iOS 14 before trying to use this book; all upgrades to the iOS are free. (You can learn how to upgrade your iPhone's iOS in Chapter 18, "Maintaining and Protecting Your iPhone and Solving Problems.")

If you're wondering what iOS stands for, it's really a term Apple used initially for marketing purposes rather than being a specific acronym. The closet direct interpretation is *internet Operating System*. (Apple has always used a lowercase "i" as a style thing.)

- **iPhone Models**—iPhones have been around since 2007, and each year Apple releases new models. Each generation of iPhone has improved, different, and more features. Therefore, each model is a bit different than the others. The good news is that there is enough common to all models that you can use this book to learn how to use whichever model you have (as long as it runs iOS 14).

  That written, understand that this book covers iPhones that can run iOS 14. Those models are 12, 12 mini, 12 Pro, 12 Pro Max, 11, 11 Pro, 11 Pro Max, Xs, Xs Max, XR, X, 8, 8 Plus, 7, 7 Plus, 6s, 6s Plus, and SE (first and second generations).

  These models can be broken down into two major groups based on whether they have a Touch ID/Home button (which is a circular button in the center of the bottom of the front of the phone) or not.

  The 12, 12 mini, 12 Pro, 12 Pro Max, 11, 11 Pro, 11 Pro Max, and all of the X models don't have a Touch ID/Home button. Instead, their screens fill the entire front of the phone (except for a gap at the top where the camera is).

  The other models have a Touch ID/Home button. These models have slightly smaller screens relative to the size of their cases than do the models without this button.

  There are some differences between these two groups, and both groups are covered in this book. These differences are detailed primarily in Chapter 2, "Getting Started with Your iPhone."

- **Multi-touch User Interface**—Most of the time, you control your iPhone by using your fingers on its screen as opposed to pressing physical buttons. You use your fingers directly on the screen to tap icons, select items, swipe, zoom, type text, and so on. If you want to get technical, this method of interacting with software is called the multi-touch user interface. (You don't really need to know this, but maybe you can use it to impress your friends at a party.)

- **Face ID**—iPhone models without a Home button support Face ID, which is Apple's facial recognition technology. These models have a camera that can "see" and recognize your face. This enables you to enter information you need for security purposes, such as passwords, by just looking at your phone.

- **Touch ID**—iPhones with a Home button support Touch ID, which is Apple's fingerprint recognition technology. You can record fingerprints and then simply touch the Touch ID/Home button to have your iPhone recognize you for security purposes, such as to enter passwords.

- **Apps**—Your iPhone can do so many great things because it can run software applications, or in the more common vernacular, apps. The iPhone comes with many apps preinstalled such as Phone, Mail, and Calendar. You can add to your iPhone's capabilities by downloading and installing more apps. (You actually use one of the iPhone's default apps, the App Store app, to download more apps.)

- **Siri**—Siri is the iPhone's voice recognition technology. It enables you to speak to your phone to do things. For example, you can read new text messages by activating Siri and saying "Read text messages." Using Siri's technology, you can also dictate text instead of typing it. Siri is one of the most useful features of your iPhone. It's not exactly like using the *Enterprise*'s main computer on *Star Trek*, but it's pretty darn close.

As you start exploring this book and your iPhone, understand that there might be different ways to do what you want to do. Some will be faster and easier for you, whereas others might take a few more taps on the screen. Over time and with practice (and with the help of this book), you'll develop your own iPhone style and learn what works best for you.

The good news is that most of the fundamental things you want to do with your iPhone, such as making phone calls, sending and receiving emails, or browsing the Web, are easy to learn. You can take your time exploring some of the more advanced areas, such as editing photos. With a little time, practice, and patience, you will truly transform *an* iPhone into *your* iPhone.

Using Your Fingers to Control Your iPhone

A lot of things you do with your iPhone involve different movements, more technically called *gestures*, of your fingers on the iPhone's screen. The following figures highlight the major ways you use your fingers to control an iPhone:

Tap an app's icon to launch it

- **Tap**—Briefly touch a finger to the iPhone's screen and then lift your finger again. When you tap, you don't need to apply pressure to the screen; simply touch your finger to the screen and immediately lift it off the screen again. For example, to open an app, you tap its icon.

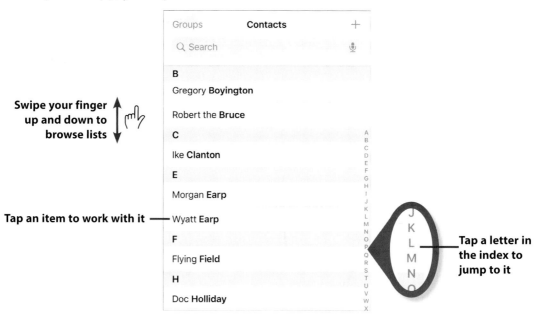

Swipe your finger up and down to browse lists

Tap an item to work with it

Tap a letter in the index to jump to it

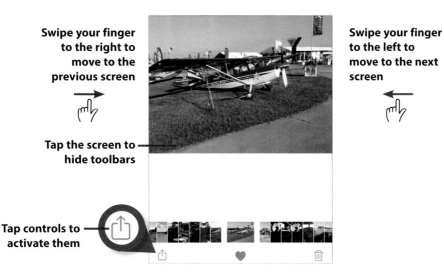

Swipe your finger to the right to move to the previous screen

Swipe your finger to the left to move to the next screen

Tap the screen to hide toolbars

Tap controls to activate them

- **Swipe**—Touch the screen at any location and slide your finger on the screen; you don't need to apply pressure, just touching the screen is enough. You use the swipe motion in many places, such as to browse a list of options or to move among Home page screens. Whatever you are swiping on moves the screen in the direction that you swipe. For example, when you swipe toward the top of the iPhone, the screen moves up. Likewise, when you swipe to the left, the screen moves to the left.

- **Drag**—Touch and hold an object and move your finger across the screen without lifting your finger; the faster you move your finger, the faster the resulting action happens. (Again, you don't need to apply pressure; just make contact.) For example, you can drag icons around the Home screens to rearrange them.

Swipe your finger up, down, left, and right to scroll

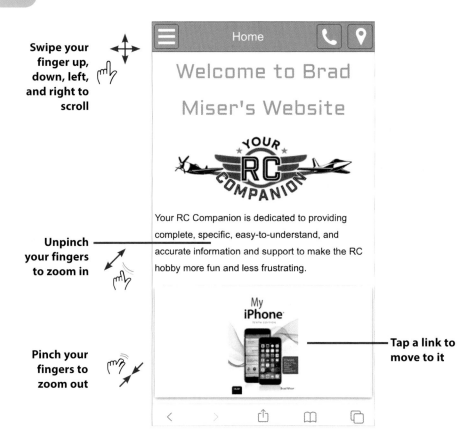

Unpinch your fingers to zoom in

Pinch your fingers to zoom out

Tap a link to move to it

- **Pinch or unpinch**—Place two fingers on the screen and drag them together (pinch) or move them apart (unpinch); the faster and more you pinch or unpinch, the "more" the action happens (such as a zoom in). For example, when you're viewing photos, you can unpinch to zoom in or pinch to zoom out again.

Rotate the iPhone to change the screen's orientation

Tap the screen to show toolbars

- **Rotate**—Rotate the iPhone to change the screen's orientation.

Touch and hold on something to perform a Peek

Swipe up on a Peek to reveal menus with actions you can select to perform them

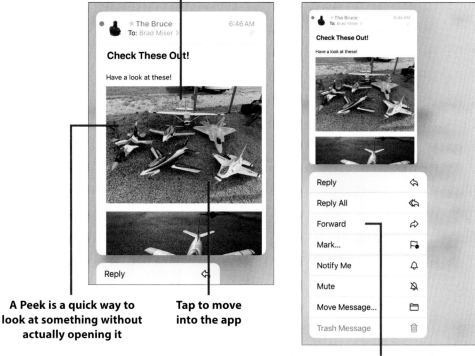

A Peek is a quick way to look at something without actually opening it

Tap to move into the app

When you swipe up on a Peek, you get a menu; tap an action to perform it

- **Peek**—When you're looking at a preview of something, such as an email, touch and hold on the screen to perform a Peek, which causes a window to open that shows a preview of the object. You can preview the object in the Peek window. Tap the Peek to move into the app to work with the object; for example, when you Peek on an email and then tap the Peek, you move into the Mail app. If you swipe up on a Peek, you get a menu of commands related to the object. For example, when you perform a Peek on an email and then swipe up on the Peek, you can tap Reply to reply to the email.

Locking/Unlocking Your iPhone

Your iPhone will probably hold information you don't want others to access, such as contact information or health data. You might also use your iPhone to work with bank, social media, or other accounts. For all these reasons, it's important that you control access to your phone to keep your account information secure.

This iPhone is in Lock mode

Tap to use the flash as a flashlight

Swipe up to unlock*

Tap to take a photo

Swipe up to unlock

To provide this security, your iPhone has a Lock mode; when it's in this mode, no one can use it. To work with the phone, it must be unlocked.

You should secure your iPhone with a passcode, which is a series of digits that must be provided to be able to unlock the phone so you can use it. To unlock one of the iPhone models without a Home button, you swipe up from the bottom and then enter the passcode. On an iPhone with a Home button, you press the Touch ID/Home button and then enter the passcode. In either case, the phone unlocks, and you can work with it.

Although entering a passcode isn't hard, you can make the process even easier by using either Face ID or Touch ID. With Face ID, your phone recognizes your face and enters your passcode for you. With Touch ID, you touch a recorded fingerprint to the Touch ID/Home button and the passcode is entered for you.

You also can use Face ID and Touch ID to enter user account information, including passwords, in many different apps. For example, when you download an app using the App Store app, you must provide an Apple ID password to complete the process. With Face ID, you can look at the screen to enter this password, or with Touch ID, you can touch the Touch ID/Home button to enter it.

Whether you can use Face ID or Touch ID depends on the specific model of iPhone you use, as does the process of unlocking your iPhone. The details of setting a password, configuring Face ID or Touch ID, and unlocking your phone are provided in Chapter 2.

When your iPhone locks, it also goes to sleep, which puts it into low power mode to extend battery life. The screen (which uses a lot of power) goes dark. Some processes keep working while your iPhone is locked/asleep, such as playing music, whereas others stop until your iPhone wakes up.

Almost all of the time, you'll lock your iPhone (which also puts it to sleep) by pressing the Side button (Sleep/Wake button on iPhone SE) once rather than turning it off because it's much faster to wake up than to turn on. Because it uses so little power when it's locked and asleep, there's not much reason to shut it down. (The steps to shut down or turn on your iPhone are model-dependent and are also covered in Chapter 2.)

Go to Sleep

Your iPhone can put itself to sleep and go into Lock mode automatically after a specified time of inactivity passes; this saves power and keeps your iPhone more secure. Configuring this is covered in Chapter 7, "Customizing How Your iPhone Looks."

When an iPhone is asleep, you need to wake it up to use it. How you do this depends on the model you're using, but you can wake up all models by pressing the Side button (Sleep/Wake button on iPhone SE) once. The Lock screen appears.

This iPhone is in Lock mode ——

Read notifications in the Notification Center ——

You can do quite a lot of things directly on the Lock screen without unlocking your phone:

- Read notifications in the Notification Center (see Chapter 2).

- Open the Control Center to use its controls (also covered in Chapter 2).

- Take photos or video (refer to Chapter 15, "Taking Photos and Video with Your iPhone").

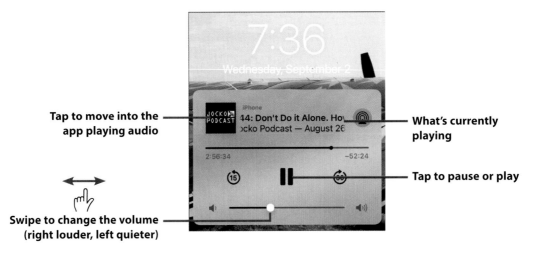

Tap to move into the app playing audio ——

What's currently playing

Tap to pause or play

Swipe to change the volume (right louder, left quieter)

- Control audio playback. When you listen to music, podcasts, or other audio, you can control the audio using the Audio Player on the Lock screen. Wake the phone and use the audio controls you see. For example, tap Pause to pause the audio or Play to play it.

- Activate the flashlight (iPhone models without a Home button). Touch and hold on the icon once to turn the flashlight on. Do the same to turn it off.

When you're ready to use apps or perform other tasks, you need to unlock your iPhone. How you unlock your iPhone depends on its state and the model you're using. The details are covered in Chapter 2.

Working with Home Screens

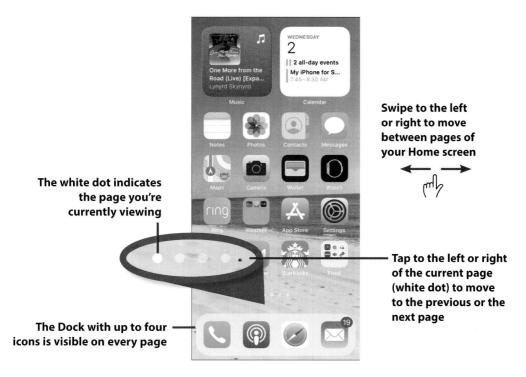

Swipe to the left or right to move between pages of your Home screen

The white dot indicates the page you're currently viewing

Tap to the left or right of the current page (white dot) to move to the previous or the next page

The Dock with up to four icons is visible on every page

When you swipe all the way to the left, you move to the App Library

When you move to a different page, you see a different set of icons and folders

Almost all iPhone activities start at the Home screen, or Home screens, to be more accurate, because the Home screen consists of multiple pages. You move to the Home screen automatically any time you restart your iPhone and unlock it. You also can move directly to the Home screen any time your iPhone is unlocked; how you do this depends on the model of phone you have. For example, if your iPhone doesn't have a Home button, you swipe up from the bottom of the screen, whereas if your iPhone has a Home button, you press that button to move to a Home screen. (The details are provided in Chapter 2.)

Along the bottom of the Home screen is the Dock, which is always visible on the Home screens. This gives you easy access to the icons it contains; you can place up to four icons on this Dock. Above the Dock are apps that do all sorts of cool things. Your iPhone comes with a number of preinstalled apps; as you install your own apps, the number of icons on the Home screens increases. To manage these icons, you can organize the pages of the Home screens in any way you like, you can place icons into folders to keep your Home screens tidy, and you can even hide pages you don't use (see Chapter 7). At the top of the screen are status icons that provide you with important information, such as whether you are connected to a Wi-Fi network and the current charge of your iPhone's battery.

As mentioned earlier, the Home screen has multiple pages. To change the page you're viewing, swipe to the left to move to later pages or to the right to move to earlier pages. The dots above the Dock represent the pages of the Home screen; the white dot represents the page being displayed. You also can change the page by tapping to the left of the white dot to move to the previous page or to the right of it to move to the next page. When you swipe all the way to the right or tap the first white dot, you move to the Widget Center (see Chapter 2 for the details). When you swipe all the way to the left, you move into the App Library (see Chapter 5, "Customizing Your iPhone with Apps and Widgets").

Working with iPhone Apps

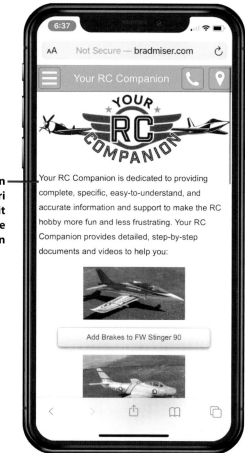

When you tap an app's icon (Safari in this example), it opens and fills the iPhone's screen

The reason an iPhone is so useful is because it can run apps. iPhones come with a number of preinstalled apps, including Phone, Mail, and Contacts; and you can download and use thousands of other apps, many of which are free, through the App Store. You learn about many of the iPhone's preinstalled apps as you read through this book. And as I mentioned earlier, to launch an app, simply tap its icon. The app opens and fills the iPhone's screen. You can then use the app to do whatever it does.

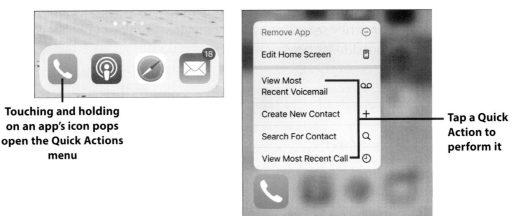

Touching and holding on an app's icon pops open the Quick Actions menu

Tap a Quick Action to perform it

You can touch and hold on an app's icon to open its Quick Actions menu; tap an action to take it. For example, when you open the Quick Actions menu for the Phone app, you can jump to the most recent voicemail. You also see the Remove App and Edit Home Screen commands at the top of all of these menus (see Chapter 7).

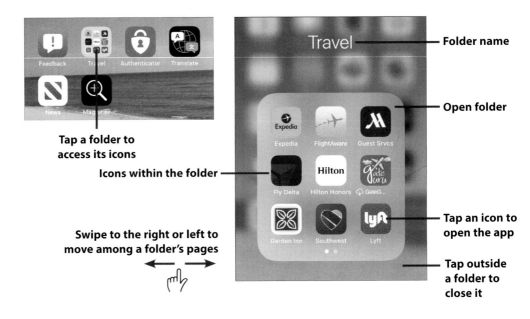

Tap a folder to access its icons

Icons within the folder

Swipe to the right or left to move among a folder's pages

Folder name

Open folder

Tap an icon to open the app

Tap outside a folder to close it

In Chapter 7, you learn how you can organize icons in folders to keep your Home screens tidy and make getting to icons faster and easier. To access an icon that is in a folder, tap the folder. It opens and takes over the screen. Under its name is a box showing the apps it contains. Like the Home screens, folders can have multiple pages. To move between a folder's pages, swipe to the left to move to the next screen or to the right to move to the previous one. Each time you "flip" a page, you see another set of icons. You can close a folder without opening an app by tapping outside its box.

To open an app within a folder, tap its icon.

To move back to a Home screen, swipe up from the bottom of the screen (models without a Home button) or press the Touch ID/Home button (models with a Home button). You return to the Home screen you were most recently using.

When you move back to a Home screen, the app you were using moves into the background but doesn't stop running (you can use the Settings app to control whether apps are allowed to work in the background). So, if the app has a task to complete, such as uploading photos or playing audio, it continues to work behind the scenes. In some cases, most notably games, the app becomes suspended at the point you move it into the background by switching to a different app or moving to a Home screen.

In addition to the benefit of completing tasks when you move into another app, the iPhone's capability to multitask means that you can run multiple apps at the same time. For example, you can run an Internet radio app to listen to music while you switch over to the Mail app to work on your email.

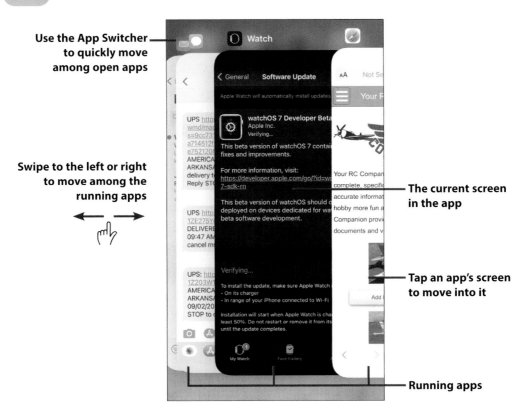

Use the App Switcher to quickly move among open apps

Swipe to the left or right to move among the running apps

The current screen in the app

Tap an app's screen to move into it

Running apps

You can control apps by using the App Switcher. To open it, swipe up from the bottom of the screen and pause toward the middle of the screen (iPhones without a Home button) or press the Touch ID/Home button twice (iPhones with a Home button). The App Switcher appears.

At the top of the App Switcher, you see icons for apps you are currently using or have used recently. Under each app's icon, you see a thumbnail of that app's current screen. You can swipe to the left or right to move among the apps you see. You can tap an app's screen to move into it. That app takes over the screen, and you can work with it, picking up right where you left off the last time you used it.

When you open the App Switcher, the app you were using most recently comes to the center to make it easy to return to. This enables you to toggle between two apps easily. For example, suppose you need to enter a confirmation number from one app into another app. Open the app into which you want to enter the number. Then open the app containing the number you need to enter. Open the App Switcher and tap the previous app to return to it quickly so that you can enter the number.

To close the App Switcher without moving into a different app, tap the top or bottom of the screen outside any app's window or icon. You move back into the app or Home screen you were most recently using.

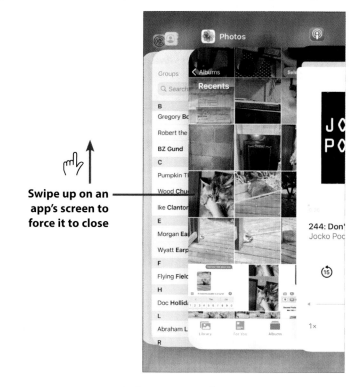

Swipe up on an app's screen to force it to close

In some cases, you might want to force an app to quit, such as when it's using up your battery too quickly or it has stopped responding to you. To do this, open the App Switcher. Swipe up on the app you want to stop. The app is forced to quit, its icon and screen disappear, and you remain in the App Switcher. You should be careful about this, though, because if the app has unsaved data, that data is lost when you force the app to quit. The app is not deleted from the iPhone—it's just shut down until you open it again (which you can do by returning to the Home screen and tapping the app's icon).

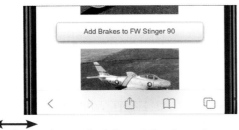

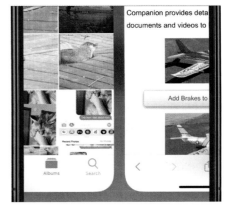

Swipe to the left or right along the bottom of the screen to move among the apps you are using

Keep swiping until the app you want to use fills the screen

Work with the app you switched to

On iPhones without a Home button, you can quickly switch apps by swiping to the left or right along the bottom of the screen. You can swipe to the right to move into apps you were using earlier or to the left to move to apps you were using more recently. (If you're on a Home screen and swipe above the Dock, swiping to the left or right moves among pages of the Home screen instead.)

Tap to return to the app you were previously using

Sometimes, a link in one app takes you into a different app. When this happens, you see a left-facing arrow with the name of the app you were using in the upper-left corner of the screen. You can tap this to return to the app you came from. For example, you can tap a link to a web page in a Mail email message to open the associated web page in Safari. To return to the email you were reading in the Mail app, tap Mail in the upper-left corner of the screen.

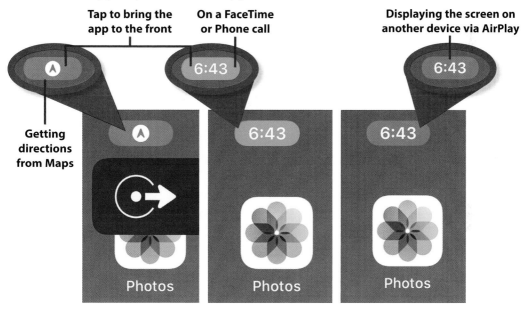

Tap to bring the app to the front

On a FaceTime or Phone call

Displaying the screen on another device via AirPlay

Getting directions from Maps

As you read, an app can continue to work in the background. For example, if you are using Maps for turn-by-turn directions, you might switch to the Music app to choose different music. Even though you don't see Maps (because it has moved into the background), it continues to provide directions to you through both audible and visual means. In many cases, you can tell if an app is actively working in the background by the presence of the active app indicator in the upper-left corner of the screen (iPhones without a Home button) or in a bar across the top of the screen (iPhones with a Home button). This appears in different colors and provides various information depending on the app it's representing. For example, if you're on a FaceTime call and move FaceTime into the background, this indicator is green. You also see the current length of the conversation.

In addition to providing you with information about the background activity, you can use this indicator to bring the background activity to the front. For example, if you're on a phone call using the Phone app and have switched to a different app, you can tap this indicator to return to the Phone screen.

Working with the Settings App

Aptly named, the Settings app is where you configure the many settings that change how your iPhone looks, sounds, and works. In fact, virtually everything you do on your iPhone is affected by settings in this app. As you use and make an iPhone into *your* iPhone, you frequently visit the Settings app.

Using the Settings App

 — **1. Tap to open the Settings app**

On the Home screen, tap Settings. The Settings app opens. The app is organized in sections starting at the top with your Apple ID information followed by Airplane Mode, Wi-Fi, Bluetooth, and Cellular.

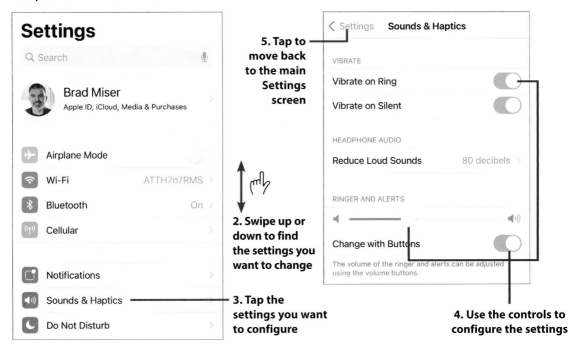

5. Tap to move back to the main Settings screen

2. Swipe up or down to find the settings you want to change

3. Tap the settings you want to configure

4. Use the controls to configure the settings

Swipe up or down the screen to get to the settings area you want to use. Then tap the area you want to configure, such as Sounds & Haptics.

Use the resulting controls to configure that area. The changes you make take effect immediately.

When you're done, you can leave the Settings app where it is and move into a different app (it remains there when you come back to it) or tap back (<), which is always located in the upper-left corner of the screen (its name changes based on where you are in the app), until you get back to the main Settings screen to go into other Settings areas.

Searching for Settings

You can quickly find settings you need by searching for them. For example, suppose you don't have this book handy and forget where to change the ringtone. You can quickly search in Settings to find what you need.

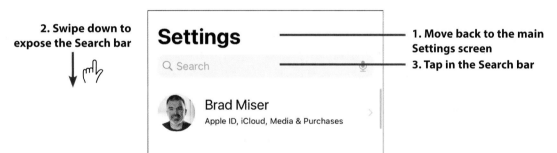

Move into the Settings app. (If you aren't on the main Settings screen, tap back (<) until you get there.) If you don't see the Search bar, swipe down from the top of the Settings screen until it appears. Tap in the Search bar.

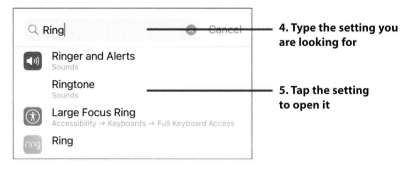

Type the setting for which you want to search. As you type, potential matches are shown on the list of results. Matches can include a settings area, such as Sounds & Haptics, or specific settings, such as the ringtone and vibrations used when you receive a call. Tap the setting you want to use.

‹ Back	Ringtone	
Vibration		Heartbeat ›
STORE		
Tone Store		
Download All Purchased Tones		
This will download all ringtones and alerts purchased using the "bradmacosx@mac.com" account.		
RINGTONES		
Carry On Wayward Son		
Kryptonite		
Oh Where Is My Cell Phone?		

6. Use the resulting screen to change the setting

Configure the setting you selected; for example, tap a ringtone to use it.

Meeting Siri

Siri is the iPhone's voice-recognition and control software. This feature enables you to accomplish many tasks by speaking. For example, you can create and send text messages, reply to emails, make phone calls, or get directions. (Using Siri is explained in detail in Chapter 13, "Working with Siri.")

When you perform actions, Siri uses the related apps to accomplish what you've asked it to do. For example, when you create an event, such as a doctor appointment, Siri uses the Calendar app.

Siri is a great way to control your iPhone, especially when you're working in handsfree mode.

Your iPhone has to be connected to the Internet for Siri to work. That's because the words you speak are sent over the Internet, transcribed into text, and then sent back to your iPhone. If your iPhone isn't connected to the Internet, this can't happen and Siri reports that it can't connect to the network or simply that it can't do what you ask right now.

Using Siri is pretty simple because it follows a consistent pattern and prompts you for input and direction.

Siri is ready to do
your bidding

Activate Siri by pressing and holding the Side button (iPhones without a Home button) or pressing and holding the Touch ID/Home button (iPhones with a Home button) until you hear the Siri chime. If so configured, you can say "Hey Siri" to activate it, too. If you are using EarPods, you can press and hold down the center part of the buttons on the right wire.

All these actions put Siri in "listening" mode, and the "What can I help you with?" text appears on the screen. This indicates Siri is ready for your command. (Although when you use "Hey Siri" to activate it, the onscreen text is skipped, and Siri gets right to work.)

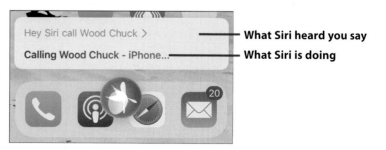

What Siri heard you say

What Siri is doing

Speak your command or ask a question. When you stop speaking, Siri goes into processing mode. After Siri interprets what you've said, it can provide two kinds of feedback to confirm what it heard (you can configure this as described in Chapter 13): It displays what it heard on the screen and provides audible feedback to you. Siri then tries to do what it thinks you've asked and shows you what it's doing. If it needs more input from you, Siri prompts you to provide it and then moves into "listening" mode automatically.

What Siri is doing for you

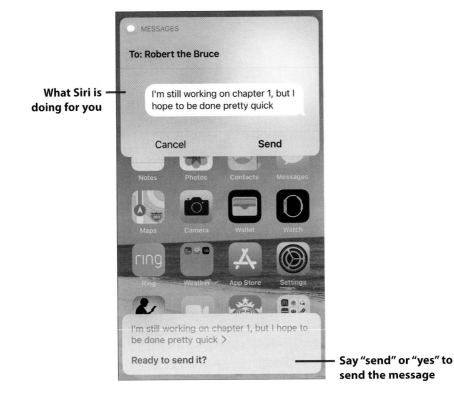

Say "send" or "yes" to send the message

If Siri requests that you confirm what it's doing or make a selection, do so. Siri completes the action and displays what it has done; it also audibly confirms the result.

Mostly, you can just speak to Siri as you would talk to a real person, and it's able to do what you want. If not, it asks you for the information it needs to do what you want.

After you get used to speaking to your iPhone, and seeing all you can do with Siri, you'll probably find yourself using it quite often.

Controlling Volume

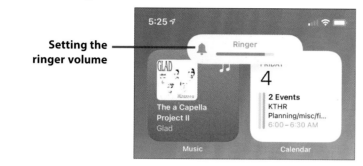

Setting the ringer volume

To change the iPhone's volume, press the up or down Volume button on the left side of the iPhone. When you change the volume, your change affects the current activity. For example, if you're on a phone call, the call volume changes, or if you're listening to music, the music's volume changes. If you're on a screen that doesn't have a Volume slider, an icon pops up to show you the relative volume you're

Drag to the left or right to change the volume level

setting and the type, such as Ringer. When the volume is right, release the Volume button.

When you're using an audio app, such as the Podcasts app, you can drag the volume slider in that app to increase or decrease the volume. Drag the slider to the left to lower volume or to the right to increase it.

When you use the iPhone's EarPods, you can change the volume by pressing the upper part of the switch on the right EarPod's wire to increase volume or the lower part to decrease it.

To mute your phone's sounds, slide the Mute switch, located at the top of the left side of the phone, toward the back of the phone. You see an on-screen indicator that the phone is muted, and you see orange within the Mute switch on the side of the iPhone; notification and other sounds won't play. To restore normal sound, slide the switch toward the front of the phone.

Unintentional Muting

If your phone suddenly stops ringing when calls come in or doesn't play notification sounds that you think it should, check the Mute switch to ensure you haven't activated it accidentally or that you forgot that you had muted your iPhone. When you mute your phone, the Silent Mode indicator appears briefly on the screen. However, there's no on-going indication on the screen that the iPhone is currently muted, so you have to look at the switch to tell.

Using Airplane Mode

When you place your iPhone in Airplane mode, its cellular transmitting and receiving functions are disabled. While it's in Airplane mode, you can't use the phone, Siri, or any other functions that require cellular communication between your iPhone and other devices or networks. You can continue to use Wi-Fi networks to access the Internet and Bluetooth to communicate with Bluetooth devices.

Swipe down from the upper-right corner of the screen to open the Control Center (iPhones without a Home button)

Tap to put your iPhone in Airplane mode

To put your iPhone in Airplane mode, open the Control Center (see Chapter 6 for the details of using the Control Center) and tap Airplane mode. All connections to the cellular network stop, and your iPhone doesn't broadcast or receive any cellular signals. Airplane mode becomes orange, and you see the Airplane mode status icon at the top of the screen.

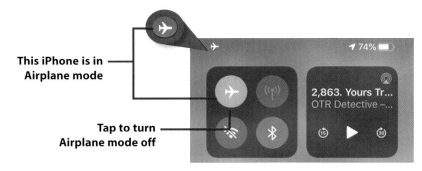

This iPhone is in Airplane mode

Tap to turn Airplane mode off

In Airplane mode, you can use apps that don't require a cellular connection. For example, you can admire and edit your pictures in the Photos app without an Internet connection, or you can connect to a Wi-Fi network, if one is available, to work with email or browse the Web.

To turn off Airplane mode, open the Control Center and tap Airplane mode; it becomes gray again and the Airplane mode status icon disappears. The iPhone resumes transmitting and receiving cellular signals, and all the functions that require a cellular connection start working again.

Locking/Unlocking Your iPhone's Orientation

Earlier, you saw how you can rotate your iPhone to change the orientation of its screen. For example, to view a photo that was taken in the landscape orientation (the long axis of the phone pointing left and right), you can rotate the phone to the same orientation. At times, you might not want the iPhone's screen to change orientation. For example, when you hand your phone to someone else to show them photos, you might want the photos to remain in the portrait (vertical) orientation, so they aren't continuously changing as you move the phone around. Or, if you're reading something while lying on your side, you might want to lock the phone's orientation so pages don't keep swapping orientation if you move the phone slightly.

Tap to lock your iPhone in portrait orientation

To lock your iPhone in portrait (vertical) orientation, open the Control Center and tap Portrait Orientation Lock. The screen's orientation won't change even when you rotate your iPhone. (If you lock the orientation when you're holding your phone horizontally, its orientation changes to portrait and then locks in the vertical orientation.)

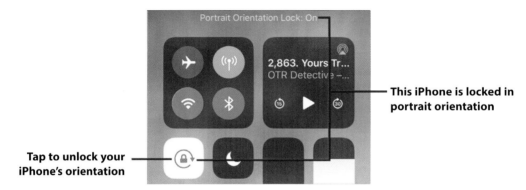

This iPhone is locked in portrait orientation

Tap to unlock your iPhone's orientation

To unlock your iPhone's orientation, open the Control Center and tap Portrait Orientation Lock again. The iPhone's screen changes orientation when you rotate it.

Understanding iPhone Status Icons

At the top of the screen, you see icons that provide you with information, such as if you are in Airplane mode, whether you are connected to a Wi-Fi or cellular data network, the time, whether the iPhone's orientation is locked, and the state of the iPhone's battery. On iPhones without a Home button, you see these icons on each side of the "notch" in the center of the screen, on the Control Center, and on the Lock screen. On iPhones with a Home button, you see these icons across the top of the screen, on the Control Center, and on the Lock screen. The following table provides a guide to the most common of these icons. You won't see all of these icons on all models. For example, you don't see the battery percentage status icon on iPhones without a Home button; you have to use the Widget or Battery screen in the Settings app to see this information.

Icon	Description	Where to Learn More
	Signal strength—Indicates how strong the cellular signal is.	Chapter 9
AT&T	**Provider name**—The provider of the current cellular network.	Chapter 9
LTE	**Cellular data network**—Indicates which cellular network your iPhone is using to connect to the Internet.	Chapter 9
	Wi-Fi—Indicates your phone is connected to a Wi-Fi network.	Chapter 3
Wi-Fi	**Wi-Fi calling**—Indicates your phone can make voice calls over a Wi-Fi network.	Chapter 9
	Do Not Disturb—Your iPhone's notifications and ringer are silenced.	Chapter 6
	Bluetooth—Indicates if Bluetooth is turned on or off and if your phone is connected to a device.	Chapter 17
79%	**Battery percentage**—Percentage of charge remaining in the battery.	Chapter 18
	Battery status—Relative level of charge of the battery.	Chapter 18
	Low Battery status—The battery has less than 20% power remaining.	Chapter 18
	Low Power mode—The iPhone is operating in Low Power mode.	Chapter 18
	Portrait Orientation Lock—Your iPhone's screen won't change when you rotate your iPhone.	Chapter 1
	Charging—The battery in the iPhone is being charged.	Chapter 18
	Location Services—An app is using the Location Services feature to track your iPhone's location.	Chapter 3
	Airplane mode—The cellular transmitting and receiving functions are disabled.	Chapter 1

Getting started with an iPhone is easy—no matter which model you have!

In this chapter, you learn fundamental skills for the major types of iPhone. Topics include the following:

→ Getting started
→ Getting started with iPhones without a home button
→ Getting Started with iPhones with a Home button
→ Using the split-screen on iPhone 12 Pro Max, iPhone 11 Pro Max, Xs Max, or Plus Models

Getting Started with Your iPhone

To get going with your iPhone, you need to understand how to use the features provided by your specific model. The differences between these models impact some fundamental tasks you do, such as moving to the Home screen. While these differences are significant, there aren't many of them. In this chapter, you learn about these fundamental tasks for the type of iPhone you have.

Getting Started

In Chapter 1, "Getting to Know Your iPhone," you learned the major ways you interact with an iPhone to do things such as opening apps, moving among Home screens, and working with the Settings app. Almost all of these tasks are performed in the same way on all models of iPhone. In that chapter, you also learned that there are some differences in models that impact how you perform certain tasks. In this chapter, you learn the details about these differences for the specific model of iPhone that you use.

In this book, iPhone models are grouped into the following major types that have corresponding sections in this chapter (you only need to read the sections that cover the model you have):

- **iPhones without a Home button**—This group includes the newest generation of iPhone models, which are the iPhone 12, iPhone 12 mini, iPhone 12 Pro, iPhone 12 Pro Max, iPhone 11, iPhone 11 Pro, and iPhone 11 Pro Max along with the older X, XR, Xs, and Xs Max. The most distinguishable difference with these models compared to prior generations is that the screen is almost as large as the frame. These models use facial recognition technology (Face ID) and generally have more advanced features in cameras, speed, and more, than do the earlier models. If you have one of these models, read the section "Getting Started with iPhones without a Home Button."

- **iPhones with a Home button**—These prior generations of iPhones have the Touch ID/Home button as their major distinguishing external feature. Because of this button, their screens are a bit smaller than those on models without this button relative to the overall size of the phones. The models in this group with higher numbers have more features and capabilities than models with lower numbers. Models in the group are the SE (first and second generations), 6s, 6s Plus, 7, 7 Plus, 8, and 8 Plus. If you have one of these models, read the section "Getting Started with iPhones with a Home Button."

- **Max and Plus Models**—These models are larger than others and include models without and with a Home button. Because of their large size, these models have the ability to use a split-screen, which can make them easier and faster to use. The models in this group are iPhone 12 Pro Max, 11 Pro Max, Xs Max, 8 Plus, 7 Plus, and 6s Plus. If you have one of these models, read "Using the Split-Screen on iPhone 12 Pro Max, iPhone 11 Pro Max, Xs Max, or Plus Models" to learn how to take advantage of its split-screen capability.

Where's My Model?

This book covers iPhones that can run iOS 14 (see Chapter 1). If you don't see your iPhone model listed in any of the groups described here, it isn't able to run iOS 14 and so isn't covered in this book. Prior editions of this book cover earlier iPhone models; you can obtain an earlier version of this book to match the phone you have.

If you aren't sure which model you have, tap the Settings icon on the Home screen, tap General, and then tap About. On the About screen, you see the model you are using.

Getting Started with iPhones without a Home Button

iPhone models without a Home button are the latest generation and feature larger screens relative to the physical size of the iPhone than prior models had. They also feature Face ID along with improved hardware that makes them work faster than older models. Except for the XR, they have dual or triple cameras on the back that provide for some amazing photographic capabilities, such as enhanced zooming and portrait photos. The iPhone 12 Pro and 12 Pro Max also have a Light Detection and Ranging (LiDAR) sensor to improve how the cameras capture images, especially in the Night mode.

Have a Button?

If your iPhone has a Home button, skip the rest of this section and jump to the section "Getting Started with iPhones with a Home Button" instead.

Getting to Know Your iPhone's External Features

iPhone models without a Home button have even fewer physical features than prior generations did. These models look like this:

- **Cameras**—All iPhone models have one camera on the front at the top near the center of the phone; this is the TrueDepth camera that you use for Face ID, selfies, and Animojis (animated emojis).

  All of these models also have cameras on their backside that you can use to take all sorts of photos and videos. Different models have different camera configurations that provide a variety of capabilities. The iPhone 12 Pro, iPhone 12 Pro Max, iPhone 11 Pro, and 11 Pro Max have three cameras that provide telephoto zoom, night mode, and wide and ultra-wide shots. The iPhone 12 Pro and 12 Pro Max also have a LiDAR sensor to further improve image quality, especially in low-light conditions. The iPhone 11 has two cameras that provide night mode, wide, and ultra-wide shots but not telephoto. The Xs, Xs Max, and X have two cameras that provide wide and telephoto capabilities. The XR has a single camera that can take wide shots.

  When you take photos or video, you can choose the cameras on either side of the phone.

- **Side button**—Press this button once to lock the iPhone's screen and put it to sleep, and press it again to wake the iPhone from Sleep mode. When you hold it down for a couple of seconds, you activate Siri. When Face ID is enabled, you press it twice to use Apple Pay, or download apps from the App Store. Press this button and either Volume button at the same time to take a screenshot. Press and hold this button and either Volume button to turn the phone off or make an emergency call. When your iPhone is turned off, press and hold this until you see the Apple logo on the screen to turn it on again.

- **Mute switch**—This switch determines whether the iPhone makes sounds, such as ringing when a call comes in or making the alert noise for notifications, such as for an event on a calendar. Slide it toward the front of the iPhone to hear sounds. Slide it toward the back of the iPhone to mute all sound. When muted, you see orange in the switch.

- **Volume buttons**—Press the upper button to increase volume; press the lower button to decrease volume. These buttons are contextual; for example, when you are listening to music, they control the music's volume, but when you aren't, they control the ringer volume. When you are using the Camera app, pressing either button takes a photo. You also use these in combination with the Side button to perform various actions (refer to the Side button description).

- **Lightning port**—Use this port, located on the bottom side of the iPhone, to plug in the EarPods or connect it to a computer or power adapter using the included USB cable. There are also accessories that connect to this port. The Lightning port accepts Lightning plugs that are flat, thin, rectangular plugs. It doesn't matter which side is up when you plug something into this port.

- **Speakers and microphone**—At the top center of the front of the phone is the speaker you use to listen when you have the phone held against your ear. There are two more speakers located along the bottom edge of the phone. When you play audio without another device (such as wireless headphones) connected to your iPhone, you hear the audio from these speakers. If you're having a hard time hearing (because of background noise for example), holding this edge to your ear can help.

There is also a microphone located along the bottom edge that captures sound, such as during a phone conversation, when you aren't using another device (for example, EarPods).

Accessing the Home Screens

To access the Home screens, swipe up from the bottom of the screen

In Chapter 1, you learned the importance of Home screens because most of the tasks for which you use your iPhone start there. On iPhones without a Home button, you move to the Home screens by swiping up from the bottom of the screen. You see a line indicating the area from which you start your swipe; you don't have to swipe very far to open the Home screens. Just a quick upward gesture from this line will do it.

Configuring a Passcode and Face ID

A passcode is a series of characters that must be entered on your phone to perform specific actions, the most important of which is unlocking your iPhone so you can use it. At other times, you need to enter a passcode to change settings.

You can use an iPhone without a passcode, but I strongly recommend that you always use a passcode to protect your iPhone's data.

When you use Face ID, you don't need to type a passcode very often. Instead, you can simply look at your iPhone's screen; when your face is recognized, the passcode is entered for you automatically.

Face ID can also be used to enter user account information, such as usernames and passwords. So, instead of having to type your passwords, you can simply look at your iPhone to sign into accounts or download apps.

When you first started your iPhone, you were prompted to configure a passcode and Face ID. Even if you have already configured a passcode and Face ID on your phone, you should know how to change your settings in the event you want to make updates and to ensure you're making the most of Face ID's capabilities.

The steps to configure a new passcode and Face ID follow; you can use very similar steps to change existing passcode and Face ID settings:

(1) Tap Settings on the Home screen to open the Settings app and then tap Face ID & Passcode.

Don't Already Have a Passcode?

If you set up your iPhone following the prompts when you turned it on for the first time, you likely set a passcode. The steps in this section assume you did that and so you have to enter that passcode to access the Face ID & Passcode Settings screen. If your iPhone doesn't currently have a passcode, you move immediately to the Face ID & Passcode screen after step 1. Tap Turn Passcode On, enter a passcode, and confirm the new passcode by entering it again. You might also be prompted to enter your Apple ID password; do so at the prompt and tap OK to create a passcode. You can then pick up these steps at step 4. See the Go Further sidebar, "Be Secure with Face ID," for some tips about updating existing settings.

---

(2) Enter your passcode.

(3) Tap Done, if necessary. Depending on your passcode's length or complexity, the passcode might be recognized as soon as you enter it, and you don't need to tap anything else.

(4) Tap Set Up Face ID.

(5) Tap Get Started.

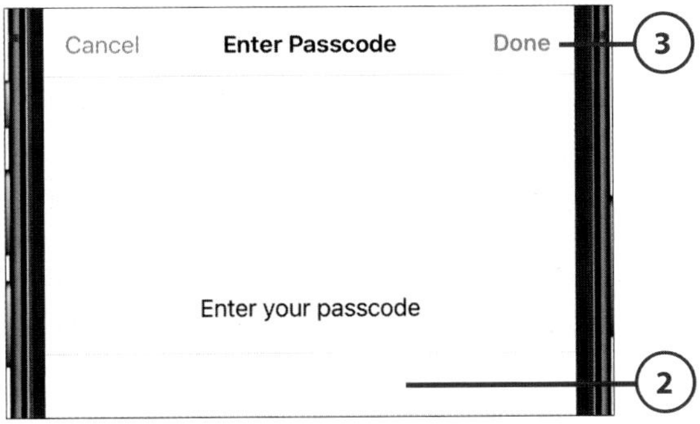

| Cancel | **Enter Passcode** | Done —(3) |

Enter your passcode

—(2)

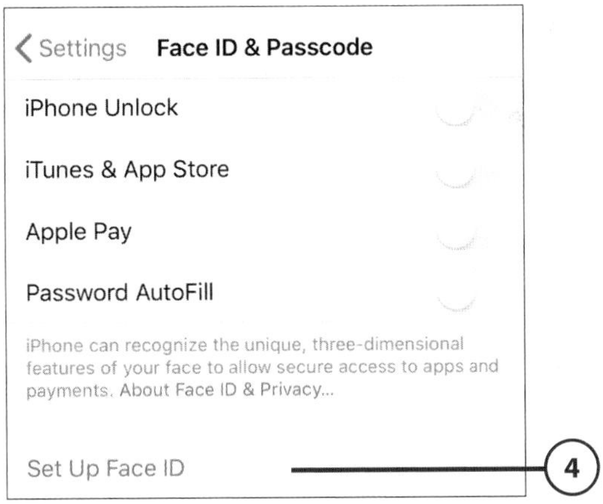

< Settings   **Face ID & Passcode**

iPhone Unlock

iTunes & App Store

Apple Pay

Password AutoFill

iPhone can recognize the unique, three-dimensional features of your face to allow secure access to apps and payments. About Face ID & Privacy...

Set Up Face ID ——————(4)

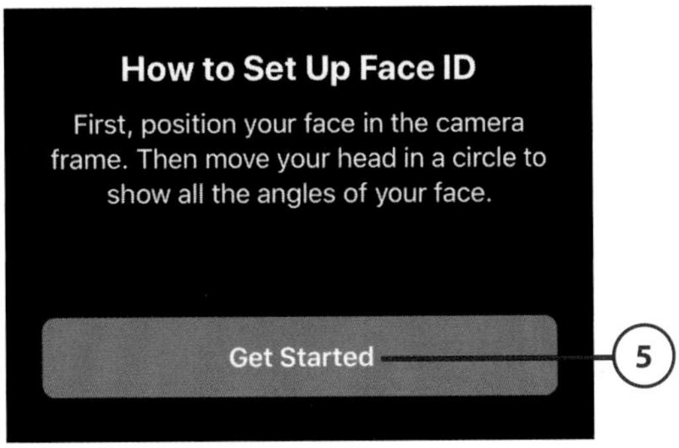

How to Set Up Face ID

First, position your face in the camera frame. Then move your head in a circle to show all the angles of your face.

Get Started ——————(5)

6. Look at the iPhone and position your face within the frame. When the iPhone recognizes a face being in the frame, it starts recording your face.

7. Move your head around in a circle. As you do, the green lines around the circle on the screen fill in, which indicates the part of the image that has been successfully recorded.

8. Continue moving your head in a circle until all of the green lines are filled in. When the process is complete, you see First Face ID scan complete appear on the screen.

9. Tap Continue. Record your face a second time.

Move your head slowly to complete the circle.

First Face ID scan complete.

Continue

10. Rotate your head until the circle is enclosed by green lines. When the second image has been recorded, Face ID is now set up appears on the screen.

11. Tap Done. You return to the Face ID & Passcode screen.

12. If you want to use Face ID for one of the options shown in the USE FACE ID FOR section, set its switch to on (green). You might need to enter the related passcode or password to complete an action. If you don't want to use Face ID for an action, set its switch to off (white). For example, when you set the iTunes & App Store switch to off (white), you'll need to enter your Apple ID password when you download apps, instead of using Face ID. I recommend you enable all of these options to make the most of Face ID.

13. Tap Other Apps (you see this only after other apps have requested and been granted permission to use Face ID).

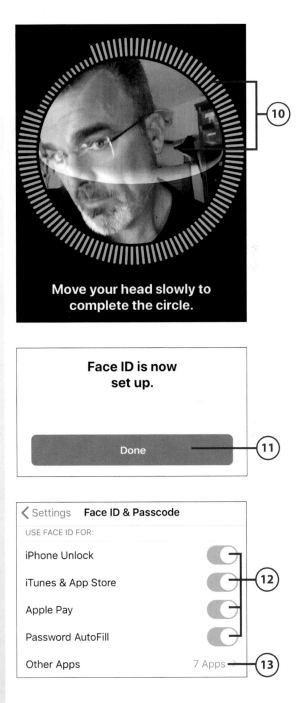

Move your head slowly to complete the circle.

Face ID is now set up.

Done

Settings   Face ID & Passcode

USE FACE ID FOR:

iPhone Unlock

iTunes & App Store

Apple Pay

Password AutoFill

Other Apps   7 Apps

(14) To allow an app to use Face ID, set its switch to on (green); to prevent one of the apps listed from using Face ID, set its switch to off (white). As you make changes, you might need to enter your user account information to work with that app.

(15) Tap Back (<).

(16) If you want to enable someone else to use Face ID with your iPhone, tap Set Up an Alternate Appearance and follow steps 5 through 11 to record the other person's face. You could also use this to capture an alternate appearance for yourself in the event your appearance changes so dramatically that it isn't recognized by Face ID. This would be unusual, however, since Face ID can handle changes in appearance due to changes in haircuts, cosmetic applications, or other similar changes.

(17) To remove the current face being used for Face ID and replace it with a new recording, tap Reset Face ID and follow steps 5 through 11.

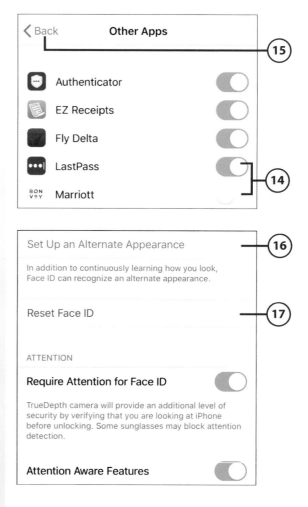

Glasses or Sunglasses? Facial Hair or Clean Shaven?

If you wear glasses or sunglasses, you don't need to create an alternate appearance. Face ID recognizes you with glasses or sunglasses on or off. Routine facial changes, such as in facial hair or different applications of makeup, won't faze it either. It's very unlikely that you'll need an alternate appearance for yourself. You typically use this to allow someone else to work with your phone.

18 If you don't want the additional security offered by the Face ID system verifying you're looking at the iPhone, set the Require Attention for Face ID switch to off (white). You should usually leave this set to on (green). However, if Face ID isn't working well, try setting this switch to off to see if Face ID works better.

19 If you don't want the TrueDepth camera to check for your attention before the display is dimmed or the alert volume is lowered, set the Attention Aware Features switch to off (white). When this is on (green), if you're looking at the iPhone, the display won't be dimmed nor will the alert volume be lowered. This is useful because the iPhone can "tell" when you are looking at it so that it won't dim the screen or lower the volume automatically.

20 Swipe up the screen until you see the Voice Dial switch.

21 To prevent Voice Dial from working, set the Voice Dial switch to off (white). (Voice Dial enables you to make calls by speaking even if you don't use Siri.)

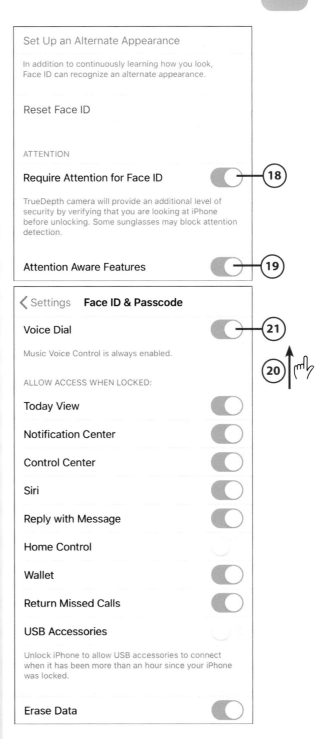

22 Use the switches in the ALLOW ACCESS WHEN LOCKED section to enable or disable the related functions used when your iPhone is locked. The options include Today View (the Today section of the Notification Center), Notification Center, Control Center, Siri, Reply with Message, Home Control, Wallet, Return Missed Calls, and USB Accessories. If you set a switch to off (white), you won't be able to access the corresponding function when your iPhone is locked.

23 If you want the iPhone to automatically erase all your data after an incorrect passcode has been entered 10 times, set the Erase Data switch to on (green) and tap Enable at the prompt (not shown in the figure).

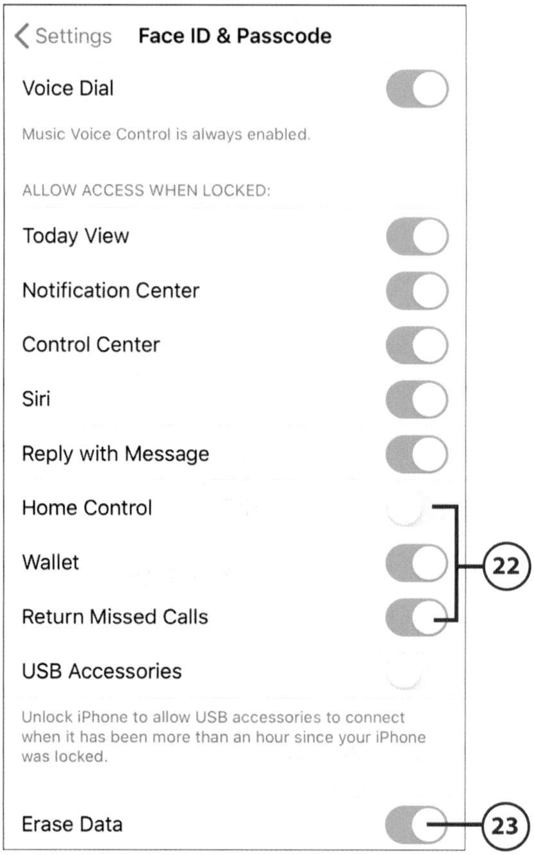

⟨ Settings   **Face ID & Passcode**

Voice Dial

Music Voice Control is always enabled.

ALLOW ACCESS WHEN LOCKED:

Today View

Notification Center

Control Center

Siri

Reply with Message

Home Control

Wallet   **22**

Return Missed Calls

USB Accessories

Unlock iPhone to allow USB accessories to connect when it has been more than an hour since your iPhone was locked.

Erase Data   **23**

>>>Go Further
BE SECURE WITH FACE ID

Here are some additional tidbits to help you with Face ID, a passcode, and your iPhone's security:

- **Face ID and Apps**—The first time you launch an app that supports Face ID, you're prompted to enable Face ID in that app. If you allow this, you can log into the associated account by looking at the screen at the prompt, just like unlocking your phone or using Apple Pay. You can change this setting by moving back to the Other Apps screen and changing the app's switch to on (green) to enable Face ID or off (white) to disable Face ID for that app.

- **Automatic Erase**—When you have enabled the Erase Data function and you (or someone else) enter an incorrect passcode when unlocking your iPhone, you see a counter showing the number of unsuccessful attempts. When this reaches 10, all the data on your iPhone is erased on the next unsuccessful attempt.

- **Making Changes**—Any time you want to make changes to your passcode or Face ID settings, move back to the Face ID & Passcode screen by tapping Face ID & Passcode, entering your passcode, and tapping Done. To change your passcode, tap Change Passcode. Enter your current passcode and enter your new passcode twice. If the passcode you enter isn't very secure, you are prompted to create a more secure passcode, such as by using more characters. You also have the option to keep the less secure version, so which you use is up to you. When the new passcode has been successfully created, you return to the Face ID & Passcode screen, and the new passcode takes effect. You can change the other settings similar to how you set them initially, as described in the steps earlier in this section.

- **Passcode Complexity**—Initially, iPhones supported only four-digit passcodes. Later, six-digit passcodes were required. Over time, more complex passcodes have been required to improve security. You also have the option to create more complex passcodes that include both letters and numbers. When you change your passcode, you can tap Passcode Options and then choose the type of passcode you want to use. You can choose Custom Alphanumeric Code, Custom Numeric Code, or 4-Digit Numeric Code. After you choose the option you want to use, you create a new passcode of that type.

  If your phone is set up with a four- or six-digit passcode, you can enter it by tapping the appropriate keys to enter the passcode's characters; you move to the next screen automatically.

  If you have a longer or more complex passcode, you have to type its characters and then tap OK or Done to enter it and move to the next screen.

- **Require Passcode**—This setting applies only if you don't use Face ID and rely on just the passcode instead. You can use it to determine how much time passes when the phone goes to sleep and when it locks automatically. For example, if you set this to After 1 minute, if you start to use the iPhone within 1 minute of the time it went to sleep, you won't need to unlock it. If more than one minute has passed, you need to enter the passcode to unlock the phone.

- **Automatic Locking**—For security purposes, you should configure your iPhone so that it locks automatically after a specific amount of idle time passes. To do this, you use the Auto-Lock setting on the Display & Brightness settings screen as explained in Chapter 7, "Customizing How Your iPhone Looks."

Using Face ID

You can use Face ID to quickly, easily, and securely provide a password or passcode in many different situations, such as unlocking your iPhone, downloading apps from the App Store, signing into an account in a banking or other app, and just about any other situation in which you need to confirm your identify.

The First Time

The first time you do something after Face ID is enabled, such as downloading an app, you might still be prompted to enter your Apple ID or other password. The next time you perform that action, you can use Face ID to complete it instead.

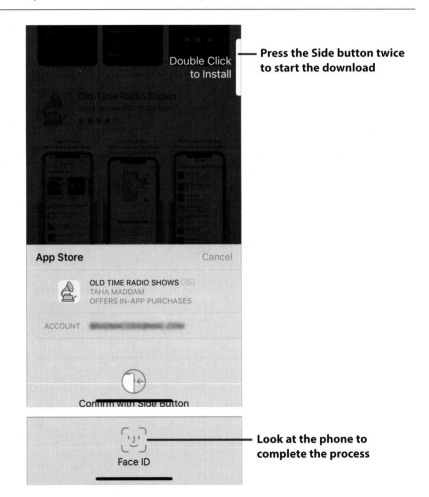

Press the Side button twice to start the download

Look at the phone to complete the process

For example, to download an app using Face ID, move to the app's screen in the App Store app and tap Get (or the price if there is a license fee). At the prompt, press the Side button twice and look at the phone. When your face is recognized, the app is downloaded and installed. Most of the time, you won't see the Face ID prompt as shown in the figure because you're already looking at the screen and so the action happens automatically. If you see this prompt, look more directly at your iPhone's screen to complete the action.

Using Face ID in other situations is similar. If you see the Face ID prompt, look at the phone. In other cases, you won't see the prompt and the action you're performing is completed as soon as your face is recognized.

The First Time, Part 2

The first time you sign into an app, such as a banking app, that supports Face ID, you're prompted to allow that app to use Face ID to enter your password. If you allow it, you can sign into the app by just looking at the phone on the log-in screen. If you don't allow it, you need to manually enter the password.

Sleeping/Locking and Waking/Unlocking Your iPhone

As you learned in Chapter 1, you can manually lock your iPhone, which also puts it to sleep. It's a good idea to do this when you aren't using your phone to both save battery power and secure your information. (You can configure this to happen automatically as explained in Chapter 7.)

To put your iPhone to sleep and lock it, press the Side button. The screen goes dark and the phone locks.

This iPhone is awake and locked

This iPhone is unlocked

Swipe up to move to the most recent screen

When an iPhone is asleep/locked, you need to wake it up to use it. You can do this in several ways: touch the screen, press the Side button, or raise the iPhone. (See Chapter 7 to learn how to enable or disable this Raise to Wake feature.) The Lock screen appears. If you are looking at the phone, it also unlocks immediately. On the Lock screen, you can tell if the phone is locked or unlocked by the Lock icon.

After your iPhone is unlocked, swipe up from the bottom of the screen. You move back to the screen you most recently used. For example, if you were using the Safari app right before your phone locked, you move back to that app.

Enter My Passcode? Why?

At times, when you try to unlock your phone, you are prompted to enter your passcode. This always happens after your phone has been restarted. After you enter the passcode correctly, the phone unlocks and Face ID becomes active again.

Most of the time, unlocking your phone is so seamless that you might not even realize that Face ID has been used. When you wake your phone, you can just swipe up the screen to unlock it. Face ID works quickly, and, as long as you are looking at the phone when you swipe up the screen, you won't even notice Face ID unlocking the phone for you.

No Face ID?

If you don't have Face ID configured, you can wake your phone in the same ways. However, your phone remains locked. To use it, swipe up from the bottom of the screen. You're prompted to enter your passcode. If you enter it successfully, you move to the screen you were most recently using.

No Passcode?

If you don't have a passcode configured, which is not recommended, you can wake your phone and swipe up from the bottom of the screen to move to the screen you were most recently using.

Face Coverings

If your face is partially or mostly covered by something, Face ID probably won't work. In that case, you have to enter your passcode or passwords manually.

Turning Your iPhone Off or On

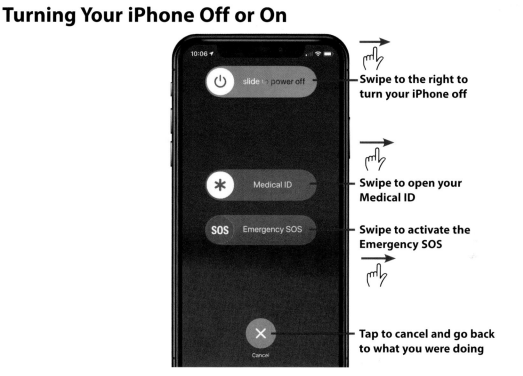

Swipe to the right to turn your iPhone off

Swipe to open your Medical ID

Swipe to activate the Emergency SOS

Tap to cancel and go back to what you were doing

You seldom need to turn your iPhone off, but you might want to shut if off if you aren't going to be using it for a while or for troubleshooting purposes.

To turn your iPhone off, press and hold the Side button and either Volume button. Several sliders appear on the screen; swipe the top slider to the right. The phone turns off.

To restart your iPhone, press and hold the Side button until the Apple logo appears on the screen, and then let go of the button.

After it starts up, you see the Lock screen. To use the phone, swipe up from the bottom of the screen. Assuming you have a passcode, you see the Enter Passcode screen. Enter your passcode and tap OK (if required) to start using your phone; once your passcode is entered correctly, you move to the Home screen. (Even if you have Face ID enabled to unlock your phone, you must enter your passcode the first time you unlock it after a restart.)

Getting Started with iPhones with a Home Button

While these models might not be the newest generation, they are certainly packed with amazing capabilities. The primary distinguishing physical feature of these models compared to later models is the presence of the Touch ID/Home button. Because of this button, the screens on these models are smaller relative to their case size than the later models that don't have a Touch ID/Home button.

This group also has a large range in physical size with the SE being the smallest and the 7 Plus and 8 Plus being the largest of this group.

Don't Have a Button?

If your iPhone doesn't have a Home button, skip the rest of this section. You should read the earlier section "Getting Started with iPhones without a Home Button" instead.

Getting to Know iPhones with a Home Button

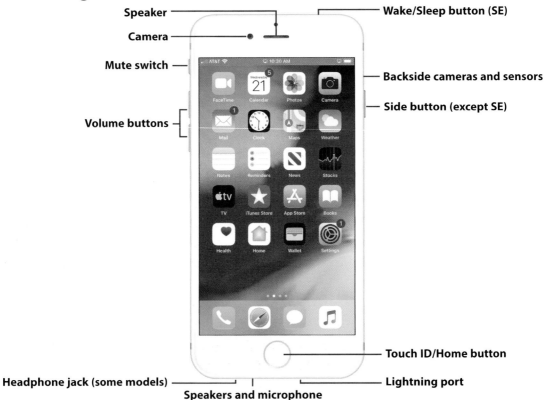

Speaker

Camera

Mute switch

Volume buttons

Wake/Sleep button (SE)

Backside cameras and sensors

Side button (except SE)

Touch ID/Home button

Headphone jack (some models)

Lightning port

Speakers and microphone

iPhones with a Home button have the following physical features:

- **Cameras**—Every iPhone has a camera on the frontside that you usually use to take selfies (you can use it for any kind of photo or video you want to take).

  Every iPhone has at least one camera on the backside too; different models have different configurations of backside cameras that provide a variety of photographic capabilities.

  The SE, 6s, 6s Plus, 7, and 8 have a single backside camera that provides wide shot capability.

  The 7 Plus and 8 Plus have two backside cameras that provide wide and telephoto zoom functions.

  You usually use the backside cameras to take photos and video of, well, anything you want.

  These models have cameras of different quality, with the later models having higher quality cameras than earlier models. For example, the 8 Plus has higher quality cameras than does the 6s Plus.

- **Side button**—Press this button (on the SE, this is called the Sleep/Wake button and is located on the top of the phone instead of on the side) to lock the iPhone's screen and put it to sleep. Press it again to wake the iPhone from Sleep mode. You also use this button to shut down the iPhone and to power it up.

- **Mute switch**—This switch determines whether the iPhone makes sounds, such as ringing when a call comes in or making the alert noise for notifications, such as for an event on a calendar. Slide it toward the front of the iPhone to hear sounds. Slide it toward the back of the iPhone to mute all sound. When muted, you see orange in the switch.

- **Volume**—Press the upper button to increase volume; press the lower button to decrease volume. These buttons are contextual; for example, when you're listening to music, they control the music's volume, but when you aren't, they control the ringer volume. When you're using the Camera app, pressing either button takes a photo.

- **Lightning port**—Use this port, located on the bottom side of the iPhone, to plug in the EarPods or connect it to a computer or power adapter using the included USB cable. Some accessories also connect to this port. The Lightning port accepts Lightning plugs that are flat, thin, rectangular plugs. It doesn't matter which side is up when you plug something into this port.

- **Headphone jack (some models)**—Some earlier models have a standard 3.5 mm jack that can be used for headphones (such as the older EarPods) and powered speakers. Most models don't have this and instead use the Lightning port to connect to other devices, including EarPods.

- **Touch ID/Home button**—This button provides multiple functions.

  The Touch ID sensor recognizes your fingerprint, so you can simply touch it to unlock your iPhone, sign in to the iTunes Store, use Apple Pay, and enter your password in Touch ID-enabled apps.

  You also use it for several other actions, such as waking and unlocking your phone, moving to the Home screens, opening the App Switcher, and opening the magnifier.

- **Speakers and microphone**—At the top center of the front of the phone is the speaker you use to listen when you have the phone held against your head. There are two more speakers located along the bottom edge of the phone. When you play audio without another device (such as wireless headphones) connected to your iPhone, you hear the audio from these speakers. If you're having a hard time hearing (because of background noise for example), holding this edge to your ear can help.

  There is also a microphone located along the bottom edge that captures sound, such as during a phone conversation, when you aren't using another device (for example, EarPods).

Accessing the Home Screens

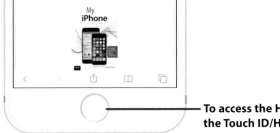

To access the Home screens, press the Touch ID/Home button

In Chapter 1, you learned the importance of Home screens because most of the tasks for which you use your iPhone start there. You move to the Home screens by pressing the Touch ID/Home button. When you press this button, you move back to the Home screen you were using most recently.

Configuring a Passcode and Touch ID

A passcode is a series of characters that must be entered on your phone to perform specific actions, the most important of which is unlocking your iPhone so you can use it. At other times, you need to enter a passcode to change settings.

Although you can use an iPhone without a passcode, I strongly recommend that you always use a passcode to protect your iPhone's data.

While typing a passcode is fairly quick and easy, when you use Touch ID, you don't need to do that very often. (A passcode is always required when you restart your iPhone.) Instead, you can simply touch the Touch ID/Home button and the passcode is entered for you automatically.

Touch ID can also be used to enter user account information, such as usernames and passwords. So, instead of having to type your passwords, you can touch the Touch ID/Home button to sign into accounts, download apps, or in just about any other situation in which you need to verify your identity.

When you first started your iPhone, you were prompted to configure a passcode and Touch ID. Even if you have already configured a passcode and Touch ID on your phone, you should know how to change your settings in the event you want to make updates and to ensure you are making the most of Touch ID's capabilities.

The steps to configure a new passcode and Touch ID follow; you can use very similar steps to change existing passcode and Touch ID settings:

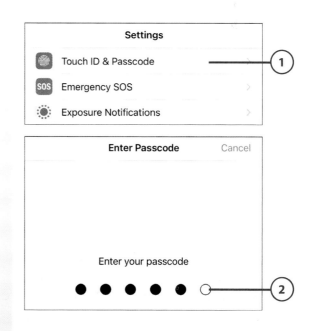

(1) Tap Settings on the Home screen to open the Settings app; then tap Touch ID & Passcode.

(2) Enter your passcode.

(3) Tap Done; if necessary (not shown in a figure). Depending on your passcode's length or complexity, the passcode might be recognized as soon as you enter it, and you don't need to tap anything else.

Don't Already Have a Passcode?

If you set up your iPhone following the prompts when you turned it on for the first time, you probably created a passcode. If you have a passcode, you have to enter it to open the Touch ID & Passcode Settings screen. If your iPhone doesn't have a passcode, you move directly to the Touch ID & Passcode screen after step 1. Tap Turn Passcode On, enter a passcode, and confirm the new passcode by entering it again. You might also be prompted to enter your Apple ID password; do so at the prompt and tap OK. This creates the passcode for your phone. You can then pick up these steps at step 4. See the Go Further sidebar, "Be Secure with Touch ID," for some tips about updating existing settings.

4 Tap Require Passcode; when you use Touch ID to unlock your iPhone, you don't have an option for when the passcode is required, so if you're going to or already use Touch ID, skip to step 7.

5 Tap the amount of time the iPhone is locked before the passcode takes effect. The shorter this time is, the more secure your iPhone is, but also the more frequently you'll have to enter the passcode if your iPhone locks frequently.

6 Tap Back (<).

7 Tap Add a Fingerprint.

Turn Passcode Off

Change Passcode

Require Passcode   Immediately   **4**

< Back   **Require Passcode**   **6**

Immediately

After 1 minute

After 5 minutes   ✓   **5**

After 15 minutes

After 1 hour

After 4 hours

FINGERPRINTS

Add a Fingerprint...   **7**

Turn Passcode Off

Change Passcode

Require Passcode   After 5 min. >

8. Touch the finger you want to record to the Touch ID/Home button, but don't press it. An image of a fingerprint appears.

9. Leave your finger on the Touch ID/Home button until you feel the phone vibrate, which indicates part of your fingerprint has been recorded. The parts of your fingerprint that are recorded are indicated by the red segments; gray segments are not recorded yet. This step captures the center part of your finger.

10. Lift your finger off the Touch ID/Home button and touch the button again, adjusting your finger on the button to record other parts that currently show gray lines instead of red ones. Other segments of your fingerprint are recorded.

11. Repeat step 10 until all the segments are red. You're prompted to change your grip so you can record more of your fingerprint.

Place Your Finger

Lift and rest your finger on the Home button repeatedly.

8

Place Your Finger

Lift and rest your finger on the Home button repeatedly.

9

(12) Tap Continue.

(13) Repeat step 10, again placing other areas of your finger to fill in more gray lines with red. This step captures the fingerprints more toward the edges of your fingers. When the entire fingerprint is covered in red lines, you see the Complete screen.

(14) Tap Continue. The fingerprint is recorded and you move back to the Touch ID & Passcode screen. You see the fingerprint that has been recorded.

(15) Tap the fingerprint you recorded.

(16) Give the fingerprint a name.

(17) Tap Touch ID & Passcode (<).

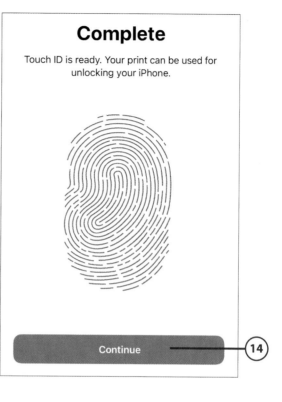

Complete

Touch ID is ready. Your print can be used for unlocking your iPhone.

Continue ────(14)

FINGERPRINTS

Finger 1 ──────(15)

Add a Fingerprint...

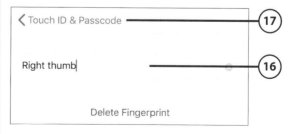

< Touch ID & Passcode ────(17)

Right thumb| ────(16)

Delete Fingerprint

(18) Repeat steps 7–17 to record up to five fingerprints. These can be yours or someone else's if you want to allow another person to access your iPhone.

(19) To be able to use Touch ID to unlock your iPhone, ensure the iPhone Unlock switch is set to on (green).

(20) If it isn't enabled already and you want to also be able to enter your Apple ID password by touch-ing your finger to the Touch ID/ Home button, set the iTunes & App Store switch to on (green). You need to enter your Apple ID password and tap Continue to complete this configuration (not shown in a figure).

(21) To use your fingerprint to make Apple Pay payments, set the Apple Pay switch to on (green).

(22) To enable your passwords for various accounts to be entered with Touch ID, set the Password AutoFill switch to on (green).

USE TOUCH ID FOR:

iPhone Unlock — (19)

iTunes & App Store — (20)

Apple Pay — (21)

Password AutoFill — (22)

FINGERPRINTS

Right thumb

Add a Fingerprint…

(18)

Using Touch ID-Aware Apps

Apps that require you to sign into accounts can also use Touch ID to make signing in easier. When you have at least one app that uses Touch ID, you see the Other Apps option on the Touch ID & Passcode screen. Tap Other Apps to enable or disable Touch ID for those other apps.

23. Swipe up the screen until you see the Voice Dial switch.

24. To prevent Voice Dial from working, set the Voice Dial switch to off (white). (Voice Dial enables you to make calls by speaking even if you don't use Siri.)

25. Use the switches in the ALLOW ACCESS WHEN LOCKED section to enable or disable the related functions when your iPhone is locked. The options include Today View (the Today section of the Notification Center), Notification Center, Control Center, Siri, Reply with Message, Home Control, Wallet, Return Missed Calls, and USB Accessories. If you set a switch to off (white), you won't be able to access the corresponding function when your iPhone is locked.

26. If you want the iPhone to automatically erase all your data after an incorrect passcode has been entered 10 times, set the Erase Data switch to on (green).

27. Tap Enable.

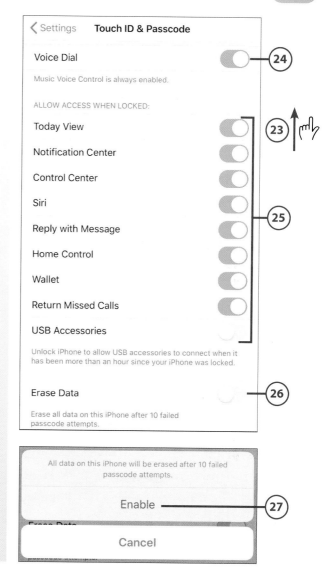

>>>Go Further

BE SECURE WITH TOUCH ID

Here are some additional tidbits to help you with your iPhone's security:

- **Touch ID and Apps**—The first time you launch an app that supports Touch ID, you're prompted to enable Touch ID in that app. If you allow this, you can log into the associated account by touching the Touch ID/Home button, just like unlocking your phone or using

Apple Pay. After you've enabled apps to use Touch ID, you can use the Other Apps option on the Touch ID & Passcode screen to enable or disable Touch ID for specific apps.

- **Automatic Erase**—When you have enabled the Erase Data function and you (or someone else) enter an incorrect passcode when unlocking your iPhone, you see a counter showing the number of unsuccessful attempts. When this reaches 10, all the data on your iPhone is erased on the next unsuccessful attempt.

- **Making Changes**—Any time you want to make changes to your passcode and fingerprint settings, move back to the Touch ID & Passcode screen by tapping Touch ID & Passcode on the Settings screen and entering your passcode (tap Done if required). To disable the passcode (not recommended), tap Turn Passcode Off, tap Turn Off, and enter the passcode. To change your passcode, tap Change Passcode. You then enter your current passcode and enter your new passcode twice. If the passcode you enter isn't very secure, you are prompted to create a more secure passcode, such as by using more characters. You also have the option to keep the less secure version, so which you use is up to you. When the new passcode has been successfully created, you return to the Touch ID & Passcode screen, and the new passcode takes effect. You can change the other settings similar to how you set them initially as described in these steps. For example, you can add new fingerprints. To remove a fingerprint, move to the Fingerprints screen, swipe to the left on the fingerprint you want to remove, and tap Delete.

- **Passcode Complexity**—Initially, iPhones supported only four-digit passcodes. Later, six-digit passcodes were required. Over time, more complex passcodes have been required to improve security. You also have the option to create more complex passcodes that include both letters and numbers. When you change your passcode, you can tap Passcode Options and then choose the type of passcode you want to use. You can choose Custom Alphanumeric Code, Custom Numeric Code, or 4-Digit Numeric Code. After you choose the option you want to use, you create a new passcode of that type.

  If your phone is set up with a four- or six-digit passcode, you can enter it by just tapping the appropriate keys to enter the passcode's characters and then move to the next screen automatically.

  If you have a longer or more complex passcode, you have to type its characters and then tap OK or Done to enter it and move to the next screen.

- **Require Passcode**—This setting applies only if you don't use Touch ID and rely on just the passcode. On the Touch ID & Passcode screen, swipe up to Require Passcode to determine how much time passes when the phone goes to sleep and when it locks automatically. For example, if you set this to After 1 minute, and you start to use the iPhone within 1 minute of the time it went to sleep, you won't need to unlock it. If more than one minute has passed, you need to enter the passcode to unlock the phone.

- **Automatic Locking**—For security purposes, you should configure your iPhone so that it locks automatically after a specific amount of idle time passes. To do this, you use the Auto-Lock setting on the Display & Brightness settings screen as explained in Chapter 7.

Using Touch ID

You can use Touch ID to quickly, easily, and securely provide a password or passcode in many different situations, such as unlocking your iPhone, downloading apps from the App Store, or signing into an account in a banking or other app.

The First Time

The first time you do something after Touch ID is enabled, such as downloading an app, you might still be prompted to enter your Apple ID or other password. The next time you perform that action, you can use Touch ID to complete it (no password required).

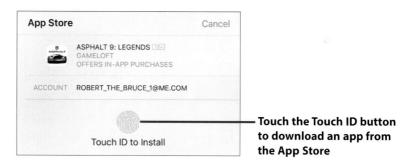

Touch the Touch ID button to download an app from the App Store

For example, to download an app using Touch ID, move to the app's screen in the App Store app and tap Get (or the price if there is a license fee). At the prompt, touch the Touch ID/Home button. When your fingerprint is recognized, the app is downloaded and installed.

Using Touch ID in other situations is similar. When you see the Touch ID prompt, simply touch the Touch ID/Home button.

The First Time, Part 2

The first time you sign into an app, such as a banking app, that supports Touch ID, you're prompted to allow that app to use your Touch ID to enter your password. If you allow it, you can sign into the app by touching the Touch ID/Home button. If you don't allow it, you need to manually enter the password to be able to access your account.

Sleeping/Locking and Waking/Unlocking Your iPhone

As you learned in Chapter 1, you can manually lock your iPhone, which also puts it to sleep. It's a good idea to do this when you aren't using your phone to both save battery power and secure your information. (You can configure this to happen automatically as explained in Chapter 7.)

To put your iPhone to sleep and lock it, press the Side button (the Sleep/Wake button on iPhone SE models). The screen goes dark and the phone locks.

This iPhone is awake and locked

To wake up and unlock the phone, press and hold on the Touch ID/Home button for a second or so. The phone wakes up and unlocks and you move to the screen you were most recently using. For example, if you were using Podcasts when the phone locked, you return to the Podcasts app when the phone unlocks.

If you don't use your iPhone for a while, it automatically goes to sleep and locks according to the preference you have set for it (this is covered in Chapter 7).

No Touch ID?

If you don't have Touch ID configured, you can wake your phone in the same ways. However, your phone remains locked. To use it, press the Touch ID/Home button. You're prompted to enter your passcode. If you enter it successfully, you move to the screen you were most recently using.

No Passcode?

If you don't have a passcode configured, which is not recommended, you can wake your phone by pressing the Touch ID/Home button to move to the screen you were most recently using.

>>>*Go Further*

THAT'S QUITE A BUTTON

Following are a number of ways to use the Touch ID/Home button:

- **Wake and Unlock**—Press and hold the button to wake and unlock your phone.
- **Go Home**—When the phone is unlocked, pressing this button takes you to the Home screen.
- **Open the App Switcher**—When the phone is unlocked, press the button twice to open the App Switcher.
- **Take a Screenshot**—Press the Side button and the Touch ID/Home button at the same time to take a screenshot.
- **Magnify**—Press the button three times to open the Magnifier. (This has to be enabled via the Accessibility settings as explained in Chapter 6, "Making Your iPhone Work for You.")

Turning Your iPhone Off or On

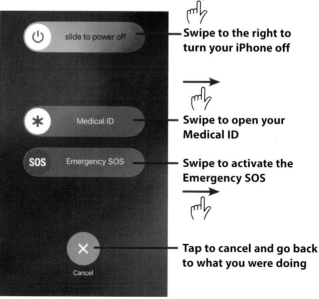

Swipe to the right to turn your iPhone off

Swipe to open your Medical ID

Swipe to activate the Emergency SOS

Tap to cancel and go back to what you were doing

You seldom need to turn your iPhone off, but you might want to shut if off if you aren't going to be using it for a while or for troubleshooting purposes.

To turn your iPhone off, press and hold the Side button (Sleep/Wake button on an SE) and either Volume button. Sliders appear on the screen; swipe the top slider to the right. The phone turns off.

To restart your iPhone, press and hold the Side button (Sleep/Wake button on an SE) until the Apple logo appears on the screen, and then let go of the button.

After it starts up, you see the Lock screen. To use the phone, press the Touch ID/ Home button. Assuming you have a passcode, you see the Enter Passcode screen. Enter your passcode to start using your phone; once your passcode is entered correctly, you move to the Home screen. (Even if you have Touch ID enabled to unlock your phone, you must enter your passcode the first time you unlock it after a restart.)

Using the Split-Screen on iPhone 12 Pro Max, iPhone 11 Pro Max, Xs Max, or Plus Models

The largest models of iPhone, including the 6s Plus, 7 Plus, 8 Plus, Xs Max, 11 Pro Max, and iPhone 12 Pro Max offer the split-screen feature in many apps (although not all apps support this function). To activate split-screen, rotate your iPhone so it's in the landscape orientation. When you're using an app that supports this feature, the screen splits automatically.

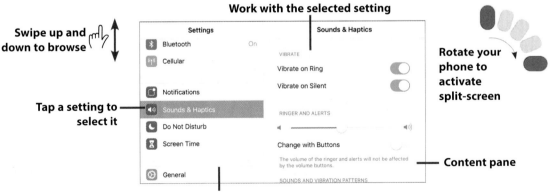

In split-screen mode, the screen has two panes. The left pane is for navigation, whereas the right pane shows the content selected in the left pane. The two panes are independent, so you can swipe up and down on one side without affecting the other.

You can browse the left pane to find what you want to work with. When you select that, its content appears in the right pane. For example, in the Settings app, you select a setting in the left pane and can use its controls in the right pane just as if you were using the phone in the single screen mode.

Swipe up and down to browse

Tap a different setting to work with it

Work with the selected setting

Settings		General	
🌙	Do Not Disturb		
⏳	Screen Time	About	>
		Software Update	
⚙️	General		
🎛	Control Center	AirDrop	>
AA	Display & Brightness	Handoff	>
♿	Accessibility	CarPlay	>
🌼	Wallpaper		
🔍	Siri & Search	Home Button	>

Split-screen makes it very easy to navigate because you can more easily and quickly jump to different areas by using the left pane (as opposed to having to tap back (<) one or more times to change what you want to work with).

In some apps that support this functionality, there is an icon you can use to open or close the split screen. This icon changes depending on the app you're using. For example, in the Mail app, you tap the Full Screen icon (two arrows pointing diagonally away from each other) to open or close the left pane.

Preinstalled apps that support this functionality include Settings, Mail, and Messages. Try rotating your phone with the apps you use to see if they support split-screen.

Dock

Home screen pages

When you hold a model that supports split-screen horizontally and move to the Home screen, the Dock moves to the right side of the screen, and you see the Home screen's pages in the left part of the window. Though this looks a bit different, it works the same as when you hold an iPhone vertically.

Connect to the Internet via
Wi-Fi or a cellular network

Use the iPhone's
great text tools in
many apps

Take advantage of an
Internet connection
wherever you are

Print email and other documents
from your phone

In this chapter, you learn to use some of your iPhone's core features. Topics include the following:

→ Getting started
→ Connecting to the Internet using Wi-Fi networks
→ Connecting to the Internet using cellular data networks
→ Securing your iPhone
→ Working with text
→ Searching on your iPhone
→ Working with Siri Suggestions
→ Printing from your iPhone

Using Your iPhone's Core Features

In this chapter, you learn fundamental skills that apply across multiple apps and functions of your phone. For example, when you are using your iPhone, you want to be connected to the Internet most of the time and you frequently work with text.

Getting Started

Here are the core features and concepts you learn about in this chapter:

- **The Internet**—Your iPhone has many functions that rely on an Internet connection; most of the apps you use either require or can use a connection to the Internet to do what they do for you. For example, to send and receive email, your iPhone has to be connected to the Internet.

- **Wi-Fi**—Wi-Fi stands for Wireless Fidelity and encompasses a whole slew of technical specifications around connecting devices together without using cables or wires. The most important thing to know is that you can use Wi-Fi networks to connect your iPhone to the Internet. This is great because Wi-Fi networks are available in many places you go. You probably also have a Wi-Fi network available in your home. (If you connect your computers to the Internet without a

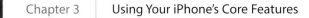

cable from your computer to a modem or network hub, you are using a Wi-Fi network.)

- **Cellular data networks**—In addition to connecting to Wi-Fi networks, your iPhone can transmit and receive data over the cellular network to which it is connected. This enables you to connect your iPhone to the Internet just about anywhere you are. You use the cellular network provided by your cell phone company. There are many different cell phone providers that support iPhones. In the United States, these include AT&T, Sprint, T-Mobile, and Verizon. You don't need to configure your iPhone to use the cellular data network, as it is set up from the start to do so.

- **Security**—Connecting your iPhone to the Internet enables you to do lots of useful, and sometimes amazing, things. But that connection comes with some risk because of the sensitive information you store on your iPhone and the tasks you perform with it. The good news is that you can protect your information with a few relatively simple precautions.

- **Text**—You enter text on your iPhone for many different purposes, including sending messages and emails and writing notes. You can type text using the iPhone's onscreen keyboards. You can also dictate text wherever you might need to create it. The iPhone has many features to help you make the text you enter "just right." For example, text is automatically checked for correct spelling, and the Predictive Text feature suggests text you might want to enter with just a tap.

- **Search**—Your iPhone has a lot of information on it. This includes apps, emails, music, and much more. The iPhone's Search tool enables you to find what you want to work with quickly and easily.

- **Siri Suggestions**—You frequently want to "go back" to something you were using recently, such as an app or a search. The Siri Suggestions tool presents these recent items so that you can return to them with a single tap. Siri can also make recommendations based on your location, such as presenting a store's app when you are close to that store. Siri also can learn from what you do and make suggestions about what you might find useful; for example, when you correct a text message that you've dictated, Siri can make suggestions about what you might have intended to say.

- **Print**—The paperless world has never become a reality—and probably never will. Fortunately, you can print emails, documents, and other content directly from your iPhone.

Connecting to the Internet Using Wi-Fi Networks

Your iPhone is designed to seamlessly connect to the Internet so apps that use the Internet to work, such as Safari to browse the Web, are always ready when you need them. Wi-Fi networks provide fast Internet connections and you usually have an unlimited amount of data to work with, so you don't have to worry about paying more based on how you are using your iPhone. Because of their speed and unlimited data (usually), Wi-Fi networks are the best way for your iPhone to connect to the Internet.

Wi-Fi networks are available just about everywhere you go, including homes, offices, hotels, restaurants, airports, stores, and other locations. Fortunately, it's very easy to connect your iPhone to the Wi-Fi networks you encounter. (And, if there isn't a Wi-Fi network available, your iPhone uses its cellular data network to connect to the Internet, which is covered later in this chapter.)

Almost all Wi-Fi networks broadcast their identifying information so that you can easily see them with your iPhone; these are called *open networks*. Anyone who is in range can attempt to join known networks because they appear on Wi-Fi devices automatically. The Wi-Fi networks you can see on your iPhone in public places (such as airports and hotels) are all open. Likewise, any Wi-Fi networks in your home or office are very likely to be open as well. To connect your iPhone to an open network, you tap its name and then enter its password (if required).

Your iPhone remembers Wi-Fi networks you've connected to previously and joins one of them automatically when available; these are called *known networks*. For example, if you have a Wi-Fi network at home and another in your office, when you change locations, your iPhone automatically changes Wi-Fi networks.

If your iPhone can't connect to a known network, it automatically searches for other Wi-Fi networks to join. If one or more are available, a prompt appears showing the networks available to your iPhone. You can select and join one of these networks by tapping its name on the list of networks and entering its password. (If a password is required, you need to obtain it from the source of the network, such as a hotel or restaurant, to be able to join that network.)

If no Wi-Fi networks are available or you choose not to connect to one, your iPhone automatically switches to its cellular data connection (covered in "Connecting to the Internet Using Cellular Data Networks" later in this chapter).

Connecting to Open Wi-Fi Networks

To connect your iPhone to a Wi-Fi network, perform the following steps:

1. On the Home screen, tap Settings. Next to Wi-Fi, you see the status of your Wi-Fi connection. It is Off if Wi-Fi is turned off, Not Connected if Wi-Fi is turned on and your phone isn't currently connected to Wi-Fi, or the name of the Wi-Fi network to which your iPhone is connected.

2. Tap Wi-Fi.

3. If Wi-Fi isn't enabled already, slide the Wi-Fi switch to on (green) and your iPhone searches for available networks. A list of available networks is displayed in the NETWORKS section (it can take a moment for all the networks available in an area to be shown). Along with each network's name, icons indicating whether it requires a password (the padlock icon) to join and the current signal strength (the radio waves icon) are displayed. As long as you see at least two "waves," the signal should be strong enough to use. You want to choose the strongest signal available because that usually means you get the best speed.

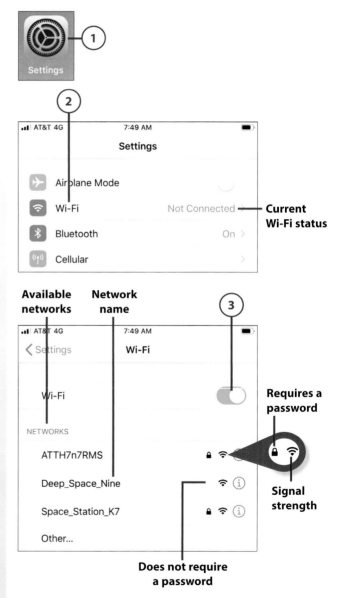

4 Tap the network you want to join. Of course, when a network requires a password, you must know that password to be able to join it.

5 At the prompt, enter the password for the network you selected. If you aren't prompted for a password, skip to step 7. (You're likely to find networks that don't require a password in public places; see the next section for information on these types of networks.)

6 Tap Join. If you provided the correct password, your iPhone connects to the network and gets the information it needs to connect to the Internet. If not, you're prompted to enter the password again. After you successfully connect to the network, you return to the Wi-Fi screen.

7 Review the network information. The network to which you are connected appears just below the Wi-Fi switch and is marked with a check mark. You also see the signal strength for that network.

.ıl AT&T 4G   7:49 AM

‹ Settings   **Wi-Fi**

Wi-Fi

NETWORKS

ATTH7n7RMS

Deep_Space_Nine

Space_Station_K7

Other...

4

6

Enter the password for "Space_Station_K7"

Cancel   **Enter Password**   Join

Password   ●●●●●●●●|

You can also access this Wi-Fi network by bringing your iPhone near any iPhone, iPad, or Mac which has connected to this network and has you in their contacts.

5

7 **The network your iPhone is using**

The current network's signal strength

✓ Sp

7:50 AM

gs   **Wi-Fi**

Wi-Fi

✓ Space_Station_K7

NETWORKS

ATTH7n7RMS

Deep_Space_Nine

Other...

Ask to Join Networks   Notify ›

8) Try to move to a web page, such as www.bradmiser.com, to test your Wi-Fi connection. (See Chapter 14, "Surfing the Web," for details.) If the web page opens, you're ready to use the Internet on your phone. If you go to a log-in web page for a Wi-Fi provider rather than the page you were trying to access, see the next task. If you see a message saying the Internet is not available, there's a problem with the network you joined. Go back to step 4 to select a different network.

9) Move back to the Wi-Fi Settings page.

10) Continue to add Wi-Fi networks that you use in various locations. Your iPhone remembers these networks (they become known networks) and connects to them automatically; these are shown in the MY NETWORKS section of the screen. Available networks that you haven't used before continue to be shown in the NETWORKS section.

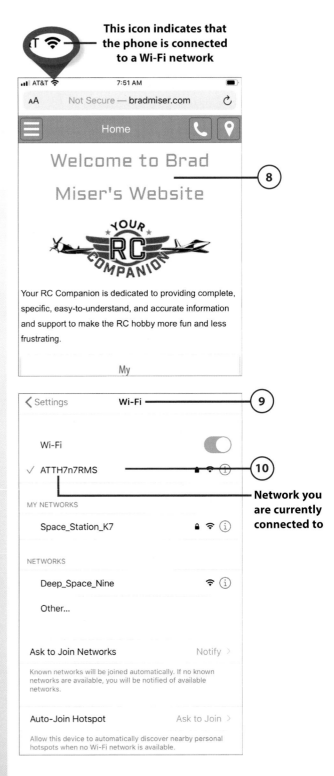

This icon indicates that the phone is connected to a Wi-Fi network

Network you are currently connected to

(11) To change how your iPhone works when it can't find a known network, tap Ask to Join Networks and select one of the three options. Off means your iPhone won't take any action when it encounters a network you haven't used. Notify means your iPhone sees that a new network is available, but you need to manually select that network to sign in. Ask pops up a dialog inviting you to join the found network. If you frequently get interrupted by network prompts, set this to Off. You can always use steps 4 through 6 to join available networks.

(12) If you want to use available hotspots, tap Auto-Join Hotspot (see the following Go Further sidebar for an explanation of hotspots). Choose one of the three options. Choose Never if you don't want to use hotspots. Choose Ask to Join if you want to be prompted when your iPhone encounters a hotspot. Or choose Automatic if you want to automatically join networks provided by a known hotspot.

< Settings   Wi-Fi

Wi-Fi

✓ ATTH7n7RMS   🔒 📶 ⓘ

MY NETWORKS

Space_Station_K7 ——————————— 🔒 📶 ⓘ **— Networks you connect to automatically**

NETWORKS

Deep_Space_Nine   📶 ⓘ

Other...

Ask to Join Networks ————————— Notify › **(11)**

Known networks will be joined automatically. If no known networks are available, you will be notified of available networks.

Auto-Join Hotspot ——————————— Ask to Join › **(12)**

Allow this device to automatically discover nearby personal hotspots when no Wi-Fi network is available.

Available networks that you haven't connected to

>>>Go Further
CONNECTING TO WI-FI NETWORKS

As you connect to Wi-Fi networks, consider the following:

- **Quick Access to Wi-Fi Controls**—You can quickly turn Wi-Fi on or off using the Control Center, which you can open by swiping down from the upper-right corner of the screen (iPhones without a Home button) or swiping up from the bottom of the screen (iPhones with a Home button). If the Wi-Fi icon (it looks like the signal strength indicator on the Wi-Fi Settings screen) is blue, Wi-Fi is on. Tap that icon to turn Wi-Fi off (the icon becomes white). Tap it again to turn Wi-Fi on and reconnect to a known network. Tap and hold on the icon to see another palette of icons. When you touch and hold on the Wi-Fi icon on that palette, you can quickly choose from available networks to join one of them. See Chapter 6, "Making Your iPhone Work for You," for more information about working with the Control Center.

- **Typing passwords**—As you type a password, each character is shown for a moment and then hidden by a dot. Keep an eye on characters as you enter them so you can fix a mistake as soon as you make it rather than finding out after you've entered the entire password and having to start over.

- **Changing networks**—You can use these same steps to change the Wi-Fi network you are using at any time. For example, if you have to pay to use one network while a different one is free, simply choose the free network in step 4. Another reason you might want to change Wi-Fi networks is if the signal strength on another available network is stronger.

- **Security recommendation**—If you are connected to a network that doesn't use what Apple considers sufficient security, you see the words "Security Recommendation" under the network's name. If you tap Info (i) for that network, you see its Info screen. At the top of that screen, you see the type of security the network is using and a recommendation about the type of security it should use. If the Wi-Fi network comes from a router or modem you own or rent, contact your Internet service provider, such as a cable company, to learn how the security provided by that router or modem can be reconfigured to be more secure. If the network is in a public place or business, you just have to use it as is (unless you can contact the administrator of that network to see if better security is available).

- **Privacy Warning**—If you are connected to a network that doesn't have sufficient privacy protection, you see the words "Privacy Warning" under the network's name. Tap Info (i) for that network to move to its Info screen. At the top of that screen, you see information about Private Wi-Fi addressing being turned off. Set the Private Address switch to on (green). Tap Disconnect to temporarily disconnect your phone from the network. Your phone then reconnects to the network. When it does, it uses a private address, which makes the phone harder to track. (I recommend you turn Private Address on whenever you have the opportunity to do so. This is especially important when you are using networks in public places.) When you return to the Wi-Fi screen, the Privacy Warning message is gone, indicating you are using a private address.

- **Have a network, but no Internet**—If you successfully connect to a network, but there is an exclamation point on top of the signal strength icon, the network you are connected to might not have a current Internet connection. Sometimes, that's because you need to provide additional information to reach the Internet (as described in the next section). At other times, it's because the network has lost its connection to the Internet. You'll need to get that connection restored (such as by resetting the modem or contacting your provider) before you can use that network to connect to the Internet. (If you are working with a network inside your home or business that you control, try resetting the modem, which usually involves unplugging the modem, waiting for about 30 seconds, and plugging it in again. This often solves the issue and should be the first thing you try, even before contacting your provider.)

- **Hotspots**— iPhones and iPads can share their cellular Internet connection with other devices by providing a Wi-Fi network to which you can connect your iPhone—the device providing a network is called a hotspot. The icons for hotspot networks consist of two connected loops rather than the wave icon that indicates a "regular" Wi-Fi network. You can select and use hotspot networks just like the other types of networks being described in this chapter. The speed of your access is determined mostly by the speed of the device's cellular data connection. Also, the data you use while connected to the hotspot's network counts against the data plan for the device to which you are connected.

You can configure your iPhone to be a hotspot. The details of configuring a hotspot and how much it costs depend on your cellular provider. To get started, contact your provider to get specific instructions to set up your phone as a hotspot and to understand any limitations or applicable fees. After your phone is a hotspot, you can connect other devices to its network using its password (which is automatically generated).

Connecting to Public Wi-Fi Networks

Many Wi-Fi networks in public places, such as hotels or airports, require that you pay a fee or provide information to access the Internet through that network; even if access is free, you usually have to accept terms and conditions for the network to be able to use it.

When you connect to one of these public networks, you're prompted to provide whatever information is required. This can involve different details for different networks, but the general steps are the same. Connect to the public network and follow the instructions that appear.

Better Safe Than Sorry

Many public Wi-Fi networks have limited or no security. This means the information being transmitted from your iPhone to the Internet (and vice versa) is susceptible to being intercepted by hackers and others who are looking for personal information. It's best practice not to use these networks for sensitive information, such as to access your bank account, to shop, or in other areas where you provide sensitive information (including user names, account numbers, or passwords). Public Wi-Fi networks are perfectly fine for browsing the Web, email, and other such activities.

Make It Private

If you see a Privacy Warning when you connect to a public network, refer to the bullet "Privacy Warning" in the previous section.

Following are the general steps to connect to many types of public Wi-Fi networks:

(1) Move to the Settings screen by tapping the Settings app on the Home screen.

(2) Tap Wi-Fi.

(3) Tap the network you want to join. You move to that network's Log In screen. Follow the onscreen prompts to complete the process. This often involves accepting terms of use for that network as this example shows.

Settings	——————————————— (1)
⬛ Airplane Mode	
🛜 Wi-Fi	——— Not Connected > — (2)

‹ Settings	**Wi-Fi**	
Wi-Fi		⬤
✓ IUH Public		🛜 ⓘ
Unsecured Network		
NETWORKS		
DIRECT-0d-HP M426 LaserJet	🔒 🛜 ⓘ	
OA Guest	🛜 ⓘ	
Otolaryngology Associates	🔒 🛜 ⓘ	——— (3)

(4) Swipe up the screen to read the terms and conditions (or browse them).

Your Experience Might Vary

You often select from different connection options, such as a free low-speed or a high-speed connection, and provide additional information before you can perform step 5. For example, when connecting from a hotel room, you probably need to provide a last name and room number. If a fee is required, you have to provide payment information. (If you're in a hotel, the fee is added to your room charges.) In most cases, you at least have to indicate that you accept the terms and conditions for using the network, which you typically do by checking a check box or reading the terms as in this example.

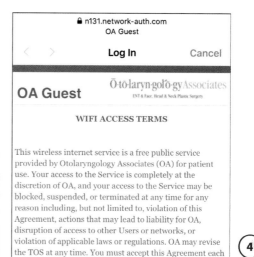

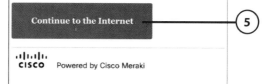

(5) Tap the icon to join the network. This icon can have different labels depending on the type of access, such as Connect, Authenticate, Done, Free Access, Login, and so on.

(6) Tap Done (if required).

(7) Try to move to a web page, such as www.wikipedia.org, to test your Wi-Fi connection (not shown). (See Chapter 14 for details.) If the web page opens, you're ready to use the Internet on your phone. If you go to a log-in web page for the Wi-Fi network's provider, you need to provide the required information to be able to use the Internet. For example, when access is free, as it is at most airports, you usually just have to accept the terms of use.

>>>Go Further
MORE ON CONNECTING TO THE NET

As you use the Internet through different networks, following are a few points to keep in mind:

- **No Prompt**—Not all public networks prompt you to log in as these steps explain. Sometimes, you use the network's website to log in instead. After you join the network (step 3 in the prior task), your iPhone is connected to the network without any prompts. When you try to move to a web page as explained in step 7, you're prompted to log in to or create an account with the network's provider on the web page that appears.

- **Forget a network**—As you learned earlier, your iPhone remembers networks you have joined and connects to them automatically as needed; these are known networks. For one reason or another, you might want to stop using that network. You can make your iPhone "forget" a known network so that it won't keep connecting to it automatically.

  To do this, tap Info (i) for the network you want to forget. Tap Forget This Network and then tap Forget at the prompt. Your iPhone stops using and forgets the network.

  You can rejoin a forgotten network at any time, just as you did the first time you connected to it.

- **Auto-Join**—On the Info screen for a network, which you open by tapping (i) for that network, you see the Auto-Join switch. If you want to stop automatically joining the network but keep its password on your iPhone, set the Auto-Join switch to off (white). Your iPhone stops automatically connecting to that network, but you can rejoin it at any time by tapping it; you don't have to re-enter the password as you do if you forget a network.

Connecting to the Internet Using Cellular Data Networks

When you don't have a Wi-Fi network available or you don't want to use one that is available (such as if it has a fee or is slow), your iPhone can connect to the Internet through a cellular data network.

The provider for your iPhone provides a cellular data connection through which your iPhone connects to the Internet automatically when you aren't using a Wi-Fi network (such as when you are in a location that doesn't have one). (Your iPhone first tries to connect to an available Wi-Fi network before using a cellular data connection because Wi-Fi is typically less expensive and faster to use.) These cellular networks cover large geographic areas and the connection to them is automatic; your iPhone chooses and connects to the best cellular network currently available. Access to these networks is part of your monthly account fee; you choose from among various amounts of data (ideally, you can choose an account with unlimited data) per month for different monthly fees.

Most providers have multiple cellular data networks, such as a low-speed network that is available widely and one or more higher-speed networks that have a more limited coverage area.

The cellular data networks you can use are determined based on your provider, your data plan, the model of iPhone, and your location within your provider's networks or the roaming networks available (when you're outside of your provider's coverage area). Your iPhone automatically uses the fastest connection available to it at any given time (assuming you haven't disabled that option, as explained later).

In the United States, the major iPhone providers are AT&T, Sprint, T-Mobile, and Verizon. There are also other smaller providers, such as Virgin Mobile. All these companies offer high-speed 5GE or Long Term Evolution (LTE) cellular networks along with the slower 4G and 3G networks. In other locations, the names and speeds of the networks available might be different.

The following information is focused on 5GE networks because I happen to live in the United States and use AT&T as my cell phone provider. If you use another provider, you're able to access your provider's networks similarly, though some details might be different. For example, the icon on the Home screen reflects the name of your provider's network, which might or might not be 5GE.

This iPhone is connected to a high-speed 5GE cellular network

5GE, 5G, LTE, or other high-speed wireless networks provide very fast Internet access from many locations. (Note: These networks might not be available every-where, but you can usually access them near populated areas.) To connect to the cellular network, you don't need to do anything. If you aren't connected to a Wi-Fi network, you haven't turned off the high-speed network, and your iPhone isn't in Airplane mode, the iPhone automatically connects to the fastest network available. If you can't access the fastest network, such as when you aren't in its coverage area, the iPhone automatically connects to the next fastest network available, such as 4G. If that isn't available, it connects to the next fastest and so on until it finds a network to which it can connect if there is one available. If it can't connect to any network, you see No Service instead of a network's name; this indicates that you currently can't connect to any network, and so you aren't able to access the Internet.

One thing to keep in mind when using a cellular network is that your cell phone plan might include a limited amount of data per month. When your data use exceeds this limit, you might be charged overage fees, which can be very expensive. Most providers send you warning texts or emails as your data use approaches your plan's limit, at which point you need to be careful about what you do while using the cellular data network to avoid an overage fee. Some tasks, such as watching YouTube videos or downloading large movie files, can chew up a lot of data very quickly and should be saved for when you are on a Wi-Fi net-work to avoid exceeding your plan's monthly data allowance. Other tasks, such as using email, typically don't use very much data.

Fortunately, most of the major providers now offer unlimited data plans for a reasonable fee. If you don't already have an account with unlimited data, check with your provider periodically to see if an unlimited data plan is available. As competition has increased among cell providers, unlimited data plans have become more common and less expensive in many areas. If other cell providers are available to you, check to see whether they offer unlimited data plans; if so, you can consider changing providers or using a competitor's plan to get your provider to lower the cost of your plan. Having an unlimited data plan is good because you don't need to worry about overage charges from using more data than your plan allows.

When you move outside your primary network's geographic coverage area, you are in roaming territory, which means a different provider might provide cellular phone or data access, or both. The iPhone automatically selects a roaming provider, if there is only one available, or allows you to choose one, if there is more than one available.

When you're outside of your primary provider's coverage area, roaming charges can be associated with calls or data use. These charges are often very expensive. The roaming charges associated with phone calls are easier to manage because it's more obvious when you make or receive a phone call in a roaming area. However, data roaming charges are much more insidious, especially if Push functionality (where emails and other updates are pushed to your iPhone from the server auto-matically) is active. And when you use some applications, such as Maps to navigate, you don't really know how much data is involved. Because data roaming charges are harder to notice, the iPhone is configured by default to prevent data roaming. When data roaming is disabled, the iPhone is unable to access the Internet when you are outside of your cellular network, unless you connect to a Wi-Fi network. (You can still use the cellular roaming network for telephone calls.)

You can configure some aspects of how your cellular network is used, as the fol-lowing task demonstrates. You can also allow individual apps to use, or prevent them from using, your cellular data network. This is especially important when your data plan has a monthly limit. (If you have an unlimited plan, you usually don't need to bother.)

In most cases, the first time you launch an app, you're prompted to allow or prevent it from using cellular data. At any time, you can use the Cellular Data options in the Settings app to enable or disable an app's access to your cellular data network.

The options you have for configuring how your iPhone uses its cellular data connection depend on the provider your iPhone is connected to and the model of iPhone you use. For example, if you live in the United States and use Sprint as your cellular provider, the Cellular screens in the Settings app look a bit different than the figures in this section (which are based on AT&T's service). Regardless of the specific options you see on your phone, the basic purpose is the same, which is to configure how your iPhone uses its high-speed network and to enable and disable roaming.

Configuring Cellular Data Use

The following steps show configuring cellular data use on an iPhone using AT&T in the United States; you can use similar steps to configure these options on an iPhone from a different provider:

1 Open the Settings app.

2 Tap Cellular.

Your Terminology Might Vary

The labels you see on the Settings screens depend on the region you are in. For example, in the United States, you see the Cellular label, whereas in the UK, the label is Mobile. Although the labels might be different, the selection process is the same.

---

3 To use a cellular Internet connection, set the Cellular Data switch to on (green) and move to step 4; if you don't want to use a cellular Internet connection, set this switch to off (white) and skip the rest of these steps. To use the Internet when the Cellular Data switch is off, you have to connect to a Wi-Fi network that provides Internet access.

4 Tap Cellular Data Options.

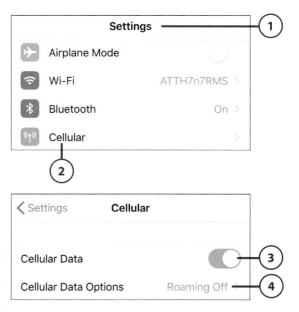

5 If you want to allow data roaming, slide the Data Roaming switch to the on (green) position. With some providers, Roaming is an option instead of a switch; tap Roaming and use the resulting switches to enable or disable roaming for voice or data and then tap Back (<). You should usually leave Data Roaming off so that you don't unknowingly start using roaming (which can lead to high fees) should you be moving around a lot. You can then enable it as needed so you know exactly when roaming is on.

6 To configure the high-speed network, tap Voice & Data.

7 Ensure *high-speed network* (where *high-speed network* is the name of your provider's high-speed network) is selected (has the check mark); if you want to limit your connection to slower speeds, choose one of the other options instead.

8 Tap Back (<).

9 If you have a limited data plan, consider enabling the Low Data Mode switch (green). This can cause specific apps to use less data. If you have an unlimited plan, you should leave this disabled (white).

10 Tap Cellular (<) to go back to the previous screen.

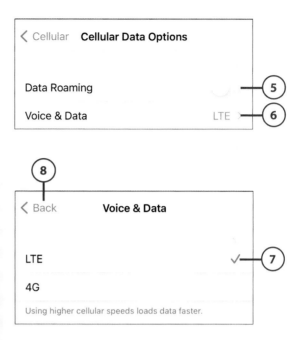

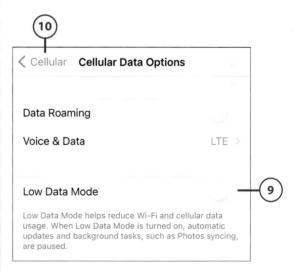

(11) Use the controls in the *PROVIDER* section, where *PROVIDER* is the name of your provider, to configure how the cellular service interacts with other services, such as to enable Wi-Fi calling, calls on other devices, and so on. Some of the options are explained in "More On Cellular Data" at the end of this section.

(12) Swipe up the screen until you see apps and services. This section enables you to allow or prevent individual apps from accessing a cellular data network. To limit the amount of data you use, it's a good idea to review this list and allow only those apps that you rely on to use the cellular data network. (Of course, if you're fortunate enough to have an unlimited cellular data plan, you can leave cellular data for all the apps enabled.) This list can be quite long if you have a lot of apps stored on your iPhone.

(13) Set an app's switch to on (green) if you want it to be able to use a cellular data network to access the Internet.

(14) Set an app's switch to off (white) if you want it to be able to access the Internet only when you're connected to a Wi-Fi network.

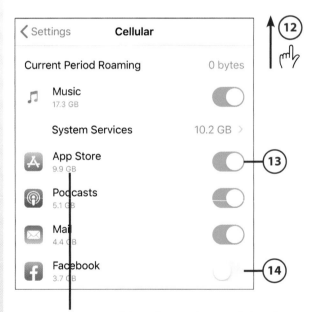

Amount of data the app has used since last reset

15 Swipe up until you reach the bottom of the screen.

16 Set the Wi-Fi Assist switch to on (green) if you want your iPhone to automatically switch to its cellular connection when the Wi-Fi connection is weak. If you have a limited cellular data plan, you might want to set this switch to off (white) to minimize cellular data use. If you have an unlimited plan, you should leave this on.

‹ Settings	**Cellular**	
	Voice Memos	⬤
✳	Walmart	⬤
◉	Wi-Fi SweetSpots	⬤
📷	WISH-TV	⬤
	Wi-Fi Assist 23.2 MB	⬤
	Automatically use cellular data when Wi-Fi connectivity is poor.	

>>>Go Further
MORE ON CELLULAR DATA

Using a cellular network to connect to the Internet means you seldom have to be without a connection unless you choose to be. Here are some things to keep in mind as you keep connected:

- **Wi-Fi Calling**—When this option is enabled, calls you make use a Wi-Fi network, which can improve the reception in certain locations. When you enable this option, you are prompted to configure an address associated with the number in case you need to call for emergency services. You also have the option to enable Wi-Fi calling for other devices.

- **Calls on Other Devices**—When enabled, you use other devices, such as an iPad or a Mac computer, to receive or place calls. When you turn the switch on, you're prompted to enable or disable this feature for each device. Be aware that the devices have to be on the same Wi-Fi network as your iPhone.

- **Control from the Control Center**— You can quickly turn cellular data on or off from the Control Center. For example, if you want to stop using cellular data (perhaps you're reaching the cap of your cellular data plan), you can open the Control Center and tap the Cellular Data icon to disable cellular data use. Tap the icon again to enable it. (When the icon is green, cellular data is enabled.)

- **Unlimited Data but Limited Speed**—Under an unlimited data plan, some providers limit the speed at which your cellular data service operates if you pass a threshold amount of data used that month. You can continue to use all the data you want, but the performance of the connection might be slower. Check your plan's details to see if it has one of these limits, and if so, how much speed reduction is applied. You probably won't hit that threshold unless you watch a lot of video on your phone, but it's good to be aware of such limitations on your account.

- **GSM versus CDMA**—There are two fundamental types of cellular networks, which are GSM (Global System for Mobile Communications) or CDMA (Code-Division Multiple Access). The cellular provider you use determines which type of network your iPhone uses; the two types are not compatible. GSM is used by most of the world whereas some very large carriers in the United States (such as Verizon and Sprint) use CDMA. There are differences between the two, which is why you might see different cellular options than shown in the figures here. (They show the options for AT&T, which uses GSM.)

- **Apps' cellular data use**—Just under each app's name in the CELLULAR DATA section, you see how much data the app has used since the counter was reset. This number can help you determine how much data a particular app uses. For example, if an app's use is shown in megabytes (MB), it's used a lot more data than an app whose use is shown in kilobytes (KB). If you have a limited data plan and you see an app uses a lot of data, you might want to disable its cellular data use.

- **Cellular data use reset**—You can reset the statistics on the Cellular screen by swiping up until you reach the bottom of the screen and see Reset Statistics. Here you can see the date of the last reset. To reset the statistics, tap Reset Statistics and then tap Reset Statistics again.

Securing Your iPhone

Even though you won't often be connecting a cable to it, an iPhone is a connected device, meaning that it sends information to and receives information from other devices, either directly or via the Internet, during many different activities. Some are obvious, such as sending text messages or browsing the Web, whereas others might not be so easy to spot, such as when an app is determining your iPhone's location. Whenever data is exchanged between your iPhone and other devices, there is always a chance your information will get intercepted by someone you didn't intend or that someone will access your iPhone without you knowing about it.

The good news is that with some simple precautions, the chances of someone obtaining your information or infiltrating your iPhone are quite small (much less than the chance of someone obtaining your credit card number when you use it in public places, for example). Following are some good ways to protect the information you're using on your iPhone:

- Always have a passcode on your iPhone so it can't be unlocked without entering the passcode. Configuring a passcode is explained in Chapter 2, "Getting Started with Your iPhone."

- Use Face ID (iPhones without a Home button) or Touch ID (iPhones with a Home button) to make entering your passcode and passwords much easier and more secure. Configuring Face ID and Touch ID also are explained in Chapter 2.

- Never let someone you don't know or trust use your iPhone, even if he needs it "just for a second to look something up." If you get a request like that, look up the information for the person and show him rather than letting him touch your iPhone.

- Learn how to use the Find My iPhone feature in case you lose or someone steals your iPhone. This is explained in Chapter 18, "Maintaining and Protecting Your iPhone and Solving Problems."

- Never click a link to verify your account in an email or text message that you aren't expecting. If you haven't requested some kind of change, such as signing up for a new service, virtually all such requests are scams, seeking to get your account information, such as username and password, or your identification, such as full name and Social Security number. And many of these scam attempts look like email from actual organizations. For example, I receive many of these emails that claim, and sometimes even look like, they're from Apple. However, Apple doesn't request updates to account information using a link in an email unless you've made some kind of change, such as registering a new email address for iMessages. Legitimate organizations never include links in an email to update account information when you haven't requested or made any changes.

  To reinforce this concept, there are two types of requests for verification you might receive via email. The legitimate type is sent to you after you sign up for a new service, such as creating a new account on a website, to confirm that the email address you provided is correct and that you are really you. If you make changes to an existing account, you might also receive confirmation request emails. You should respond to these requests to finish the configuration of your account.

  If you receive a request for account verification, but you haven't done anything with the company or organization from which you received the request, don't respond to it. For example, if you receive a request that appears to be from Apple, PayPal, or other organization, but you haven't made any changes to your account, the email request is bogus and is an attempt to scam you. Likewise, if you've never done anything with the organization apparently sending the email, it's also definitely an attempt to scam you.

If you have any doubt, contact the organization sending the request directly (not by responding to the email) before responding to the email.

- If you need to change or update account information, you can go directly to the related website using an address that you type in or have saved as a bookmark using the Safari app. This protects you because it ensures you can move directly to the legitimate website rather than clicking a link in an email that might take you to a fraudulent website.

- Be aware that when you use a Wi-Fi network in a public place, such as a coffee shop, hotel, or airport, there's a chance that the information you send over that network might be intercepted by others. The risk of this is usually quite small, but you need to be aware that there's always some level of risk. To have the lowest risk, don't use apps that involve sensitive information, such as online banking or shopping apps, when you're using a public Wi-Fi network.

- If you don't know how to do it, have someone who really knows what they're doing set up a wireless network in your home. Wireless networks need to be configured properly to be secure. Your home's Wi-Fi network should require a password to join.

- For the least risk, only use your home's Wi-Fi network (that has been configured properly) or your cellular data connection (you can turn Wi-Fi off when you aren't home) for sensitive transactions, such as shopping, accessing bank accounts, or accessing other financial information.

- Never accept a request to share information from someone you don't know. In Chapter 17, "Working with Other Useful iPhone Apps and Features," you learn about AirDrop, which enables you to easily share photos and lots of other things with other people using iOS or iPadOS devices. If you receive an AirDrop request from someone you don't recognize, always decline it. In fact, if you have any doubt, decline such requests. It's much easier for someone legitimate to confirm with you and resend a request than it is for you to recover from damage that can be done if you inadvertently accept a request from someone you don't know.

Passwords

Secure your information by using complex passwords. You can use apps, such as LastPass, or built-in capabilities, such as Safari's suggested passwords, to create passwords for you. Since passwords can be saved on your device, you don't have to type them so it doesn't matter how complicated they are. And don't use the same password for multiple accounts.

Reality Check

Internet security is a complex topic, and it can be concerning to think about. It's best to keep in mind the relative level of risk when you use your iPhone compared to other risks in the physical world that most of us don't think twice about. For example, every time you hand your credit card to someone, there is a chance that that person will record the number and use it without your knowledge or permission. Even when you swipe a credit card in a reader, such as at a gas station, that information is communicated across multiple networks and can be intercepted. (For example, there have been numerous compromises of credit card information at a number of well-known retailers.) If you take basic precautions like those described here, the risks to you when you're using your iPhone are similar to or less than the other risks we all face in everyday life.

My recommendation is to take the basic precautions, and then don't worry about it overly much. However, you might want to consider signing up with an identity theft protection company, which is a company that looks for possible threats against you like someone trying to get a new credit card using your data, and it alerts you to this happening. (Of course, this can happen whether you use an iPhone or don't use one.) Try to find an identity theft protection company that assigns someone to do the work of recovering for you should your identity be stolen because the recovery process can be very time-consuming and difficult. You can research available companies by performing a web search for "identity theft protection services." Be cautious, however; some of these companies charge high fees without providing recovery services, which is the most important part of the service.

Working with Text

You can do lots of things with an iPhone that require you to provide text input, such as writing emails, sending text messages, or taking notes. There are a couple of ways you can enter text, the most obvious of which is by typing. The iPhone's keyboard pops up automatically whenever you need it, whether it's for emailing, messaging, entering a website URL, performing a search, or in any other typing situation.

You can also dictate text, which can often be faster and easier than typing.

Typing Your Way

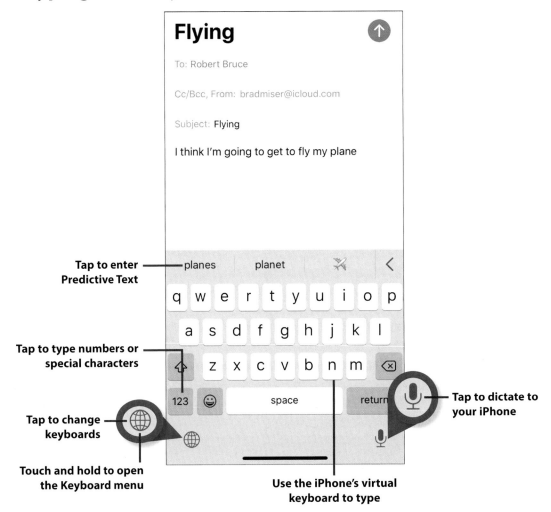

Tap to enter Predictive Text

Tap to type numbers or special characters

Tap to change keyboards

Touch and hold to open the Keyboard menu

Use the iPhone's virtual keyboard to type

Tap to dictate to your iPhone

To type, just tap the keys. As you tap each key, you hear audio feedback (you can disable this sound in the Settings app by going to Sounds then turning off the Keyboard Clicks switch) and the key you tapped pops up in a magnified view on the screen (this can also be disabled using the Settings app). The keyboard includes all the standard keys, plus a few for special uses. To change from letters to numbers and special characters, just tap the 123 key. Tap the #+= key to see more special characters. Tap the 123 key to move back to the numbers and special characters or the ABC key to return to letters. The keyboard also has contextual keys that appear when you need them. For example, when you enter a website address, the .com key appears so you can enter those four characters with a single tap.

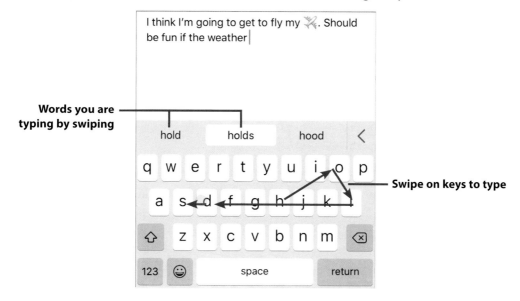

Words you are typing by swiping

Swipe on keys to type

You can also type by swiping on the keys without taking your finger off the screen. For example, to type "holds," touch "h" and then swipe to "o," "l," "d," and "s." Pause your finger when you get to the last letter in the word, and it's entered for you followed by a space. Swipe the letters of the next word, pause when you reach the end of it, and so on. As you swipe, an arrow traces your path on the keyboard so you can see what you're entering.

You'll probably either love swiping to type or hate it. Give it a good try to see which is the case for you. It can be a faster way to type, but it definitely requires some practice.

Keyboard Settings

You can change how you work with text through Keyboard Settings. These are explained in Chapter 6.

Working with Predictive Text

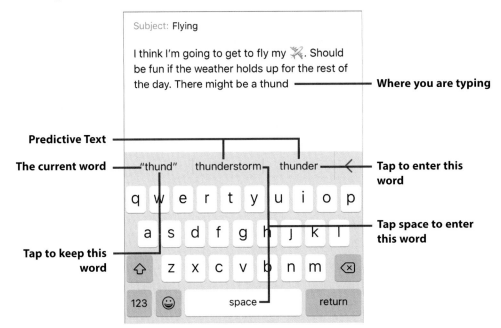

Where you are typing

Predictive Text

The current word

Tap to enter this word

Tap space to enter this word

Tap to keep this word

You can also use Predictive Text, which tries to predict text you want to enter based on the context of what you are currently typing and what you have typed before. The choices that Predictive Text provides appear in the bar between the text and the keyboard and present you with three or more options. The option on the far left in quotes is what you're currently typing. The other options are words or phrases that you might want to type based on context and what you've typed before. The second option from the left is the one that Predictive Text has selected as the most likely word. To the right of that you might see other text options or one or more emojis (more on these later).

Don't Like Predictive Text?

If you don't want to use the Predictive Text feature, open the Settings app, tap General, tap Keyboard, and then set the Predictive switch to off (white).

If you want to enter the most likely option (the second from the left), tap the space key; that word or phrase is entered followed by a space so you can keep typing. If the option to the right is what you want to enter, tap it to add it to the text at the current location of the cursor. To keep what you have currently typed, tap the word in quotes that appears on the left.

If you don't see an option you want to enter, keep typing; the options change as the text you are typing changes. You can tap an option on the Predictive Text bar at any time to enter it.

The nice thing about Predictive Text is that it gets better at predicting your text needs over time. In other words, the more you use it, the better it gets at predicting what you want to type. And it can even suggest phrases based on what you're typing; tap a phrase to enter it.

Predictive Text Need Not Apply

When you're entering text where Predictive Text doesn't apply, such as when you're typing email addresses, the Predictive Text bar is hidden. This makes sense because there's no way text in things such as email addresses can be predicted. When you move back into an area where it does apply, the Predictive Text bar appears again.

Using Other Writing Tools

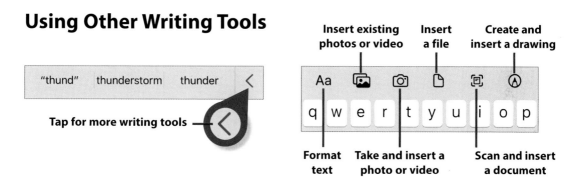

When you tap the left-facing arrow (<) on the end of the Predictive Text bar, you see a palette of tools you can use to format text and to add other types of content to it. For example, you can insert a photo you've taken or even take a new photo and insert it into the current location. These tools work similarly, and a quick example explaining how to format text will help you use the other options.

Position the cursor where you want to start formatting text as you type it. Then tap the left-facing arrow to open the tool palette. Tap Format text. The Formatting tools appear.

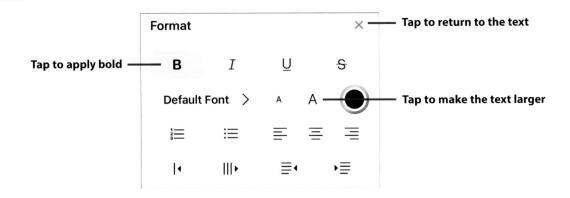

Tap the options you want to apply to the text you're going to type. For example, tap **B** to apply bold or the large A to increase the size of the text. When you're done selecting options, tap close (x).

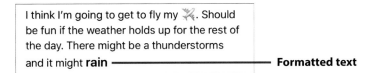

Continue typing. As you type, you see the current formatting options applied to the text. When you want to change the formatting, such as to go back to the prior format, open the Formatting tools and "undo" the selections you made earlier.

Later in this chapter, you learn how to select text to work with it. If you want to format a portion of your text, it's usually easier to apply formatting after you've typed it because you can quickly select the part that you want to format and apply the options only to that text . If you format as you type, you have to open the formatting tool, choose the formats you want to use, close the formatting tool, type the text you are formatting, and then reset the format tool to stop formatting the next text you type.

The other tools enable you to insert various types of content at the current position of your cursor. For example, in addition to using the Camera tool to take and insert a photo, you can use the Photo tool to insert a photo you already have on your iPhone or use the Drawing tool to insert drawings.

You probably aren't likely to be writing complex documents on your iPhone, but these additional tools are nice to have when you want to enhance what you write in emails, text messages, or notes.

Working with Keyboards

The great thing about a virtual keyboard like the iPhone's is that it can change to reflect the language or symbols you want to type. As you learn in Chapter 6, you can install multiple keyboards, such as one for your primary language and more for your secondary languages. You can also install third-party keyboards to take advantage of their features.

By default, two keyboards are available for you to use. One is for the primary language configured for your iPhone (for example, mine is U.S. English). The other is the Emoji keyboard (more on this shortly). How you change the keyboard depends on whether you've installed additional keyboards and the orientation of the iPhone.

If you haven't installed additional keyboards, you can change keyboards by tapping the Emoji key, which has a smiley face on it. You see the Emoji keyboard. Tap ABC to return to letters and numbers.

If you've installed other keyboards, you change keyboards by tapping the Globe key.

Each time you tap this key, the keyboard changes to be the next keyboard installed; along with the available keys changing, you briefly see the name of the current keyboard in the space key. When you have cycled through all the keyboards, you return to the one where you started.

Orient Yourself

As you type, try holding the phone in landscape orientation. The keyboard is much wider and can be easier to use, especially when you're typing a lot. The downside to this is that the area you are typing in is smaller. Try typing in both orientations to see which works best for you.

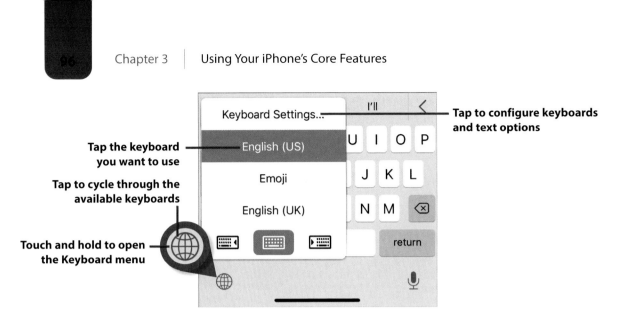

Tap to configure keyboards and text options

Tap the keyboard you want to use

Tap to cycle through the available keyboards

Touch and hold to open the Keyboard menu

You also can select the specific keyboard you want to use and access keyboard and text options by touching and holding on the Globe key (or the Emoji key if you don't see the Globe key). The Keyboard menu appears. Tap a keyboard to switch to it. Tap Keyboard Settings to jump to the Keyboards screen in the Settings app where you can configure keyboards and enable or disable text options (these settings are covered in Chapter 6).

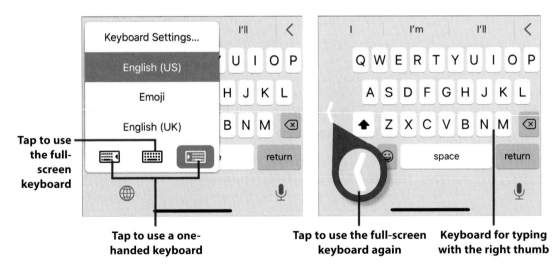

Tap to use the full-screen keyboard

Tap to use a one-handed keyboard

Tap to use the full-screen keyboard again

Keyboard for typing with the right thumb

Because you often type on your iPhone while you're moving around, it has a one-handed keyboard (this needs to be enabled via the Keyboard settings covered in Chapter 6). This keyboard "squishes" all the keys to the left or right side of the screen to suit typing with a thumb. To use a one-handed keyboard, touch and hold the Globe or Emoji key to open the Keyboard menu (this only works when the iPhone is held vertically). Tap the left or right keyboard; the keyboard compresses toward the side you selected and you can more easily tap its keys with one thumb. To return to the full-screen keyboard, tap the right- or left-facing arrow that appears in the "empty" space on the side of the screen not being used for the keyboard or open the Keyboard menu and tap the full-screen keyboard.

Using Emojis

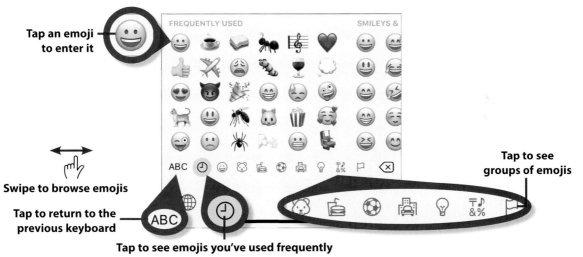

Tap an emoji to enter it

Swipe to browse emojis

Tap to return to the previous keyboard

Tap to see emojis you've used frequently

Tap to see groups of emojis

Emojis are icons you insert into your text to liven things up, communicate your feelings, or just have some fun. You can open the Emoji keyboard by tapping its key (the smiley face), by tapping the Globe until it appears, or by selecting it on the Keyboard menu. You see a palette containing many emojis, organized into groups. You can change the groups of emojis you're browsing by tapping the icons at the bottom of the screen. Swipe to the left or right on the emojis to browse the emojis in the current group. Tap an emoji to enter it at the cursor's location in your message, email, or other type of document. To use an emoji you've used often, tap the Clock icon to see emojis you've used frequently; you'll probably find that you use this set of emojis regularly so this can save a lot of time. To return to the mundane world of letters and symbols, tap the ABC key.

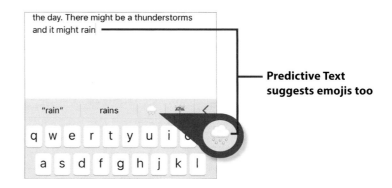

Predictive Text suggests emojis too

The Predictive Text feature also suggests one or more emojis when you type certain words; just tap an emoji before you tap the space key to replace the current word with it. Or, if you've already tapped the space key, tap the emoji to add it after the word.

Emoji Options

If you tap and hold on some emojis, you see options. For example, if you tap and hold on the thumbs-up emoji, you see a menu with the emoji in different flesh tones. Slide your finger over the menu and tap the version you want to use. The version you select becomes the new default for that emoji. You can go back to a previous version by opening the menu and selecting it.

Correcting Spelling as You Type

If you type a word that the iPhone doesn't recognize, that word is flagged as a possible mistake and suggestions are made to help you correct it. How this happens depends on whether Predictive Text is enabled.

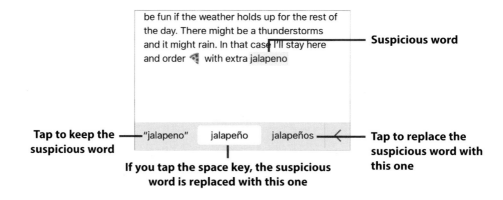

Suspicious word

Tap to keep the suspicious word

Tap to replace the suspicious word with this one

If you tap the space key, the suspicious word is replaced with this one

If Predictive Text is enabled, suspicious words are highlighted in blue. Potential replacements for suspicious words appear in the Predictive Text bar. When you tap the space key, the suspicious word is replaced with the most likely word toward the center of the Predictive Text bar. Tap the word on the far left to keep what you've typed (because it isn't a mistake) or tap another word or emoji to enter it instead of what you've typed.

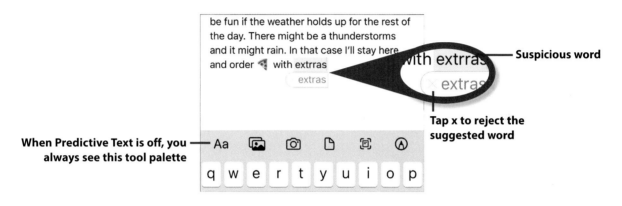

Suspicious word

Tap x to reject the suggested word

When Predictive Text is off, you always see this tool palette

If Predictive Text isn't enabled, a suspicious word is highlighted and a suggestion about what might be the correct word appears in a pop-up box. To accept the suggestion, tap the space key. To reject the suggestion, tap the x in the pop-up box to close it and keep what you typed. You can also use this feature for short-hand typing. For example, to type "I've" you can simply type "Ive" and iPhone suggests "I've," which you can accept by tapping the space key.

Typing Tricks

Many keys, especially symbols and punctuation, have additional characters. To see a character's options, touch and hold on it. If it has options, a menu pops up after a second or so. To enter one of the optional characters, drag over the menu until the one you want to enter is highlighted, and then lift your finger off the screen. The optional character you selected is entered. For example, if you tap and hold on the period when you're writing text, you can select an ellipsis (…). If you tap and hold on the period when you're typing a web or email address, you can select .com or .edu.

By default, the iPhone attempts to correct the capitalization of what you type. It also automatically selects the Shift key when you start a new sentence, start a new paragraph, or in other places where its best guess is that you need a capital letter. If you don't want to enter a capital character, simply tap the Shift key before you type. You can enable the Caps Lock key by tapping the Shift key twice. When the Caps Lock key is highlighted (the upward-facing arrow is black), everything you type is in uppercase letters.

No Caps Lock?

You can enable or disable the Caps Lock key by opening the Settings app, tapping General, tapping Keyboard, and then setting the Enable Caps Lock to on (green) or off (white).

Editing Text

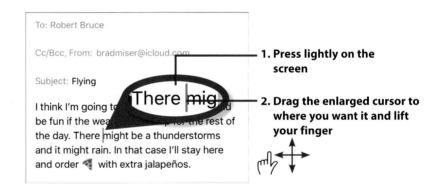

1. Press lightly on the screen

2. Drag the enlarged cursor to where you want it and lift your finger

To edit text you've typed, touch and hold the screen. You see a large cursor. Drag your finger until the cursor is placed where you want to start editing. Then lift your finger from the screen. The cursor remains in that location, and you can use the keyboard to make changes to the text or to add text at that location.

Selecting, Copying, Cutting, Pasting, or Formatting Text

Many times, you'll want to select text before editing or formatting it. You also can select text or images to copy and paste the selected content into a new location or to replace that content.

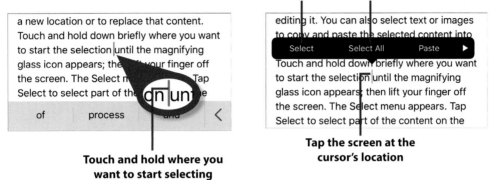

Tap Select to choose a portion of what's in the window

Tap Select All to choose everything in the window

Tap the screen at the cursor's location

Touch and hold where you want to start selecting

Touch and hold down briefly where you want to start the selection to position the cursor there. Tap the screen. The Select menu appears. Tap Select to select part of the content on the screen, or tap Select All to select everything in the current window.

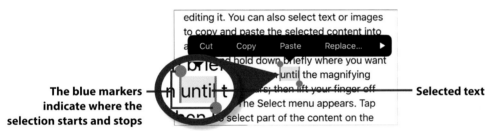

The blue markers indicate where the selection starts and stops

Selected text

You see markers indicating where the selection starts and stops. (The iPhone attempts to select something logical, such as the word or sentence.) New commands appear on the menu; these provide actions for the text currently selected.

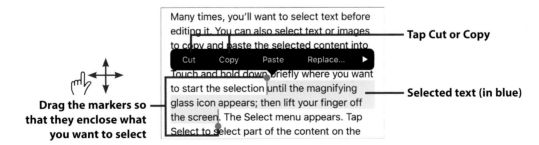

Tap Cut or Copy

Selected text (in blue)

Drag the markers so that they enclose what you want to select

Drag the two markers so that the content you want to select is between them; the selected portion is highlighted in blue. When the selection markers are located correctly, lift your finger from the screen. (If you tapped the Select All command, you don't need to do this because the content you want is already selected.)

Tap Cut to remove the content from the current window, or tap Copy to just copy it.

Tap where you want to paste

Tap Paste

Pasted content

Move to where you want to paste the content you selected; for example, use the App Switcher to change to a different app. Tap where you want the content to be pasted. The menu appears. Then tap Paste.

The content you copied or cut appears where you placed the cursor.

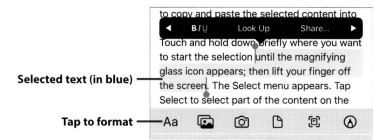

Selected text (in blue)

Tap to format

You can use similar steps to format text using the Formatting tools you learned about earlier. Select the text you want to format as described previously. Tap Format (Aa).

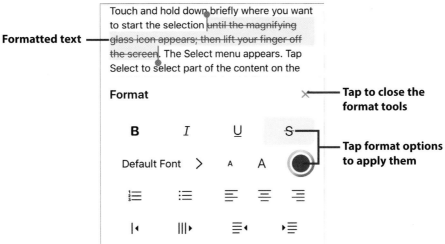

Formatted text

Touch and hold down briefly where you want to start the selection until the magnifying glass icon appears; then lift your finger off the screen. The Select menu appears. Tap Select to select part of the content on the

Format

Tap to close the format tools

B   *I*   U̲   S̶

Tap format options to apply them

Default Font >   ᴀ   A

Tap the format options you want to apply such as strikethrough or tap the color wheel and choose a color. The formatting options you choose are applied to the selected text. Tap Close (x) to close the Format tools.

More Commands

Some menus that appear when you're making selections and performing actions have a right-facing arrow at the right end. Tap this to see a new menu that contains additional commands. These commands are contextual, meaning that you see different commands depending on what you're doing at that specific time. You can tap the left-facing arrow to move back to a previous menu.

Correcting Spelling After You've Typed

Suspicious word

The iPhone also has a spell-checking feature that comes into play after you have entered text (as opposed to the Predictive Text and autocorrect/suggests features that change text as you type it). When you've entered text the iPhone doesn't recognize, it's underlined in red.

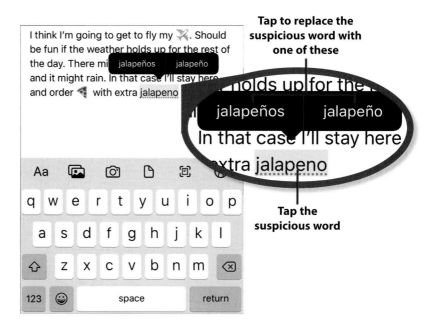

Tap to replace the suspicious word with one of these

Tap the suspicious word

Tap the underlined word. It's shaded in red to show you what's being checked, and a menu appears with one or more replacements that might be the correct spelling. If one of the options is the one you want, tap it. The incorrect word is replaced with the one you tapped.

Undo

The iPhone has a somewhat hidden undo command. To undo what you've just done, such as typing text, gently shake your phone back and forth a couple of times. An Undo prompt appears on the screen. Tap Undo to undo the last thing you did, or tap Cancel if you activated the undo command accidentally.

Speaking Text Instead of Typing

You can also enter text by dictating it. This is a fast and easy way to type, and you'll be amazed at how accurate the iPhone is at translating your speech into typed words. Dictation is available almost anywhere you need to enter text.

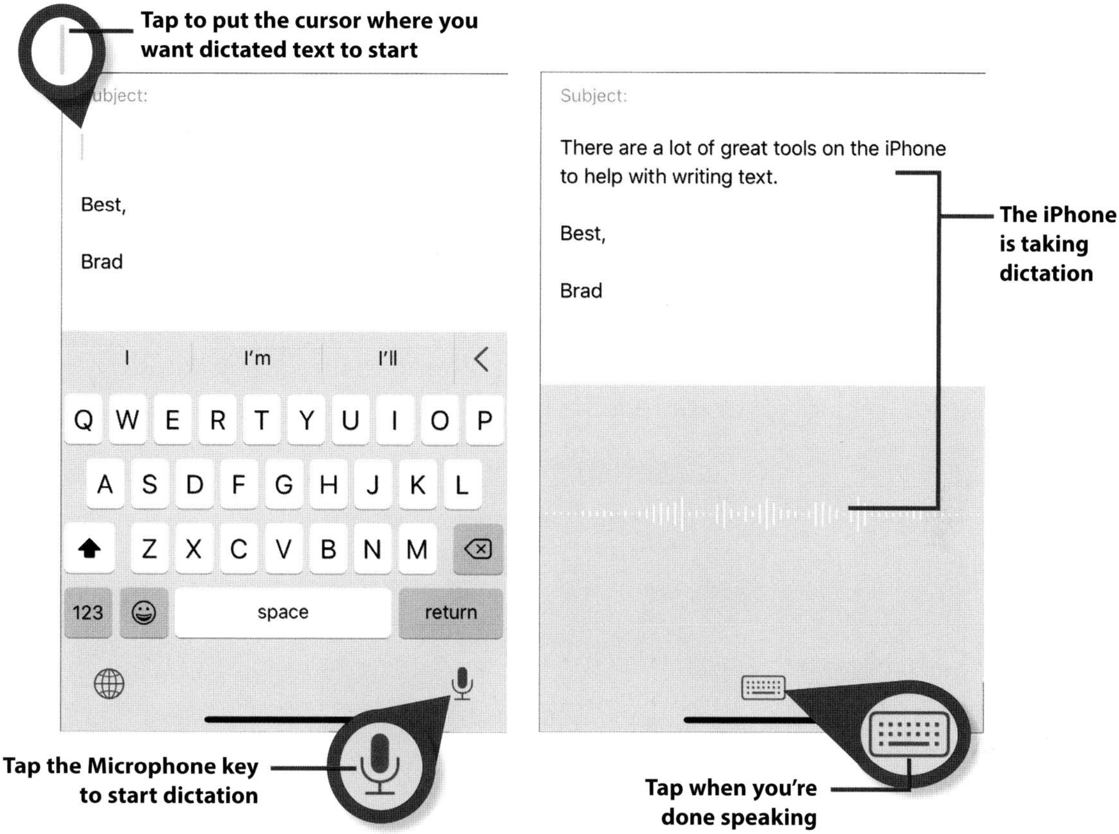

Tap to put the cursor where you want dictated text to start

Subject:

Best,

Brad

| I | I'm | I'll | < |

Q W E R T Y U I O P

A S D F G H J K L

⬆ Z X C V B N M ⌫

123 ☺ space return

🌐 🎤

Tap the Microphone key to start dictation

Subject:

There are a lot of great tools on the iPhone to help with writing text.

Best,

Brad

The iPhone is taking dictation

⌨

Tap when you're done speaking

To start dictating, tap where you want the dictated text to start and then tap the Microphone key. The iPhone goes into Dictation mode. A gray bar appears at the bottom of the window. As the iPhone "hears" you, the line oscillates.

Start speaking the text you want the iPhone to type. As you speak, the text is entered starting from the location of the cursor. Speak punctuation when you want to enter it. For example, when you reach the end of a sentence, say "period," or to enter a colon say "colon." To start a new paragraph, say "new paragraph."

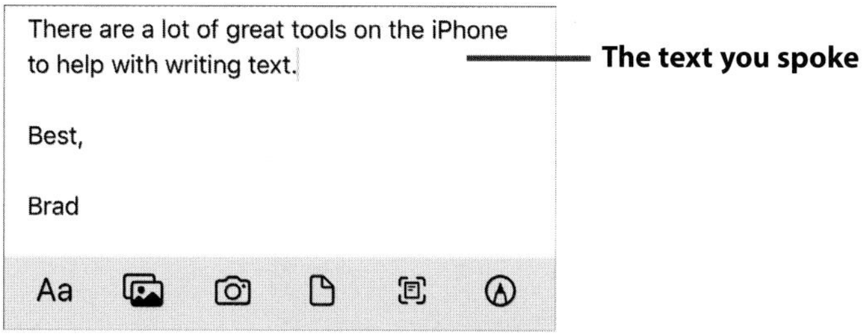

There are a lot of great tools on the iPhone to help with writing text.

Best,

Brad

The text you spoke

Aa 🖼 📷 📄 🔲 ✒

When you've finished dictating, tap Keyboard. The keyboard reappears and you see the text you spoke. This feature is amazingly accurate and can be a much faster and more convenient way to enter text than typing it.

You can edit or format text you dictated just like text you typed using the keyboard.

Searching on Your iPhone

You can use the Search tool to search your iPhone to find many different types of information, including locations, emails, messages, or apps.

Swipe down from the center part of the screen to search your iPhone

There are a couple of ways you can start a search:

- On a Home screen, swipe down from the center of the screen. The Search bar appears at the top of the screen.

- Swipe to the right to open the Widget Center. The Search bar is at the top of the screen.

Tap what you want to search for ———

Recent or suggested searches ———

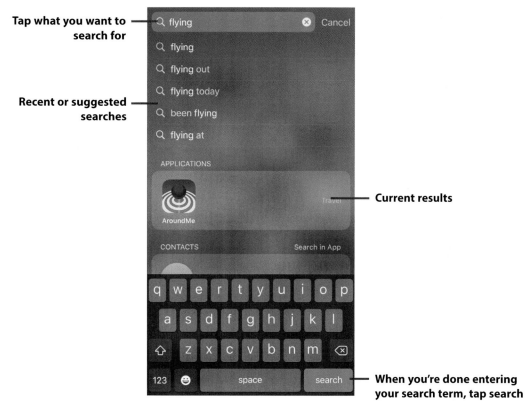

——— **Current results**

——— **When you're done entering your search term, tap search**

To perform a search, tap in the Search bar and type the search term using the onscreen keyboard. As you type, recent or suggested searches appear just below the Search bar; tap a search to perform it. Under the search list, you see the current items that match your search. If you don't tap one of the recent or suggested searches, when you finish typing the search term, tap search to see the full list of results.

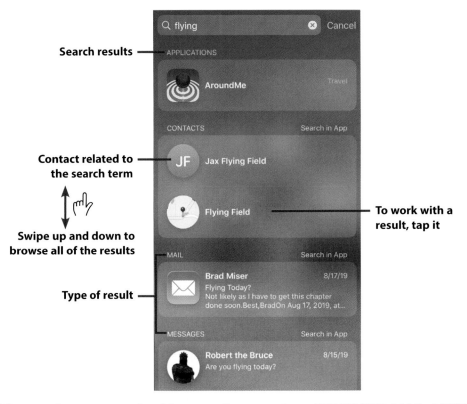

Search results

Contact related to the search term

Swipe up and down to browse all of the results

Type of result

To work with a result, tap it

The results are organized into sections, such as CONTACTS, MAIL, MESSAGES, APPLICATIONS, MAPS, MUSIC, and so on. Swipe up and down the screen to browse all of the results. To work with an item you find, such as to view a location you found, tap it; you move to a screen showing more information or into the associated app and see the search result that you tapped.

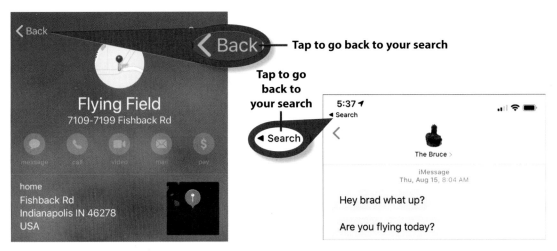

Tap to go back to your search

Tap to go back to your search

The results remain in the Search tool as you work with them. To move back to the search results, tap Back (<) in the upper-left corner of the screen or tap Search. (The option you see depends on the result you tapped on.)

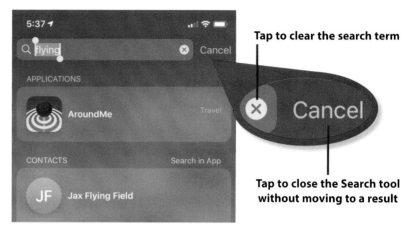

Tap to clear the search term

Cancel

Tap to close the Search tool without moving to a result

The results of the most recent search are still listed. To clear the search term, tap Clear (x). To close the Search tool without going to one of the results, tap Cancel.

Tell Me More

If one of the categories you find in a search has a lot of entries, you see the Show More command. Tap this to show more of the results for that category. Tap Show Less to collapse the category again. When you can search within an app, you see the Search in App text on the right side of the screen aligned with the results section; tap this to open the app and perform the search within that app.

Working with Siri Suggestions

Siri Suggestions can make it easy to get back to apps, searches, or other items you've used recently. These can also be tasks that Siri thinks you might want to do, such as making a call to someone. Using Siri Suggestions can lead you to useful things you weren't necessarily looking for. These suggestions show up in many different areas on your iPhone and you can access them directly at any time.

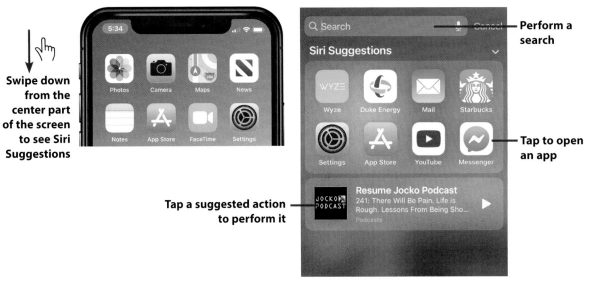

Swipe down from the center part of the screen to see Siri Suggestions

Perform a search

Tap to open an app

Tap a suggested action to perform it

To access Siri Suggestions, swipe down from the center of a Home screen. Just under the Search bar, you see the Siri Suggestions panel. This panel shows you apps you've used recently or apps that might be useful to you based on your location. For example, the Starbucks app may be suggested when you are near a Starbucks location. Tap an app to open it.

You might also see actions Siri has created for you. These can be actions you recently performed or actions Siri has created based on your activity. Tap a suggested action to perform it.

Siri Suggestions can also appear in other apps. Siri monitors your activity and "learns" from what you do to improve the suggestions it makes. These suggestions can appear in many different places, such as when you're entering email addresses in a new email, dealing with new contact information, performing searches, or editing a text message. When you see a list of Siri Suggestions, you can tap the suggestion you want to use. For example, if it's an email address, that address is entered for you. If it's a search, the search is performed.

Printing from Your iPhone

You can print directly from your iPhone to AirPrint-compatible printers.

First, set up and configure your AirPrint printer (see the instructions that came with the printer you use).

AirPrint?

AirPrint is an Apple technology that enables an iOS device to wirelessly print to an AirPrint-compatible printer without installing any printer drivers on the iOS device. To be able to print directly to a printer via Wi-Fi, the printer must support AirPrint (a large number of them do). When an iOS device, such as your iPhone, is on the same Wi-Fi network as an AirPrint printer, it automatically detects that printer and is able to print to it immediately.

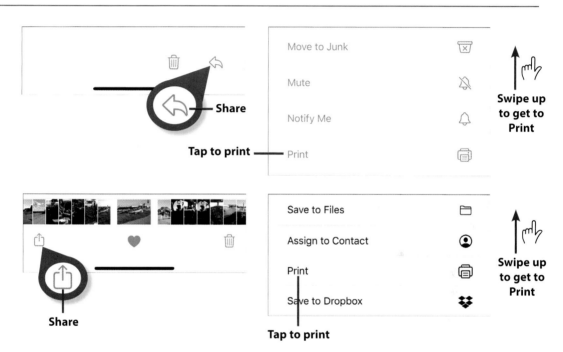

When you are in the app from which you want to print, tap Share. You see the Share menu with a number of actions. Swipe up the menu and tap Print. (If you don't see Share or the Print command on the Share menu, the app you're using doesn't support printing.)

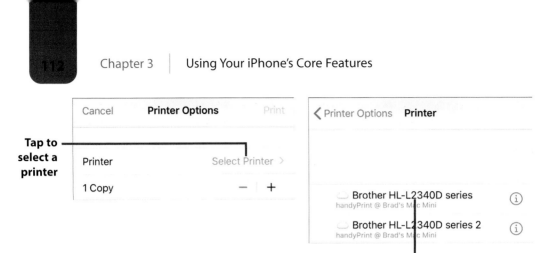

Tap to select a printer

Tap the printer you want to use

The first time you print, you need to select the printer you want to use. On the Printer Options screen, tap Select Printer. Then tap the printer you want to use. As long as your AirPrint-compatible printer is turned on and the Wi-Fi has been enabled, you should see it there. You move back to the Printer Options screen and see the printer you selected.

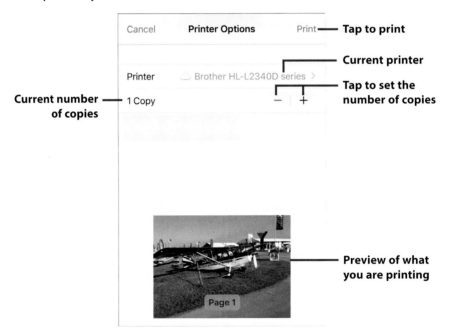

Tap to print

Current printer

Tap to set the number of copies

Current number of copies

Preview of what you are printing

Tap – or + to set the number of copies; the current number of copies is shown to the left of the controls. You can use other controls that appear to configure the print job, such as the Black & White switch to print in black and white on a color printer; the controls you see depend on the capabilities of the printer you selected and the type of content you are printing. Tap Print to print the document.

The next time you print, if you want to use the same printer, you can skip the printer selection process because the iPhone remembers the last printer you used. To change the printer, tap Printer and then tap the printer you want to use.

With your calendars on the cloud, you'll always know where you're supposed to be and when to be there

Storing your photos on the cloud protects them and makes them more accessible

Storing your contacts on the cloud makes them easy to use in all of your apps

Go here to configure and manage your iCloud, Google, and other online accounts

In this chapter, you learn how to configure an Apple ID on your iPhone and to set up various types of accounts, such as iCloud and Google, so that apps on your iPhone can access data stored on the Internet cloud. Topics include the following:

→ Getting started
→ Configuring an Apple ID
→ Configuring and using iCloud
→ Setting up other types of online accounts on your iPhone
→ Setting how and when your accounts are updated

4

Setting Up and Using an Apple ID, iCloud, and Other Online Accounts

Connecting your iPhone to the Internet enables you to share and sync a wide variety of content using popular online accounts such as iCloud and Google. Using iCloud, you can put your email, contacts, calendars, photos, and more on the Internet so that multiple devices—most importantly your iPhone—can connect to and use that information. There are other online accounts you might also want to use, such as Google for email, calendars, and contacts as well as email accounts provided by your Internet Service Provider (for example, a cable company). And, many people like to access social media, such as Facebook, with their iPhone; you can easily configure those accounts on your phone, too.

Getting Started

To access the cloud-based services provided by Apple, Google, and others, you need to have an account for each service that you want to use. Then, you sign into your account on your iPhone and configure which of the account's services you use. For example, if you have an iCloud account, you can choose to use it for email, calendars, contacts, and other information.

To use iCloud and access other services provided by Apple, such as the App Store, iTunes Store, and iMessage, you need to have an Apple ID and have that account configured on your iPhone. Similarly, to use Google, Facebook, or basically any other type of cloud-based service, you obtain an account for that service.

This chapter includes sections explaining how to configure several different online accounts you might want to use. Of course, you need to refer only to the sections related to the accounts you actually use. And, after you've configured a couple of these types of accounts, adding more is simple because you configure them similarly.

This chapter also explains how to configure how and when your cloud-based information is updated. Having the most current information on your phone at all times can be helpful, but it also causes your iPhone to use more power, which drains the battery faster. Updating your phone less frequently saves power, but the information you see isn't necessarily current.

Here are some of the key terms for this chapter:

- **Apple ID**—You'll need an Apple ID to access many Apple services, especially iCloud, and make purchases from the App Store, iTunes Store, and Apple's online store. An Apple ID also enables you to use iMessage to send and receive messages using Apple's iMessage service. Think of an Apple ID as the "connector" between all your devices; you can start tasks on one device, such as writing an email on your iPhone, and finish them on another, for example, an iPad. Similarly, if you subscribe to Apple's Music Library, your Apple ID provides access to that music on each of your devices (iPhones, iPads, or computers).

- **iCloud**—This is Apple's online service that offers lots of great features that you can use for free; if you store a lot of information (especially photos and video) online, you might need to add storage to your account for an additional fee. iCloud includes email, online photo storage and sharing, backup, calendars, Find My iPhone, and much more.

- **Google account**—A Google account is similar to an iCloud account except it's provided by Google instead of Apple. It also offers lots of features, including email, calendars, and contacts. You can use iCloud and a Google account on your iPhone at the same time.

- **Social media**—This term refers to various services that provide "social" connections between people or organizations. Popular examples include Facebook, Instagram, and Twitter. You can configure these accounts on your iPhone, and then use the associated app, such as the Facebook app, to access the service.

- **Push, Fetch, or Manual**—Information has to get from your online account onto your iPhone. For example, when someone sends an email to you, it goes to an email server, which then sends the message to devices that are configured with your email account. You can choose how and when new data is provided to your phone. The three ways data gets moved onto your iPhone (Push, Fetch, Manual) are explained in "Setting How and When Your Accounts Are Updated" later in this chapter.

Configuring an Apple ID

An Apple ID is required to access Apple's online services, including iCloud, the App Store, iTunes, and Apple's online store. You can access all these services with one Apple ID.

An Apple ID has two elements. One is the email address associated with your account; this can be an iCloud address provided by Apple, or an address from a different service (such as Google Gmail). The other element is a password.

In addition to your email address and password, your contact information (a physical address and phone number) and payment information (if you make purchases through your account, such as for apps or storage upgrades) is also part of your Apple ID.

If you have used Apple technology or services before, you probably already have an Apple ID. If you don't already have one, obtaining an Apple ID is simple and free.

If you have any of the following accounts, you already have an Apple ID:

- **iTunes Store**—If you've ever shopped at the iTunes Store, you created an Apple ID.

- **Apple Online Store**—As with the iTunes Store, if you made purchases from Apple's online store, you created an account with an Apple ID.

- **Find My iPhone**—If you obtained a free Find My iPhone account, you created an Apple ID.

Another way you might have already obtained an Apple ID is during the initial iPhone startup process when you were prompted to sign in to or create an Apple ID.

If you don't have an Apple ID, read the next section to obtain one. If you already have an Apple ID, move to "Signing In to Your Apple ID." If you both have an Apple ID and have already signed in to it on your iPhone, skip to the section "Configuring and Using iCloud."

Obtaining an Apple ID

If you don't have an Apple ID, you can use your iPhone to create one by performing the following steps:

1. On the Home screen, tap Settings.

2. Tap Sign in to your iPhone.

3. Tap Don't have an Apple ID or forgot it?.

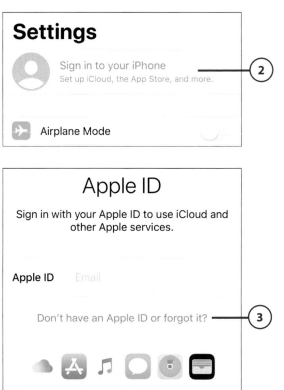

Tap Create Apple ID.

(5) Enter the information required on the following screens; after you've completed all the required information on a screen, tap Next. (If Next is gray, that means you haven't provided all the required information on that screen. After you have provided all the needed information, Next turns blue.) The first step is to provide your name and birthday. You are guided through each step in the process.

During the process, you're prompted to use an existing email address or to create a free iCloud email account. You can choose either option. The email address you use becomes your Apple ID that you use to sign in to iCloud. If you create a new iCloud email account, you can use that account from any email app on any device, just like other email accounts you have.

You also create a password, enter a phone number and verification method, verify the phone number you entered, and agree to license terms. When your account has been created, you sign in to iCloud, which you do by entering your iPhone's passcode.

You might then be prompted to merge information already stored on your iPhone, such as Safari bookmarks, onto iCloud. Tap Merge to copy the information that currently is stored on your iPhone to the cloud or Don't Merge to keep it out of the cloud.

When you've worked through merging your information, you're prompted to sign in to the iTunes and App Stores to make sure your new account can work with those services, too. You can choose to review your account information now or skip it and configure it at another time.

When the process is complete, you're signed in to your new Apple ID; skip ahead to "Configuring and Using iCloud" to learn how to change its settings.

Signing In to Your Apple ID

You can sign in to an existing Apple ID on your iPhone by doing the following:

(1) On the Home screen, tap Settings.

(2) Tap Sign in to your iPhone.

(3) Enter your Apple ID (email address).

(4) Tap Next.

(5) Enter your password.

(6) Tap Next. If you are using the same Apple ID on other devices, you're prompted to enter the verification code sent to those devices. See the sidebar "Two-Factor Authentication" for more information.

Don't Forget Your Password!

Make sure you know your Apple ID password or use a secure place to store it. If you forget your password or can't sign into your account, tap Don't have an Apple ID or forgot it? or similar link. This takes you to Apple's account restore process. You should be able to work through the process to restore access to your account.

You can also manage your Apple ID on a computer using the appleid.apple.com website.

Be aware that if you make multiple unsuccessful login or recovery attempts, your account can be locked. In this case, the recovery process can take weeks. Apple uses a fully automated process that works by strict rules. You won't be able to call Apple to get your account restored, you have to wait for whatever recovery time is set by the automated system.

---

7. Enter your verification code; if you weren't prompted to enter a code, skip this step.

8. Enter your iPhone's passcode. You might be prompted to merge existing information on the iPhone with your iCloud storage.

9. Tap Merge to copy information from your iPhone onto the cloud or Don't Merge if you don't want that information copied to the cloud. You might be prompted to perform this merge step more than once depending on the kind of information already stored on your iPhone. If you don't have any information that can be merged, you skip this step entirely.

When you've finished these steps, you are signed in to your Apple ID and can configure it further using the information in the next section.

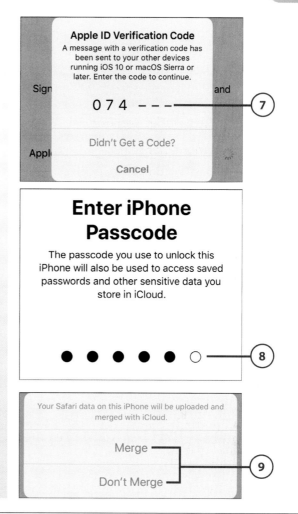

Two-Factor Authentication

For added security, you can use two-factor authentication if you have at least one other device that is configured with the same Apple ID as your iPhone, such as an iPad or Mac. When you need to authenticate a device, you use two sets of information. One is your Apple ID email and password. The other is a verification code sent to other devices on which your Apple ID is also configured (called trusted devices). This makes using Apple services more secure because even if someone was able to get your Apple ID and password, he would also need the verification code to access your account (meaning he would also have to be able to access another device on which your Apple ID is configured). You should use two-factor authentication if you have at least one other device configured with your Apple ID.

You can configure two-factor authentication for your Apple ID in the Password & Security settings on the iCloud Settings screen. There, you can enable or disable two-factor authentication, change your password, edit trusted phone numbers, or get a verification code.

Confirmations Galore

When you make changes to your Apple ID, such as adding a new email address or changing security settings, you receive confirmation emails and notifications from Apple on all the devices tied to your account. These are helpful because they confirm the actions you're taking. If you ever receive such a notification when you haven't changed your account, carefully review it. In many cases, especially email notifications, these unexpected notifications are attempts to get your information for nefarious purposes. Don't respond to emails you don't expect even if they appear to be from Apple (if you make a change to your account and then receive an email, that is expected). Instead, log in directly to your account on your iPhone or other device (such as a computer to log into the iCloud website) to make sure it hasn't been changed by someone else.

Signing Off

If you don't want to continue accessing your Apple ID on your iPhone or you want to sign into a different Apple ID, tap Sign Out at the bottom of the Apple ID screen. If Find My iPhone is enabled, you need to enter your password and tap Turn Off to continue the process. You might be prompted to keep some information on your iPhone; if you choose to keep that information, it remains on your iPhone but is no longer synchronized to the information stored on the cloud. After you have completed the process, you can sign in to the same or a different Apple ID using the information in "Signing In to Your Apple ID" earlier in this chapter.

Configuring and Using iCloud

iCloud provides you with storage space on the Internet. You can store content from your iPhone, computer, or other devices here, and because iCloud is on the Internet, all your devices are able to access that information at the same time. This means you can easily share your information on your iPhone, a computer, and iPad, so that the same information and content is available to you no matter which device you are using at any one time.

Although your iPhone can work with many types of online/Internet accounts (such as Google), iCloud is integrated into the iPhone like no other (not surprising because the iPhone and iCloud are both Apple technology).

There are many types of content you can store in and access from your iCloud account, including the following:

- **Photos**—Storing your photos in iCloud protects them and makes them easy to share.

- **Email**—An iCloud account includes an @icloud.com email address (unless you choose to use an existing account instead). You can configure any device to use your iCloud email account, including an iPhone, an iPad, or a computer.

- **Contacts**—You can store your contact information in iCloud.

- **Calendars**—Putting your calendars in iCloud makes it much easier to manage your time.

- **Reminders**—Through iCloud, you can be reminded of things you need to do or anything else you want to make sure you don't forget.

- **Notes**—With the Notes app, you can create text notes, draw sketches, and capture photos for many purposes; iCloud enables you to use these notes on any iCloud-enabled device.

- **Messages**—This puts all of your messages in the Messages app on the cloud so you can access them from all your devices.

- **Safari**—iCloud can store your bookmarks, letting you easily access the same websites from all your devices.

- **News**—iCloud can store information from the News app online, making reading news on multiple devices easier.

- **Stocks**—If you use the Stocks app to track investments, iCloud ensures you have the same investments in the Stocks app on all your devices.

- **Home**—If you use your iPhone for home automation, using iCloud ensures you have the same automation controls on all your devices.

- **Health**—This allows the information stored using the Health app to be available to multiple devices. For example, you might track information on your Apple Watch and want to be able to analyze it on your iPhone.

- **Wallet**—The Wallet app stores coupons, tickets, boarding passes, and other documents so you can access them quickly and easily. With iCloud, you can ensure that these items are available on any iCloud-enabled device.

- **Game Center**—This capability stores information from the Game Center app on the cloud.

- **Siri**—It can be helpful to manage Siri information on multiple devices; this setting puts that information on the cloud.

- **Keychain**—The Keychain securely stores sensitive data, such as passwords, so that you can easily use that data without having to remember it.

- **iCloud Backup**—You can (and should) back up your iPhone to the cloud so that you can recover your data and your phone's configuration should something ever happen to the iPhone itself (if you loose it, for example).

- **iCloud Drive**—iCloud enables you to store documents and other files on the cloud so that you can seamlessly work with them using different devices.

- **App Data**—When iCloud Drive is enabled, you can allow or prevent individual apps from storing data there. For example, if you use the Books app on an iPhone and an iPad, you can store Books information on the cloud so that you always pick up reading where you left off when you change devices.

Configuring iCloud to Store Photos

Storing your photos on the cloud provides many benefits, not the least of which is that the photos you take with your iPhone are automatically saved on the cloud so that you can access them from computers and other devices, and your photos remain available even if something happens to your iPhone, such as if you drop it in water and it stops working. To configure your photos to be stored in iCloud, do the following:

1. In the Settings app, open the Apple ID screen.

2. Tap iCloud.

3. On the iCloud screen, tap Photos.

4 To store your entire photo library on the cloud, set the iCloud Photos switch to on (green). This stores all of your photos and video in iCloud, which both protects them by backing them up and makes them accessible on other iOS/iPadOS devices (iPhones or iPads) and computers (including Macs, Windows PCs, and via the Web).

5 If you enable the iCloud Photos feature, tap Optimize iPhone Storage to keep lower-resolution versions of photos and videos on your iPhone (this means the file sizes are smaller so that you can store more of them on your phone), or tap Download and Keep Originals if you want to keep the full-resolution photos on your iPhone. In most cases, you should choose the Optimize option so that you don't use as much of your iPhone's storage space for photos. (You can still access the full-resolution versions on the cloud; for example, to download them to a computer.)

6 Ensure the Upload to My Photo Stream switch is on (green) (if you aren't using iCloud Photos, this switch is called My Photo Stream); if you're not enabling this, skip to step 8. Any photos you take with the iPhone's camera are copied onto iCloud, and from there they're copied to your other devices on which the Photo Stream is enabled. Note that Photo Stream affects only photos that you take with the iPhone from the time you enable it, whereas the iCloud Photos feature uploads all of your photos—those you took in the past and will take in the future.

⟨ iCloud   **Photos**

ICLOUD

iCloud Photos ⟶ **4**

Automatically upload and safely store all your photos and videos in iCloud so you can browse, search, and share from any of your devices.

Optimize iPhone Storage ✓

Download and Keep Originals ⟶ **5**

If your iPhone is low on space, full-resolution photos and videos are automatically replaced with smaller, device-sized versions. Full-resolution versions can be downloaded from iCloud anytime.

Upload to My Photo Stream ⟶ **6**

(7) If you want all of your burst photos (photos taken in sequence, such as for action shots) to be uploaded to iCloud, set the Upload Burst Photos switch to on (green). In most cases, you should leave this off (white) because you typically don't want to keep all the photos in a burst. When you review and select photos to keep, the ones you keep are uploaded automatically.

(8) To be able to share your photos and to access photos other people share with you, set the Shared Albums switch to on (green).

Enabling iCloud to Store Information on the Cloud

As you learned earlier, one of the best things about iCloud is that it stores email, contacts, calendars, reminders, bookmarks, notes, and other data on the cloud so that all your iCloud-enabled devices can access the same information. You can choose the types of data stored on the cloud by performing the following steps:

(1) In the Settings app, move to the iCloud screen. Just below the Photos section are the iCloud data options. A few of these have a right-facing arrow that you tap to configure options, whereas most have a two-position switch. The types of data that have switches are Mail, Contacts, Calendars, Reminders, Notes, Messages, Safari,

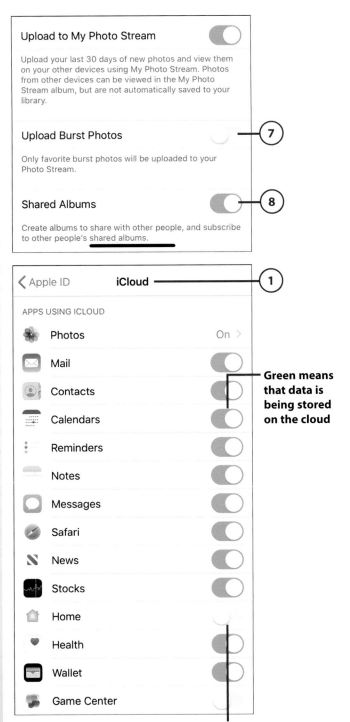

Upload to My Photo Stream

Upload your last 30 days of new photos and view them on your other devices using My Photo Stream. Photos from other devices can be viewed in the My Photo Stream album, but are not automatically saved to your library.

Upload Burst Photos — (7)

Only favorite burst photos will be uploaded to your Photo Stream.

Shared Albums — (8)

Create albums to share with other people, and subscribe to other people's shared albums.

‹ Apple ID   iCloud — (1)

APPS USING ICLOUD

Photos   On ›
Mail
Contacts
Calendars   **Green means that data is being stored on the cloud**
Reminders
Notes
Messages
Safari
News
Stocks
Home
Health
Wallet
Game Center

White means the data is not being stored on the cloud

News, Stocks, Home, Health, Wallet, Game Center, and Siri. When a switch is green, it means that switch is turned on and the related data is stored to your iCloud account and kept in sync with the information on the iPhone.

2 To store data on the cloud, set an app's switch to on (green). You might be prompted to merge that information with that already stored on the cloud. For example, if you have contacts information on your iPhone already and want to merge that with the contacts already on the cloud, tap Merge. If you don't want the contacts currently stored on your iPhone copied to the cloud, tap Don't Merge instead.

3 If you don't want a specific type of data to be stored on the cloud and synced to your iPhone, tap its switch to turn that data off (the switch becomes white instead of green). That data is no longer stored on the cloud. If you have data of that type stored on the phone, such as contacts, it remains there but won't be accessible using other devices.

4 Swipe up the screen until you see the list of apps below the iCloud Drive switch.

5 Enable or disable the apps to use iCloud to store documents and data. Just like the other options, when you enable an app, its data and documents are stored on the cloud and can be shared with other devices. If you don't allow this, the data is stored only on your iPhone.

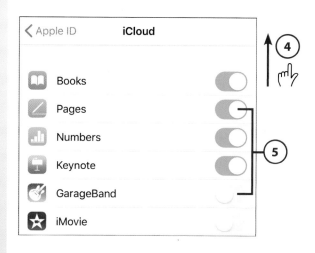

To Keep or Not to Keep?

When you turn off a switch because you don't want that information stored on the cloud any more, you might be prompted to keep the associated information on your iPhone or delete it.

If you choose Keep on My iPhone, the information remains on your iPhone but is no longer connected to the cloud; this means any changes you make exist only on the iPhone. If you choose Delete from My iPhone, the information is erased from your iPhone. Whether you choose to keep or delete the information, any information of that type that was previously stored on the cloud remains available there; the delete action affects only the information stored on the iPhone.

>>>Go Further

CONFIGURING FIND MY iPHONE

With Find My iPhone, you can locate and secure your iPhone if it is lost or stolen. This feature is enabled by default when you sign in to your iCloud account. You should leave it enabled so that you have a better chance of locating your iPhone should you need to—or in case you need to delete its data in the event that you won't be getting the iPhone back. There are a couple of configuration tasks you can do for Find My iPhone. To make changes, open the Apple ID screen and tap Find My. You move to the Find My screen. To disable or configure Find My iPhone, tap Find My iPhone. You move to the Find My iPhone screen and can perform the following actions:

- To disable Find My iPhone (not recommended), set the Find My iPhone switch to off (white) and enter your Apple ID password at the prompt. You can no longer access your iPhone via the Find My iPhone feature.

- When you turn Find My network on (green), Find My iPhone attempts to locate your iPhone even when it isn't connected to the Internet via Wi-Fi or a cellular connection. This is done via signals from Bluetooth devices for which their locations are known. Enabling this could potentially increase the chances of recovering your phone even if it doesn't have an Internet connection.

- To send the last known location of the iPhone to Apple when power is critically low, set the Send Last Location switch to on (green). When your iPhone is nearly out of power, its location is sent to Apple. You can contact Apple to try to determine where your iPhone was when the battery was almost out of power.

Configuring Your iCloud Backup

As with other digital devices, it's important to back up your iPhone's data so that you can recover it should something bad happen to your iPhone, such as if it's lost, stolen, or just isn't working. You can back up your iPhone's data and settings to iCloud, which is really useful because that means you can recover the backed-up data using a different device, such as a replacement iPhone. Configure your iCloud backup with the following steps:

(1) On the iCloud settings screen, tap iCloud Backup.

(2) Set the iCloud Backup switch to on (green). Your iPhone's data and settings are backed up to the cloud automatically.

Back Me Up on This

You can manually back up your iPhone's data and settings at any time by tapping Back Up Now on the Backup screen. This can be useful to ensure recent data or settings changes are captured in your backup. For example, if you know you're going to be without a Wi-Fi connection to the Internet for a while, back up your phone to ensure that your current data is saved.

Managing Your iCloud Storage

Your iCloud account includes storage space that you can use for your data such as photos, documents, and so on. By default, your account includes 5 GB of free storage space. For many people, that's enough, but if you take a lot of photos and video and use iCloud Photos, you might find that you need more space. It's easy (and relatively inexpensive) to upgrade the amount of room you have on your iCloud Drive. You can use the STORAGE section on the iCloud settings screen to manage your storage space as follows:

(1) Move to the STORAGE section located at the top of the iCloud settings screen. Here you see a gauge that displays the amount of space you have and how that space is currently being used. The gray portion of the bar indicates how much free space you have; if this portion of the bar is very small, you might want to consider upgrading your storage space. The colored bars show the data being used by various types of data or apps; for example, Photos.

(2) Tap Manage Storage. At the top of the resulting iCloud Storage screen, you see the same storage information as on the prior screen. Under that, you see the tools you can use to manage your storage.

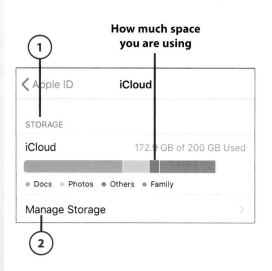

How much space you are using

(3) To change the amount of storage space available to you, tap Change Storage Plan and follow the onscreen prompts to upgrade (or downgrade) your storage.

(4) Swipe up the screen to review all the apps that are currently using iCloud storage space and to see how much space they're using.

(5) Tap an app to get details about how it's using iCloud space.

(6) If the app works with documents, you see the list of documents and how much space each is using (if it doesn't use documents, you only see a total).

Delete or Disable?

When you delete documents or data from your iCloud storage, that content is removed from the cloud as well as from every device using that data on the cloud. If you want the app to stop storing data on the cloud, but keep the data on the devices currently using it, prevent it from using iCloud storage as described previously in "Enabling iCloud to Store Information on the Cloud" instead.

(7) To remove documents and data from iCloud, tap Delete Documents & Data, Disable & Delete, or Delete Data. After you confirm the deletion, the app's data is deleted from your iCloud storage. (Of course, make sure you have this information stored in another location if you are going to need it again.)

< iCloud   iCloud Storage

iCloud	172.9 GB of 200 GB Used

● Docs   ● Photos   ● Others   ● Family

Family Usage	50.9 GB >
Change Storage Plan	200 GB —— **(3)**

☁ iCloud Drive	87.6 GB >
✿ Photos	23.8 GB > **(4)**
💬 Messages	4.2 GB >
✉ Mail	2.9 GB
🔄 Backups	2.1 GB >
🎤 Keynote	1.1 GB >
📄 Pages	115.7 MB —— **(5)**

< iCloud Storage   Pages

📄 **Pages**
    Apple Inc.

Documents & Data	80.4 MB

Delete Documents & Data ——————— **(7)**

This will delete all app data from iCloud and all connected devices. This action can't be undone.

DOCUMENTS	80.4 MB
Newbie Freewing F-15 Build Log_wo...	22 MB —— **(6)**
plane_storage.pages	14.4 MB
Plane Test Stand.pages-tef	5 MB
bike_flyer.pages	4.7 MB
Sodexo-Tulsa School Dist- VFA Pro...	3.8 MB
Oshkosh 2013.pages	3.3 MB

Setting Up Other Types of Online Accounts on Your iPhone

Many types of online accounts provide different services, including email, calendars, contacts, or social networking. To use these accounts, you need to configure them on your iPhone. The process you use for most types of accounts is similar to the steps you used to set up your iCloud account. In this section, you learn how to configure a Google account and an account that you might have through your Internet provider, such as a cable company.

Configuring a Google Account

A Google account provides syncing for email, contacts, calendar, and notes that's similar to iCloud. To set up a Google account on your iPhone, do the following:

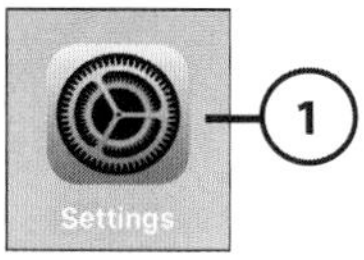

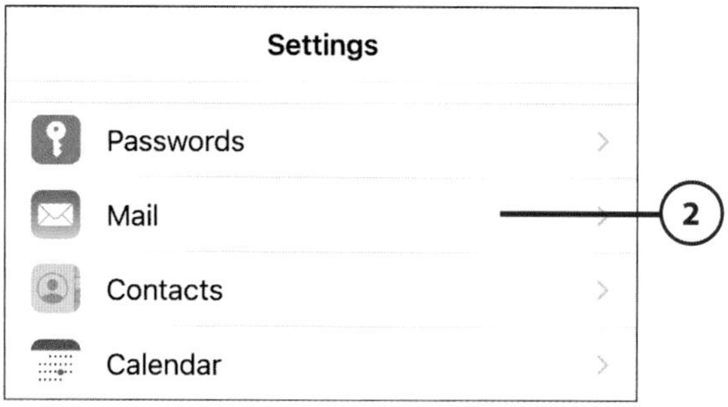

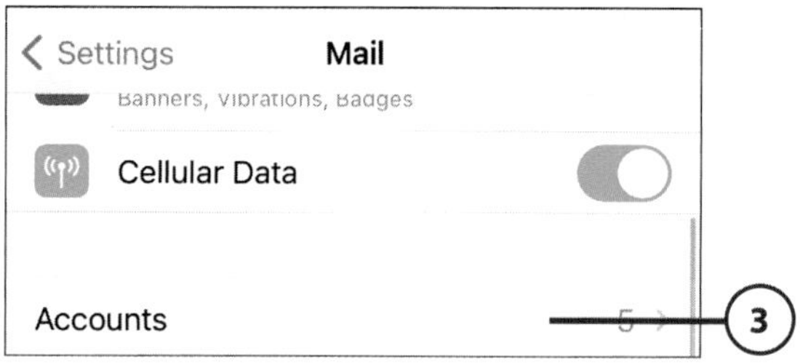

(1) On the Home screen, tap Settings.

(2) Tap the type of account you want to configure, such as Mail to set up an account for email.

Where to Begin?

It doesn't really matter what type of account you tap in step 2 because you access the same account tools. For example, if you tap Contacts or Calendar instead, you can perform the same steps.

---

(3) Tap Accounts.

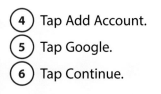 Tap Add Account.

 Tap Google.

 Tap Continue.

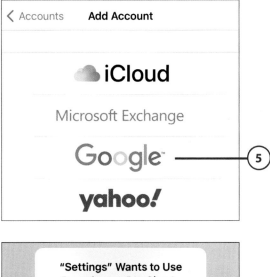

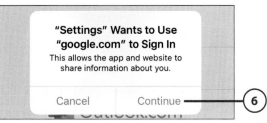

(7) Enter your Google email address.

Got Google?

This section assumes you already have a Google account. If you don't, you can tap Create account on the Google Sign in screen (shown for step 7) to create an account and sign into it at the same time.

(8) Tap Next.

(9) Enter your Google account password.

(10) Tap Next.

(11) Enable the features of the account you want to access on the iPhone—which are Mail, Contacts, Calendars, and Notes—by setting the switch to on (green) for the types of data you do want to use or to off (white) for the types of data you don't want to use. Just like iCloud, when you enable a type of data to be stored on the cloud, it's available to your iPhone and to any device signed into your Google account.

(12) Tap Save. The account is saved, and the data you enabled becomes available on your iPhone.

Google Password

You can't change the password for a Google account in the Settings app on your iPhone. The most common way to change your Google password is by accessing your Google account via the Google website. After you've changed your Google password, you're prompted to enter the new password the next time your iPhone attempts to access your account.

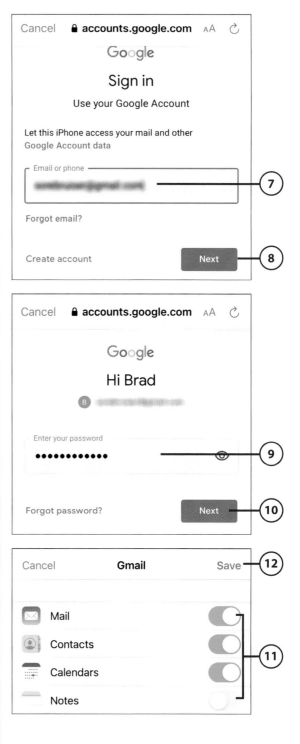

Setting Up an Online Account that Isn't Built In

You can access many types of online accounts on your iPhone. These include accounts that are "built in," which include AOL, Exchange, Google, iCloud, Outlook.com, and Yahoo! Setting up an AOL, Exchange, Outlook.com, or Yahoo! account is similar to configuring a Google or iCloud account on your iPhone. Just select the account type you want to use and provide the information for which you are prompted.

There are other types of accounts you might want to use that aren't "built in." For example, if you have an account with a cable Internet service provider, you probably also have one or more email accounts from that provider. Support for these accounts isn't built in to the iOS; however, you can usually set up such accounts on your iPhone fairly easily.

When you obtain an account, for example, email accounts that are part of your Internet service, you should receive all the information you need to configure those accounts on your iPhone. If you don't have this information, visit the provider's website and look for information on configuring the account in an email application. You need to have this information to configure the account on the iPhone.

With the configuration information for the account you want to use on your iPhone in hand, you're ready to set it up:

1. On the Home screen, tap Settings.

2. Tap the type of account you want to configure, such as Mail to set up an account for email.

Where to Begin?

It doesn't really matter what type of account you tap in step 2 because you access the same account tools. For example, if you tap Contacts or Calendar instead, you can perform the same steps.

 3. Tap Accounts.

(4) Tap Add Account.

(5) Tap Other.

(6) Tap the type of account you want to add. For example, to set up an email account, tap Add Mail Account.

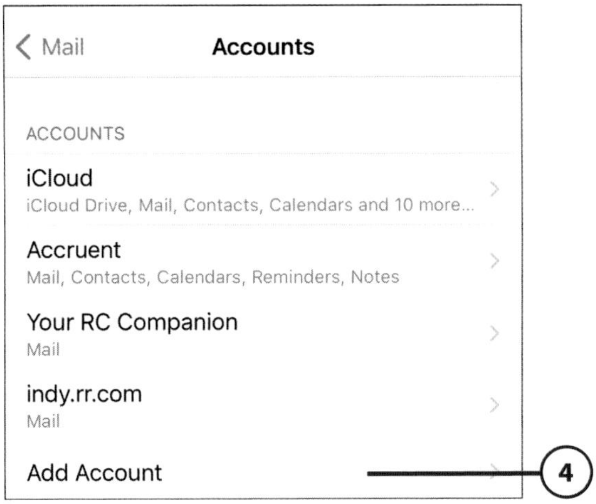

< Mail   **Accounts**

ACCOUNTS

iCloud
iCloud Drive, Mail, Contacts, Calendars and 10 more...   >

Accruent
Mail, Contacts, Calendars, Reminders, Notes   >

Your RC Companion
Mail   >

indy.rr.com
Mail   >

Add Account   > ———————— (4)

Google
yahoo!
Aol.
Outlook.com

Other ———————— (5)

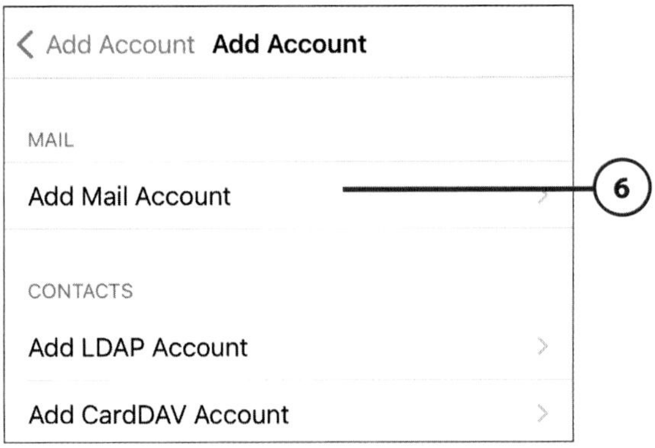

< Add Account   **Add Account**

MAIL

Add Mail Account ———————— (6)

CONTACTS

Add LDAP Account   >

Add CardDAV Account   >

7 Enter the information by filling in the fields you see; various types of information are required for different kinds of accounts. You just need to enter the information you received from the account's provider.

8 Tap Next. If the iPhone can set up the account automatically, its information is verified, and it's ready for you to use. (If the account supports multiple types of information, you can enable or disable the types with which you want to work on your iPhone.) If the iPhone can't set up the account automatically, you're prompted to enter additional information to complete the account configuration. When you're done, the account appears on the list of accounts and is ready for you to use.

9 Configure the switches for the data sync options you see. For example, to use the account for email, set the Mail switch to on (green).

10 Tap Save. The account you configured is available in the related app; if you set up an email account, you access it with the Mail app.

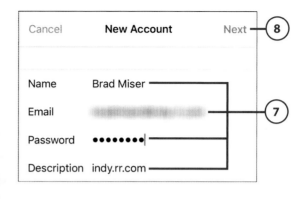

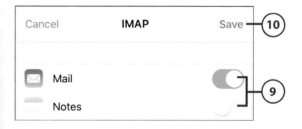

Multiple Accounts

There is no limit (that I have found so far) on the number of online accounts (even of the same type, such as Gmail or email accounts from an Internet provider) that you can access on your iPhone. (However, you can only have one iCloud account configured on your iPhone at the same time.)

Configuring Social Media Accounts on Your iPhone

Social media apps are useful for keeping in touch with others, sharing your opinions and reading the opinions of others, and exchanging photos. Examples of these types of social media include Facebook, Instagram, and Twitter. You can easily download and configure social media apps to work on your iPhone.

Unlike Google, iCloud, and other accounts that are configured through the Settings app, you configure your social media accounts directly in their apps.

To use a social media app, you perform the following three general steps:

1. Download and install the app you want to use.

2. Configure the app to access your social media account.

3. Configure other settings for the social media app.

The details of downloading and installing apps (step 1) are provided in Chapter 5, "Customizing Your iPhone with Apps and Widgets." Examples of steps 2 and 3 for Facebook follow. You can configure other social media apps similarly.

Configuring Facebook on Your iPhone

Facebook is one of the most popular social media channels you can use to keep informed about other people and inform them about you. Use these steps to download and configure Facebook on your iPhone:

1. Use the App Store app to download and install the Facebook app on your iPhone (see Chapter 5 for the details of working with the App Store app).

2 Tap the Facebook icon on a Home screen to open the app.

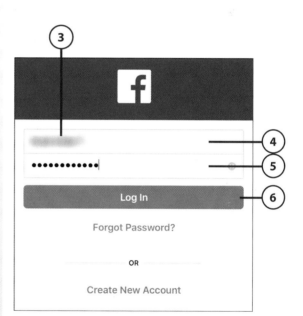

Already Signed In?

If you've previously signed into Facebook on the phone, you see that account on the opening screen. Tap the account shown. You're prompted to enter your password; when you do, you sign in and can jump to step 10.

Don't Have a Facebook Account?

If you don't already have a Facebook account, you can create one by tapping Create New Account on the opening screen in the app or on the Log In screen. Follow the onscreen instructions to create a new Facebook account and log into it.

3 Tap in the Email or phone number field.

4 Enter the email address, phone number, or Facebook account name associated with your account.

5 Enter your password. (If you don't know your password, tap Forgot Password? and follow the onscreen instructions to reset it.)

6 Tap Log In. The first time you sign into the account on your iPhone, you're prompted to configure several settings associated with how the app works. After the first time, you don't have to do steps 7 through 11.

(7) If you want to receive notifications from Facebook, such as when someone posts on your Timeline, tap Allow; if you don't want these notifications, tap Don't Allow.

(8) To use your photo to sign in instead of typing your account name, tap OK.

(9) If you want your account's password saved so you don't need to enter it when you log into your account, tap Save Password; if you prefer to have to enter your password, tap Not Now.

To Save or Not to Save?

Saving your password is convenient because you don't have to enter it each time you sign into Facebook. It also means that anyone who can access the Facebook app on your phone is able to sign into your account. If you have your phone protected with a passcode (which you should), it's probably fine to save your password. If you don't use a passcode or you let other people use your iPhone, you might want to choose Not Now and enter your password each time.

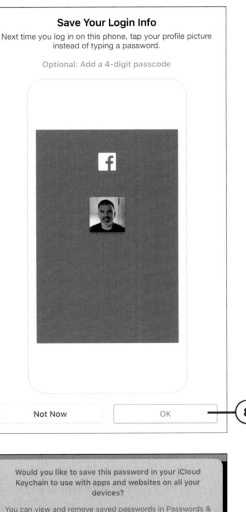

10 If you want the app to be able to access your iPhone's location, tap OK. This enables it to use features that change the information you see based on where you're using your iPhone. If you tap Not Now, Facebook can't use your location information.

Location, Location, Location

Allowing Facebook to access your location means the app can tailor your experience based on where you are, such as identifying events close to you. However, this can also expose your location to other Facebook users, which you might not want to allow. If you're concerned about making your location available to Facebook, tap Not Now or Don't Allow. The app still works fine, but it can't provide any information to you based on your location.

11 If you tapped OK in step 10, tap Allow While Using App to limit Facebook to using your location only while you're using the app, Allow Once to enable the app to use your location only one time, or Don't Allow to prevent it from using your location.

12 Use the Facebook app to post comments, add photos, and so on.

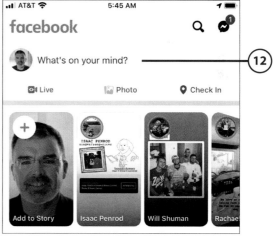

Facebook Settings

Like most of the apps that you install on your iPhone, the Facebook app has settings you can use to change how it works. To configure these settings, open the Settings app and tap Facebook. For example, you can prevent Facebook from updating its information when you're connected to the Internet with a cellular connection by setting the Cellular Data switch to off (white). You might want to do this if you have a limited data plan so that the Facebook app only transmits or receives information when your iPhone is connected to a Wi-Fi network.

Setting How and When Your Accounts Are Updated

The great thing about online accounts is that their information can be updated any time your iPhone can connect to the Internet. This means you have access to the latest information, such as new emails, changes to your calendars, and so on. There are three basic ways information gets updated:

- **Push**—When information is updated via Push, the server pushes (thus the name) updated information onto your iPhone whenever that information changes. For example, when you receive a new email, that email is immediately sent (or pushed) to your iPhone. Push provides you with the most current information but uses more battery than Fetch or Manual do.

- **Fetch**—When information is updated via Fetch, your iPhone connects to the account and retrieves the updated information according to a schedule; for example, every 15 minutes. Fetch doesn't keep your information quite as current as Push does, but it uses much less battery than Push does.

- **Manual**—You can cause an app's information to be updated manually. This happens whenever you open or move into an app or by a manual refresh. For example, you can get new email by moving onto the Inboxes screen in the Mail app and swiping down from the top of the screen. With this method, you have to take action to get updated information, but it uses the least amount of power (assuming you aren't constantly manually updating the information, of course).

You can configure the update method that is used globally, and you can set the method for specific accounts. Some account types, such as iCloud, support all three options, whereas others might support only Fetch and Manual. The global option for updating is used unless you override it for individual accounts. For example, you might want your personal email account to be updated via Push so your information there is always current, whereas configuring Fetch for email associated with a club of which you're a member might be frequent enough.

Configuring How New Data Is Retrieved for Your Accounts

To configure how your information is updated, perform the following steps:

(1) On the Home screen, tap Settings.

(2) Tap an account type that uses cloud data, such as Mail or Calendar.

Where to Begin?

It doesn't really matter what type of account you tap in step 2 because you access the same account tools. For example, if you tap Contacts instead of Mail or Calendar, you can perform the same steps.

(3) Tap Accounts.

(4) Tap Fetch New Data.

(5) To enable data to be pushed to your iPhone, slide the Push switch to on (green). To disable push to extend battery life, set it to off (white). This setting is global, meaning that if you disable Push here, it's disabled for all accounts even though you can still configure Push to be used for individual accounts. For example, if your iCloud account is set to use Push but Push is globally disabled, the iCloud account's setting is ignored.

(6) To change how an account's information is updated, tap it. The account's screen displays. The options on this screen depend on the kind of account it is. You always have Fetch and Manual; Push is displayed only for accounts that support it.

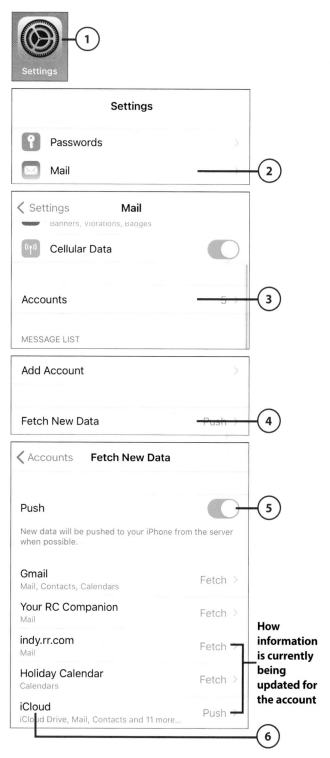

How information is currently being updated for the account

(7) Tap the option you want to use for the account: Push, Fetch, or Manual.

If you choose Manual, information is retrieved only when you manually start the process by opening the related app (such as Mail to get your email) or by using the refresh gesture, regardless of the global setting.

If you choose Fetch, information is updated according to the schedule you set in step 11.

(8) If you choose the Push option in step 7 and are working with an email account, choose the mailboxes whose information you want to be pushed by tapping them so they have a check mark; to prevent a mailbox's information from being pushed, tap it so that it doesn't have a check mark. (The Inbox is selected by default and can't be unselected.)

(9) Tap Fetch New Data (<).

(10) Repeat steps 6 through 9 until you have set the update option for each account. (The current option is shown to the right of the account's name.)

(11) Tap the amount of time when you want the iPhone to fetch data when Push is turned off globally or for those accounts for which you have selected Fetch or that don't support Push; tap Manually if you want to manually check for information for Fetch accounts or when Push is off. Information for your accounts is updated according to your settings.

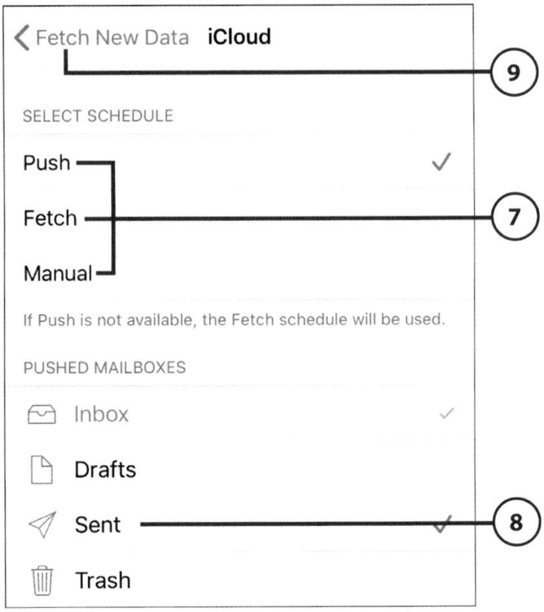

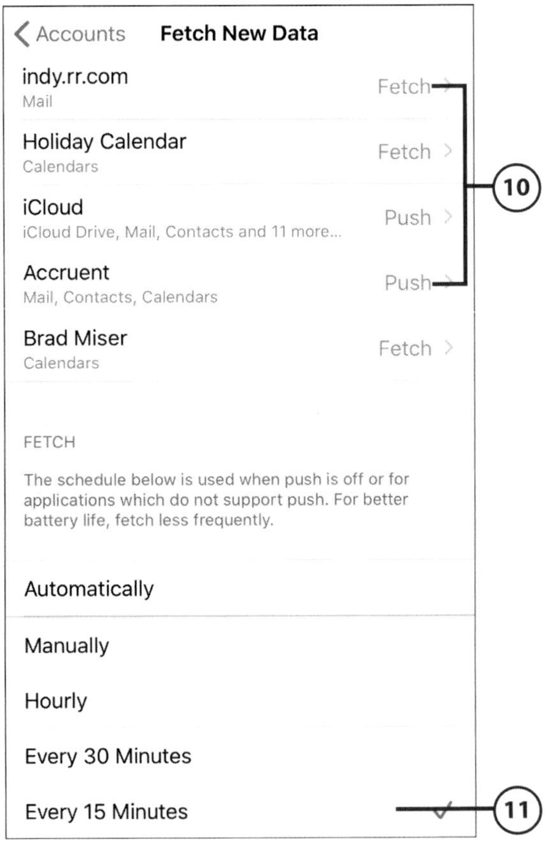

< Fetch New Data   iCloud

SELECT SCHEDULE

Push   ✓

Fetch

Manual

If Push is not available, the Fetch schedule will be used.

PUSHED MAILBOXES

Inbox   ✓

Drafts

Sent   ✓

Trash

< Accounts   Fetch New Data

indy.rr.com   Fetch
Mail

Holiday Calendar   Fetch >
Calendars

iCloud   Push >
iCloud Drive, Mail, Contacts and 11 more...

Accruent   Push
Mail, Contacts, Calendars

Brad Miser   Fetch >
Calendars

FETCH

The schedule below is used when push is off or for applications which do not support push. For better battery life, fetch less frequently.

Automatically

Manually

Hourly

Every 30 Minutes

Every 15 Minutes   ✓

>>>*Go Further*

TIPS FOR MANAGING YOUR ACCOUNTS

As you add and use accounts on your iPhone, keep the following points in mind:

- You can temporarily disable data for any account by moving to the Accounts screen and tapping that account. Set the switches for the data you don't want to use to off (white). You might be prompted to keep or delete that information; if you choose to keep it, the data remains on your iPhone but is disconnected from the account and is no longer updated. If you delete it, you can always recover it again by simply turning that data back on. For example, suppose you're going on vacation and don't want to deal with club-related email or meeting notifications. Move to your club email account and disable all its data. That data disappears from the related apps; for example, the account's mailboxes no longer appear in the Mail app. When you want to start using the account again, simply re-enable its data.

- If you want to completely remove an account from your iPhone, move to its configuration screen, swipe up the screen, and tap Delete Account. Tap Delete from My iPhone in the confirmation dialog box and the account is removed from your iPhone. (You can always sign in to the account to start using it again.)

- You can have different notifications for certain aspects of an account, such as email. See Chapter 6, "Making Your iPhone Work for You," for the details about configuring and using the notifications that your online accounts use. For example, you might want to hear a different sound when you receive personal emails versus those sent to a club email account.

- Although using Push is great because you always have the most current information on your iPhone, it has one certain drawback—it uses more power, draining your battery faster than Fetch or Manual—and one potential drawback—it uses more data. As long as you keep an eye on your battery's status, using more power probably won't be a big deal. But if you have a plan that has a limited amount of data each month, you need to consider whether using Push is worth the additional data it consumes. If you have an unlimited data plan, the extra data Push uses doesn't matter at all.

Configure your iPhone to make installing and maintaining apps easy

Install apps so you can do all kinds of useful and fun things with your iPhone

Use widgets for quick information and easy access to apps

In this chapter, you learn how to make an iPhone into your iPhone by making it work the way you want it to. Topics include the following:

→ Getting started
→ Customizing how your iPhone works with apps
→ Customizing how your iPhone works with widgets

5

Customizing Your iPhone with Apps and Widgets

Your iPhone comes with a lot of very useful apps that can do all sorts of amazing things. But that's just for starters. An iPhone gets even more useful as you add apps that enable you to do all types of useful and fun things.

Getting Started

The iPhone is a powerful computing platform that software developers can use to deliver amazing apps for you to use. And, because of the App Store (which is an app in and of itself), installing and maintaining apps on your iPhone is very simple.

Following are some key concepts you'll use as you explore the wonderful world of iPhone apps:

- **App Store app**—This app connects your iPhone to Apple's App Store, which is where you find apps that you can explore and download to your phone— some free, some for a fee. In addition to being very easy to use, the App Store also protects you from harmful apps. That's because before an app can be posted in the App Store, it must pass a rigorous set of tests by Apple to ensure that it can't do any damage to your iPhone or to your information.

- **Apple ID**—To access the Apple App Store, you need to have an Apple ID. See Chapter 4, "Setting Up and Using an Apple ID, iCloud, and Other Online Accounts," for the details of obtaining and configuring an Apple ID on your iPhone. As part of that process, you likely provided payment information to Apple, such as a credit or debit card. If you did, it's very convenient to pay for purchases from the App Store. If you didn't set up a payment method, you'll need to provide that information if you download any apps that have a price. Even without payment information, you can still shop for, download, and use thousands of apps for free.

- **Face ID**—As discussed in Chapter 2, "Getting Started with Your iPhone," your iPhone's Face ID feature enables you to sign into accounts or enter passwords by simply looking at your iPhone. You can configure the App Store to accept Face ID so that you don't have to enter your Apple ID password every time you download an app.

- **Touch ID**—Also covered in Chapter 2, Touch ID enables you to enter pass-words or sign into accounts by touching a finger to the Touch ID/Home but-ton. Using Touch ID makes downloading apps easier and faster because you only have to touch your phone rather than typing in a password each time.

- **Widgets**—Some apps provide widgets, which are mini versions of those apps that provide key information or functionality in a small window. You can move into the full app by tapping its widget.

- **Widget Center**—The Widget Center provides a collection of widgets that are ready for you to use. You can determine the widgets that are in the Widget Center and how they are organized.

- **Smart Stacks**—You can "stack" widgets on top of each other so that you can access multiple widgets from one location.

Installing apps on your iPhone enables it to do so much more than it can "out of the box." You'll want to explore and download apps to completely customize how you use your iPhone; the possibilities of what your phone can do with apps are limitless!

Widgets make using apps even easier and more convenient. You can configure widgets to suit your preferences, and you might find that widgets are just as, or more, useful than their associated apps.

Customizing How Your iPhone Works with Apps

Installing apps on your iPhone enables you to add more functionality than you can probably imagine. As the old Apple ad used to proclaim, "There's an app for that." And in all likelihood, there probably is an app for a lot of what you would like to use your iPhone for. The App Store app enables you to find, download, and install apps onto your iPhone.

Before you jump into the App Store, take a few moments to ensure your iPhone is configured for maximum ease and efficiency of dealing with new apps.

First Things First

The rest of this chapter assumes you have configured an Apple ID on your iPhone and that you have Face ID or Touch ID set up. If you don't have an Apple ID yet, see Chapter 4. If you haven't configured your iPhone for Face ID or Touch ID, go back to Chapter 2, "Getting Started with Your iPhone," and do so. Then come back here.

Configuring Your iPhone to Download and Maintain Apps

Use the following steps to configure your iPhone so that it's extremely easy to download and install apps and to keep them current:

1. Open the Settings app.

2. Tap App Store.

Settings	
![A] App Store	> 2
![Wallet] Wallet & Apple Pay	>

(**3**) Ensure the Apps switch is on (green); this causes any apps you download onto other devices using the same Apple ID to be installed on your iPhone automatically. This helps you have the same apps available on all your devices.

(**4**) Ensure the App Updates switch is on (green); this causes any updates to apps you have installed on your iPhone to be downloaded and installed automatically. I recommend you use this option so you can be sure you are always running the most current versions of your apps.

(**5**) If you don't have an unlimited cellular data plan, you might want to set Automatic Downloads to off (white) so apps and other content are downloaded only when you're connected to a Wi-Fi network. If this is enabled (green), apps and content are downloaded to your iPhone when you're using a cellular network, which can consume significant amounts of your data plan.

(**6**) If you have Automatic Downloads enabled, tap App Downloads. If not, skip to step 9.

(**7**) Choose how you want apps and updates to be downloaded when you're using a cellular connection. The options are Always Allow, Ask If Over 200 MB, or Always Ask. If you choose one of the latter two options, you're prompted to allow downloads in the corresponding situations.

(**8**) Tap App Store (<).

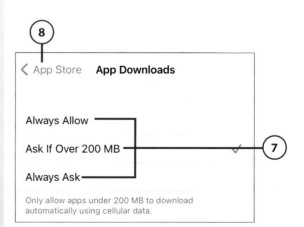

9 Tap Video Autoplay.

10 Some content in the App Store has a video preview; if you want this video to play automatically, leave Video Autoplay set to On. If you want it to play only when you're using Wi-Fi, tap Wi-Fi Only; this is a good option if you have a limited cellular data plan. If you don't want video previews to play at all, tap Off.

11 Tap App Store (<).

12 If you want to be able to provide feedback about apps you use, set In-App Ratings & Reviews to on (green). If you set this to off (white) instead, you won't be prompted to rate and review apps as you use them.

13 If you want apps that you don't use to be removed from your iPhone, set the Offload Unused Apps to on (green). This is handy because you likely won't end up using all the apps you download or you might use some for only a short time and then not need them again. When Offload Unused Apps is on, these unused apps are removed automatically so they aren't cluttering up your phone or taking up storage space. (Any data or documents are saved and reinstalled if you ever install the app again.)

14 Tap Settings.

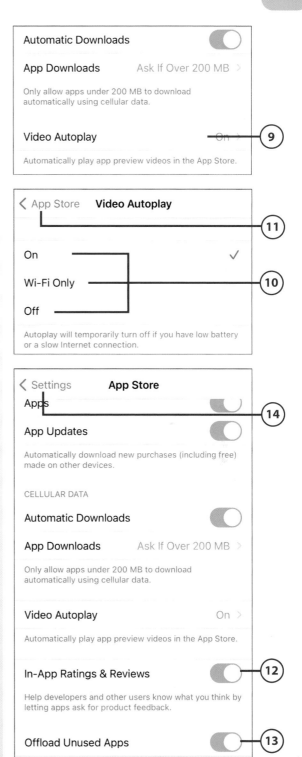

(15) Tap Face ID & Passcode.

You've Got the Touch

These steps show setting up an iPhone for the App Store on an iPhone that uses Face ID. If your iPhone uses Touch ID instead, the steps are nearly identical, but you see references to Touch ID on your screens rather than to Face ID as shown in the figures.

(16) Enter your passcode.

(17) Ensure the iTunes & App Store switch is on (green). If it isn't enabled, tap the switch, enter your Apple ID password, and tap OK to enable it.

(18) Tap Settings (<).

(19) Tap Home Screen.

(20) Tap Add to Home Screen if you want apps you download to be added directly to a Home screen or App Library Only if you don't want them added directly to a Home screen (they are only available via the App Library). Refer to "Working with the App Library" later in this chapter for an explanation of the App Library.

(21) If you want badges to appear on app icons when those apps are in the App Library, set the Show in App Library switch to on (green). With this switch off, app badges only appear when an app is on a Home screen.

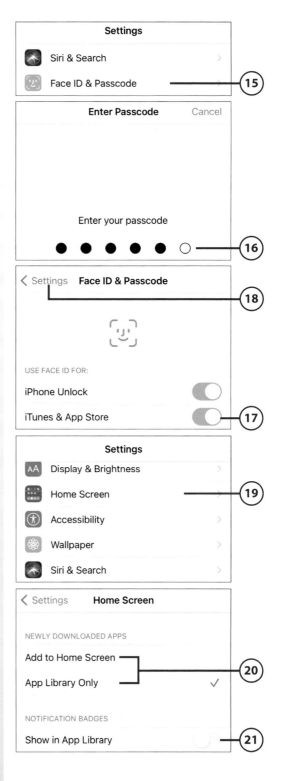

Using the App Store App to Find and Install iPhone Apps

The App Store app enables you to quickly and easily browse and search for apps, view information about them, and then download and install them on your iPhone with just a few taps.

When you use the App Store app, you can find apps to download using any of the following options:

- **Today**—This tab takes you to apps featured in the App Store on the day you visit the store. When you tap any of the items on the Today screen, you move into the group or app on which you tapped.

- **Games**—Easily the most popular category of apps, Games enables you to find those critical games you need to prove your skills and pass the time. Games are grouped in various ways, such as What to Play This Week, Popular Games, New Games We Love, and More Games You Might Like.

- **Apps**—This category leads you to apps that aren't games. On this screen, you see a number of lists including New to iPhone?, Popular Apps, Apps We Love Right Now, Top Free Apps, and Top Paid Apps. You can use the Top Categories list to see apps organized by category, which is a useful way to find apps you want to use for specific purposes. You can browse apps in the top categories directly from the list or tap See All to browse all categories.

- **Arcade**—This option enables you to subscribe to Apple's Arcade service that provides many games you can play without ads or in-game charges. Apple regularly adds to the games in the Arcade so you can always find new games to try. You can try the Arcade service for a month for free; thereafter, a monthly fee is required.

- **Search**—This tool enables you to search for apps. You can search by name, developer, and other keywords.

Finding and downloading any kind of app follows this same pattern:

1. **Find the app you're interested in.** You can use the options described in the previous list to find apps by browsing for them, or you can use the search option to find a specific app quickly and easily.

2. **Evaluate the app.** The information screen for apps provides lots of information that you can use to decide whether you want to download an app (or not). The information available includes a text description, screenshots, and ratings and reviews from users.

3. **Download and install the app**. This usually requires only a tap, two presses of the Side button, and a glance at your iPhone (Face ID) or a tap and then a touch on the Touch ID/Home button (Touch ID). If you don't have Face ID or Touch ID configured, you have to enter your Apple ID password to download apps.

The following tasks provide detailed examples for each of these steps.

Searching for Apps

If you know something about an app, such as its name, its developer, or its purpose you can quickly search the App Store to find the app. Here's how to search for an app:

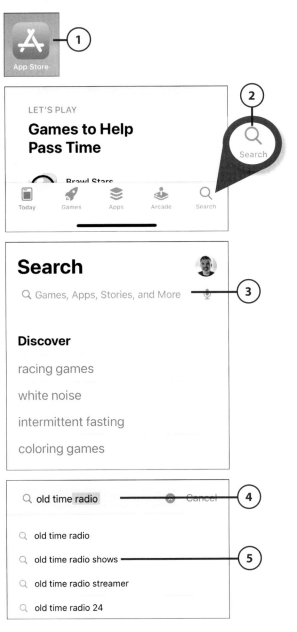

1. Move to the Home screen and tap App Store.

2. Tap Search.

3. Tap in the Search box.

Follow the Trends?

Before you enter a search term on the Search screen, you see the Trending section, which shows the searches that are being performed most frequently. You can tap one of these to use it to search for apps.

---

4. Type a search term. This can be the type of app you're looking for based on its purpose (such as *Travel*) or the name of someone associated with the app, its title, its developer, or even a topic. As you type, the app suggests searches that are related to what you're typing.

5. Tap the search you want to perform or tap the Search key on the keyboard to search for the term you entered in step 4. The apps that meet your search term appear.

6 Swipe up and down on the screen to review the apps in the search results.

7 If none of the apps are what you are looking for, tap Clear (x) in the Search box and repeat steps 4–6.

8 When you find an app of interest to you, tap it. You move to the app's information screen.

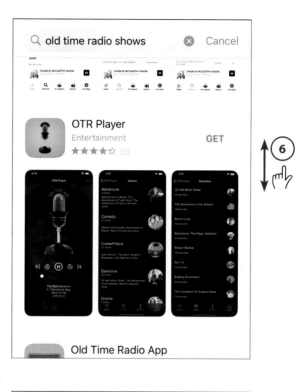

9 Use the app's information on the information screen to evaluate the app and decide if you want to download it. You can read about the app, see screenshots (tap them to make them larger), play video previews, and read other people's reviews to help you decide. If you want to download the app, see "Downloading Apps" later in this chapter for the details.

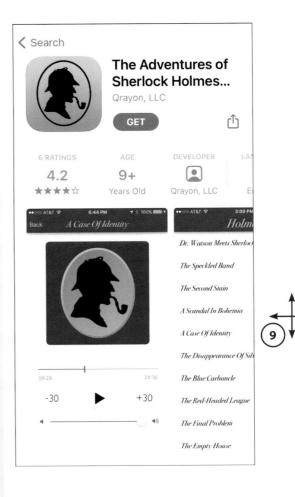

Browsing for Apps

If you don't know of a specific app you want, you can browse the App Store. To browse, you can tap any graphics or links you see in the App Store app. One of the most useful ways to browse for apps is by using categories.

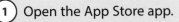

1 Open the App Store app.

2 Tap Apps (browsing for games or using the Today option works very similarly).

3 Swipe up the screen until you see the Top Categories section.

4 Tap See All to browse all available categories. (If you see a category on the Top Categories list that interests you, tap it instead and skip to step 6.)

5 Swipe up and down to browse the categories.

6 Tap a category in which you are interested.

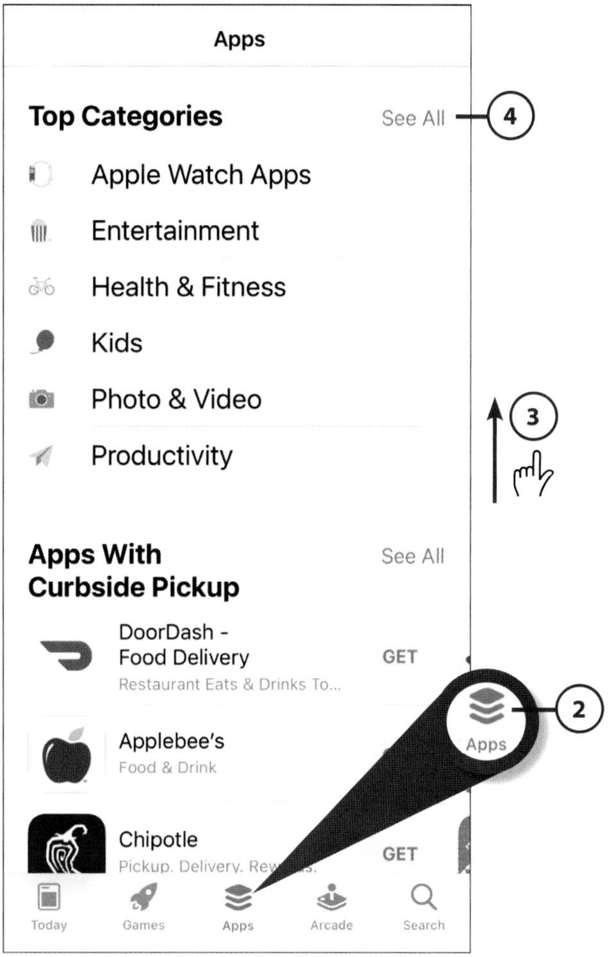

Apps

Top Categories — See All — **4**

- Apple Watch Apps
- Entertainment
- Health & Fitness
- Kids
- Photo & Video
- Productivity

3

Apps With Curbside Pickup — See All

DoorDash - Food Delivery — GET
Restaurant Eats & Drinks To...

Applebee's
Food & Drink

2

Chipotle — GET
Pickup. Delivery. Rew...

Today   Games   Apps   Arcade   Search

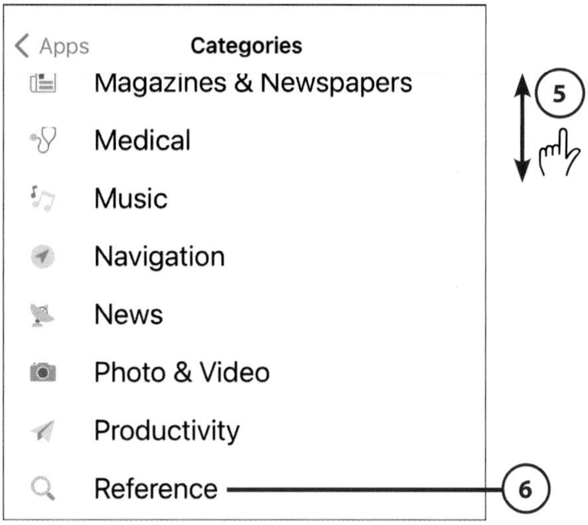

‹ Apps   **Categories**

- Magazines & Newspapers
- Medical
- Music
- Navigation
- News
- Photo & Video
- Productivity
- Reference —— **6**

5

(7) Swipe up and down to browse the groupings of apps, such as Top Free, Apps We Love, or Top Paid.

(8) Tap See All for a grouping to browse the apps it contains.

(9) Swipe up and down to browse the apps in the group you selected in step 7.

(10) Tap an app in which you are interested. You move to that app's information screen.

(11) Use the information on the information screen to decide whether you want to download the app. You can read about the app, see screenshots, play video previews, and read other people's reviews to help you decide. If you want to download the app, see "Downloading Apps" later in this chapter for the details. Or you can continue browsing by tapping Back (<) in the top-left corner of the screen to return to the previous list.

>>>Go Further
MORE ON FINDING APPS

Following are a few pointers to help you use the App Store:

- Some apps include video previews indicated by the Play icon on an image. Tap the Play icon to watch it. Tap Done in the upper-left corner of the screen to move back to the screenshots.

- After you have used an app, you can add a review by moving back to its Reviews tab and tapping Write a Review. You move to the Write a Review screen where you have to enter your App Store account information before you can write and submit a review. If the In-App Ratings & Reviews setting is enabled, you're prompted to provide feedback for apps that you use.

- You can read user reviews of the apps in the App Store. You should take these with a grain of salt. Some people have an issue with the developer or the type of app, are reviewing an older version of the app, or are commenting on issues unrelated to the app itself; these issues can cause an unfairly low rating. The most useful individual user reviews are very specific, as in "I wanted the app to do x, but it only does y." Sometimes, it can be more helpful to look at the number of reviews and the average user rating than reading the individual reviews.

Downloading Apps

Downloading and installing apps is about as easy as things get, as you can see:

① In the App Store, view the app you want to download.

② Tap GET (for free apps) or the price (for apps that have a license fee). You're prompted to download and install the app.

___

Why Apple ID? Why?

At times, you might be prompted to enter your Apple ID password to download an app even if you have Face ID or Touch ID set up to work with the App Store. This can happen for a variety of reasons so don't be too surprised if it happens to you. After you've used Face ID or Touch ID for a while, you might find it annoying to have to actually type a password... at least I do.

___

More Interruptions?

Periodically, you might have to review and update payment information, license agreements, and the like. When this happens, follow the onscreen directions to complete whatever you need to do. When that process is complete, you go back to downloading the app. Fortunately, these interruptions don't happen that often.

3 Press the Side button twice and look at the screen to confirm that you want to download and install the app (Face ID) or touch the Touch ID/Home button (Touch ID) (Touch ID not shown in the figures).

After you see the Done message, the app starts to download, and you see the progress of the process.

Wait for It... Or Not

You can do something else while apps are downloading; for example, look for and download other apps or move into a different app. The download process completes on its own.

When the process is complete, the status information is replaced by the OPEN button. You can tap OPEN to start using the app.

3

Double Click to Install

App Store   Cancel

Merriam-Webster Dictionary
Merriam-Webster, Inc.
Offers In-App Purchases

ACCOUNT

Confirm with Side Button

Merriam-Webster Dictionary
Merriam-Webster, Inc.

The app is being downloaded

‹ Back

Merriam-Webster Dictionary
Merriam-Webster, Inc.

OPEN

This app is ready to use

84K RATINGS   AGE   CHART

>>>Go Further

MORE ON APPS

As you use the App Store app to install apps on your iPhone, keep the following hints handy:

- When you install an app, it is placed in the first "open" position on the Home screens if you selected the Add to Home Screen option on the Home Screen settings page. You can move it to a more convenient location as described in Chapter 7, "Customizing How Your iPhone Looks." If you selected the App Library option instead, apps you download are available only in the App Library.

- Like other software, apps are updated regularly to fix problems, add features, or make other changes. If you set the Updates setting to on (green) as described earlier in this chapter, updates to your apps happen automatically in the background. Your apps are always current so you don't have to update them manually. (More information on updating apps is in Chapter 18, "Maintaining and Protecting Your iPhone and Solving Problems.")

- If you tap the icon for your user account located in the upper-right corner of the App Store app's screen, you access a number of other features on the Account screen. You can tap your account to update it. You can see the apps you've downloaded (called Purchased regardless of if they are free or required a fee), see your subscriptions, redeem gift cards, and more. At the bottom of the Account screen, you see a list of apps that have been updated recently.

- If you see the Download icon next to an app rather than GET or BUY, that means you have previously downloaded (and paid for, if it isn't free) the app but it's not currently installed on your iPhone. Tap the icon to download and install it; if it's a paid app, you won't have to pay for it again.

- To let someone else know about an app, tap the Share icon on the app's page and then tap how you want to let him know; options include AirDrop, Messages, and Mail.

- Apps can work in the background to keep their information current, such as Weather and Stocks. To configure this, open the Settings app, tap General, and tap Background App Refresh. Tap Background App Refresh again and tap Off to disable this, Wi-Fi to enable it only when your iPhone is connected to a Wi-Fi network, or Wi-Fi & Cellular Data to allow it any time your iPhone is connected to the Internet. Tap Back (<), located in the upper-left corner of the screen, to see the list of apps installed on your iPhone. To enable an app to work in the background, set its switch to on (green). To disable background activity for an app, set its switch to off (white).

- After you install an app, move to the Settings screen and look for the app's icon. If it's there, the app has additional settings you can use to configure the way it works. Tap the app's icon in the Settings app and use its Settings screen to configure it. If you don't see an app you have installed on the Settings screen, move back to a Home screen and tap the app's icon to launch it. (You might have to had opened an app once for it to appear.) Move back to the Settings screen and look for the app again. If it isn't there, it doesn't have settings.

Working with the App Library

It is likely you will have a lot of apps installed on your iPhone. When you install an app, it is placed on the next available slot on your Home screen (unless you chose the App Library Only option, in which case new apps are not added to a Home screen). Over time, you can end up with many pages of the Home screen. Swiping through all of these pages to find a specific app can be time consuming and annoying. Of course, you can quickly search for a specific app (see "Searching on Your iPhone," in Chapter 3, "Using Your iPhone's Core Features") to locate and use it.

You can also use the App Library to find apps that you want to use. On the App Library screen, you can search for apps or browse apps in categories that are created for you automatically. For example, apps you use to listen to podcasts or watch television shows are grouped into the Entertainment category.

Keeping Things Tidy

You can organize your iPhone such that the first page or two of the Home screens contain the apps you use most frequently. You can use the App Library to get to those you don't use that often. You can hide all the pages of the Home screens that you don't use that often to make accessing the App Library even faster. (See Chapter 7, "Customizing How Your iPhone Looks," for details.)

To keep your Home screens even tidier, use the App Library Only setting so that new apps only appear in the App Library. You can move apps you end up using frequently to a Home screen for faster access.

The App Library provides several ways to quickly find apps you want to use.

Using the App Library's Categories to Find Apps

The App Library automatically groups apps by category. These categories make it easy to get to apps.

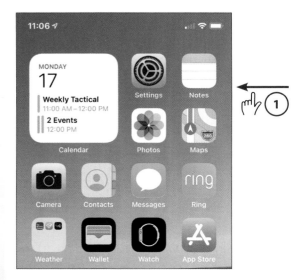

(1) Move to a Home screen and swipe to the left as far as you can. The App Library appears.

(2) Swipe up or down the screen to see the categories of apps available. Within the categories, you see two sizes of app icon. Larger icons lead you directly into an app. Smaller icons lead you to a group of apps.

(3) To open an app, tap its large icon. The app launches, and you can skip the rest of these steps.

(4) To use a small-icon app, tap the group in which it is contained. The group expands to show all of the apps in that group.

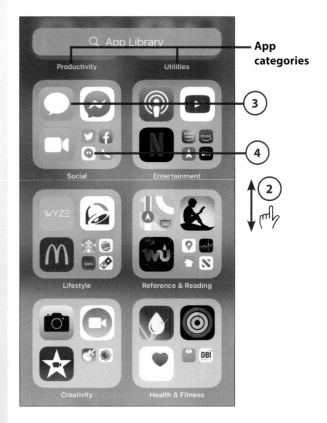

App categories

(5) Tap the app you want to use.

Using the App Library to Search for Apps

You can use the App Library to search for apps as follows:

(1) Move to a Home screen and swipe to the left as far as you can. The App Library appears.

(2) Tap in the Search bar.

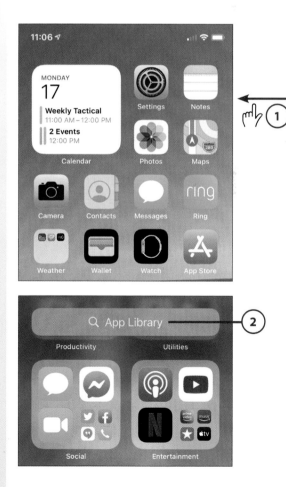

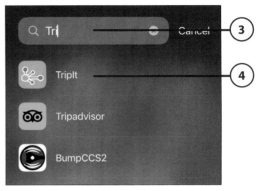

(3) Type your search term. As you type, the list of apps shown narrows to just those that match your search.

(4) When you see the app you want to use, tap it.

Using the App Library to Browse for Apps

Browsing the App Library is fast and can sometimes help you find an app when you don't remember what it was called or you might have even forgotten it is on your phone.

(1) Open the App Library and tap in the Search bar. You see a list of all the apps installed on your phone.

(2) Swipe up the screen. The keyboard closes and the list of apps fills the screen.

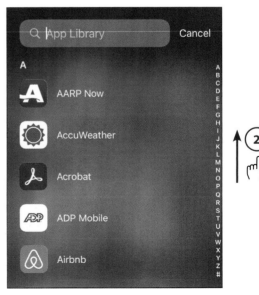

3 Swipe up and down the screen to browse the list of apps.

4 Swipe up and down on the index to browse rapidly.

5 When you see the app you want to use, tap it.

Deleting Apps

As you accumulate apps, it is likely that you'll end up with a number of apps on your phone that you never or seldom use. These clutter your screen and consume some of your iPhone's memory. It's a good idea of get rid of apps you don't use.

The Phone Does It for You

If you enabled the Offload Unused Apps setting as described earlier in the chapter, unused apps disappear from your phone automatically.

To get rid of an app whose time has come, perform the following steps:

1 Move to a Home screen containing the app you want to remove.

2 Touch and hold on the app's icon until the menu appears.

3 Tap Remove App.

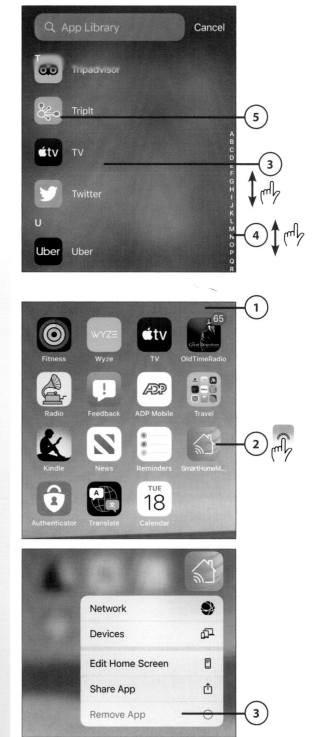

(4) To remove the app from the Home screen but keep it on your phone, tap Move to App Library. The icon disappears from the Home screen, but you can still access the app through the App Library. This is useful if you aren't sure that you won't want the app or its data again. Skip the rest of these steps.

(5) To delete the app from your phone, tap Delete App.

(6) If you are sure you won't need any data stored by the app, tap Delete. The app and its data are deleted from your phone.

When It's Gone, It's Gone

When you delete an app from your phone, any data associated with that app that is only stored on your phone is deleted, too. So, make sure you don't need that data any more before you delete the app.

Pay Once, Download Forever

Don't be concerned about deleting apps that you have paid for because you can download and install them again for no additional fee.

>>>Go Further
APP CLIPS

App clips enable you to use an app's functionality quickly and easily without needing to install the app on your phone. For example, suppose you want to park in a lot that uses an app to collect payment but you don't have that app installed on your phone. Instead of taking the time to find and download the app, you can scan a QR code, an App Clip code, or Near Field

Communication (NFC) tag. The associated app clip appears automatically, and you can complete whatever you are trying to do, such as to pay for a parking spot.

QR Code   *App Clip Code*   *NFC Code*

And in some cases, you might want to use the app only once so it's not worth installing on your phone. App clips can help prevent your phone from being overloaded with unused or seldom-used apps.

To use app clips, look for a sign or placard providing an app clip code that you can scan. Point your phone at the code; if the code isn't recognized and scanned automatically, press the Side button at the prompt to scan it. When the clip appears, tap Open and follow the onscreen prompts to use the app's functionality. Any identification or payment you need to provide comes from your Apple ID and Apple Pay, so you don't have to create additional accounts or provide payment information to complete the action you are performing.

App clips that you have used recently can be found in the App Library. When you want to get back to a clip, swipe all the way to the left on a Home screen to open the App Library. Browse or search for the clip you want to use. Tap it to launch it.

Customizing How Your iPhone Works with Widgets

Widgets are "mini" versions of apps installed on your iPhone that provide you with information at-a-glance and enable you to move into the associated app. You can access widgets from three locations:

- **Widget Center**—The Widget Center provides a collection of widgets that you can access and use quickly. You can determine which widgets are installed and how those widgets are organized.

- **Smart stacks**—A smart stack enables you to access multiple widgets that share the same space on the screen. All the widgets appear within the same window; you can swipe through them to change the active widget. Smart stacks can also be configured to "flip" through widgets automatically.

- **Widgets on Home screens**—You can put widgets and smart stacks on Home screens, which makes them even easier to access. They appear alongside app and folder icons and work similarly.

Working with the Widget Center

Swipe to the right to open the Widget Center

You can open the Widget Center in a number of ways:

- Wake your iPhone and swipe to the right on the Lock screen.

- Move to a Home page and swipe all the way to the right.

- Open the Notification Center and then swipe to the right.

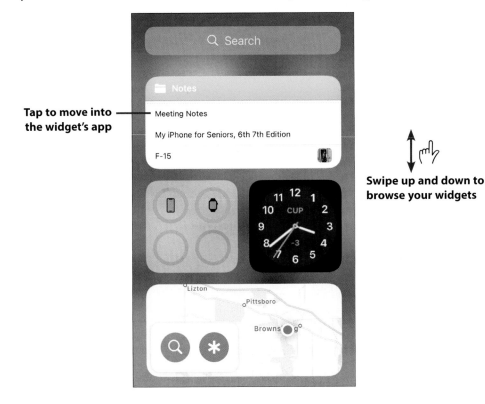

Tap to move into the widget's app

Swipe up and down to browse your widgets

At the top of the Widget Center, you see the Search tool. Beneath that, you see widgets for apps installed on your iPhone. Swipe up and down the screen to browse your widgets.

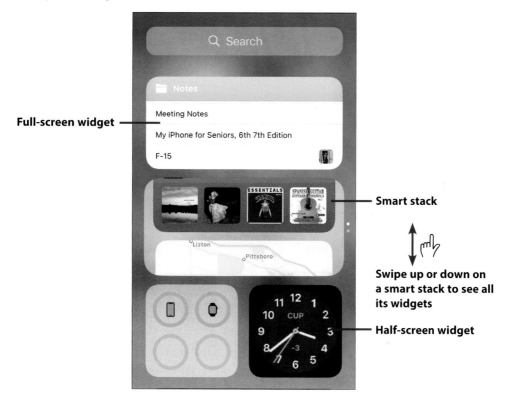

Full-screen widget

Smart stack

Swipe up or down on a smart stack to see all its widgets

Half-screen widget

Widgets come in two different sizes: half-screen or full-screen. Half-screen widgets can be placed side by side so two of them appear on the same row. Full-screen widgets take up an entire row.

Each widget provides information or functions based on its app. For example, you can use the Notes widget to see your recent notes or notes folder. You can see your daily calendar in the associated widget, get news in the News app's widget, or control music in the Music widget.

You can interact with widgets in several ways. Some widgets provide information that you can view within the widget, such as Calendar, Stocks, or Weather. When you tap something within some widgets (Notes is an example), the associated app opens and either performs the task you indicated or shows more information about what you selected.

Smart stacks contain multiple widgets. You can move among the widgets in a stack by swiping up and down on the stack. As you swipe, the next widget in the stack appears. You can keep swiping until you return to the first widget in the stack. Smart stacks can also be configured to rotate between widgets automatically.

If you don't move into an app from a widget, you can close the Widget Center by swiping to the left. You move back to the screen you came from, such as a Home screen. If you do move into an app from a widget, you work with that app just as if you moved into it from a Home screen.

Be aware that there are two basic types of widgets: core iPhone widgets or app widgets. Core iPhone widgets are installed automatically and behave in specific ways. Some apps that you install on your phone also have widgets; these behave and are configured slightly differently than the core widgets. App widgets always appear at the bottom of the Widget Center.

You can determine which widgets appear in the Widget Center and the order in which those widgets appear on the screen; for example, you might want your most frequently used widgets to be at the top of the screen.

Organizing the Widget Center

To organize the Widget Center, perform the following steps:

1. Open the Widget Center by moving to a Home screen and swiping all the way to the right.

2. Swipe all the way up the screen.

3. Tap Edit. The widgets start jiggling and you see the delete (–) symbol in the upper-left corner of the widgets.

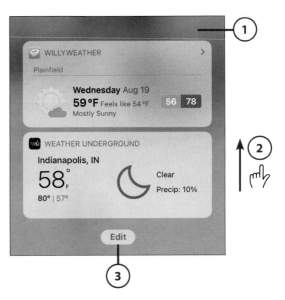

4 To remove a widget, tap remove (–).

5 Tap Remove at the prompt. This removes the widget from the Widget Center (you can add it back if you change your mind).

6 Touch and hold on a widget and drag it up or down the screen to change its location. If you drag a half-screen widget that has another one in the same row, both widgets move at the same time. You can't change the position of app widgets, which always appear at the bottom of the Widget Center.

7 When the widget is where you want it to be, take your finger off the screen. The widget stays in its new location.

8 When you're done removing or moving widgets, tap Done. The widgets are locked in place.

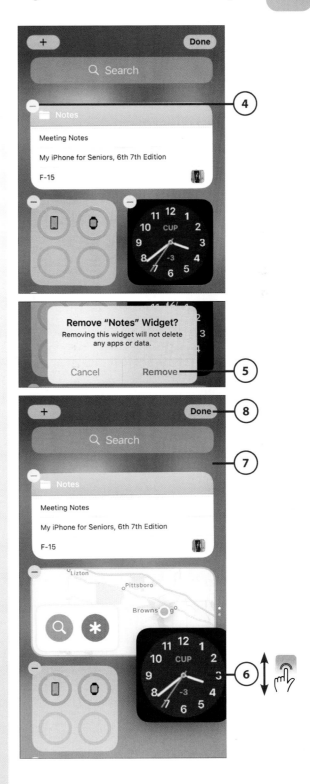

Adding Core Widgets to the Widget Center

You add core widgets to the Widget Center with the following steps:

(1) Open the Widget Center by moving to a Home screen and swiping all the way to the right.

(2) Swipe all the way up the screen.

(3) Tap Edit. The widgets start jiggling and you see the delete (–) symbol in the upper-left corner of the widgets.

(4) Tap Add (+). The palette of available widgets appears.

(5) Find the widget you want to add by swiping up or down the screen to browse for widgets or by searching for a specific widget.

(6) Tap the widget you want to add.

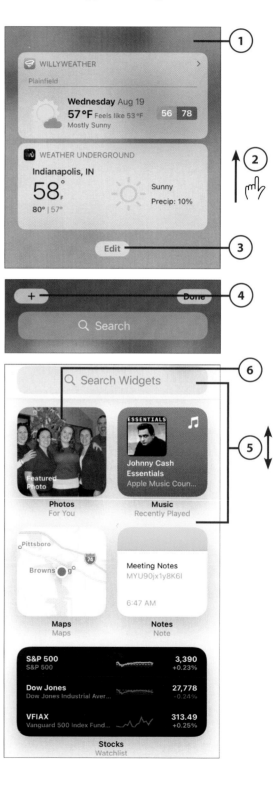

(7) Swipe left and right to see the widget's options.

(8) When you see the option you want to add, tap Add Widget. The widget is added to the Widget Center.

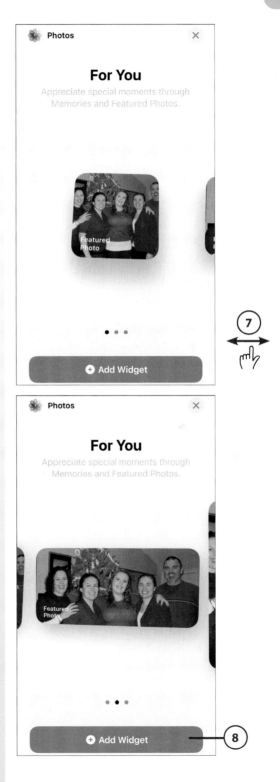

9 Drag the widget to where you want it to appear on the Widget Center.

10 Tap Done.

Adding and Configuring a Default Smart Stack

You can add and configure a smart stack that contains default widgets as follows:

1 Perform steps 1 through 4 in the previous task to put the Widget Center in Edit mode and move to the Add screen.

2 Tap the Smart Stack widget.

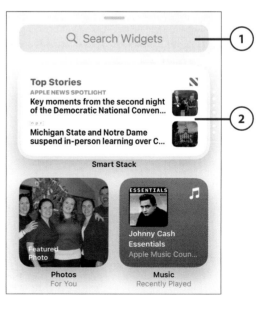

3 Swipe left and right to see the smart stack options.

4 When you see the option you want to add, tap Add Widget. The smart stack is added to the Widget Center.

5 Tap twice on the smart stack. The configure sheet appears.

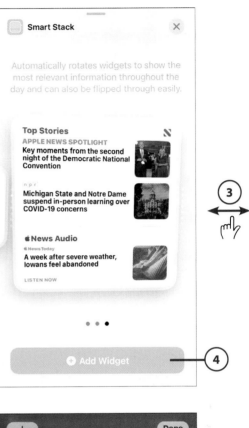

(6) If you don't want the widgets inside the smart stack to rotate automatically, set the Smart Rotate switch to off (white).

(7) To remove a widget from the stack, swipe to the left on it.

(8) Tap Delete. The widget is removed from the stack.

(9) To change the order in which the widgets appear in the stack, drag widgets up or down by their list order icon.

(10) When you're done configuring the stack, tap close (x).

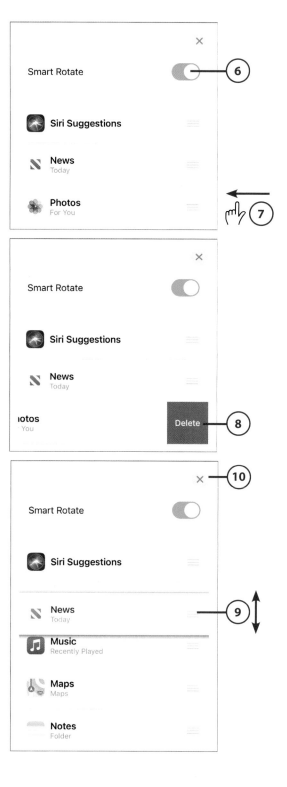

11 Drag the stack up or down the Widget Center.

12 When you're done configuring the stack, tap Done.

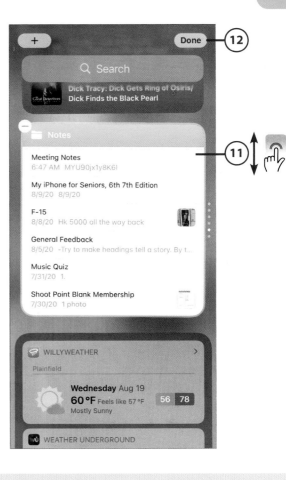

>>>Go Further

APP WIDGETS

App widgets (which come with apps you install on your phone) behave a bit differently than core widgets that are associated with apps preinstalled on your phone. App widgets always appear in a section at the bottom of the Widget Center. Like core widgets, you can view their information or tap them to move into the associated app.

You can also configure app widgets: Put the Widget Center in Edit mode and swipe all the way up the screen. Tap Customize in the app widget section at the bottom of the Widget Center. You see the Add Widgets screen.

At the top of the screen, you see the app widgets currently installed on the Widget Center. Tap Delete (–) to remove them or drag them up or down to change the order in which they appear.

In the MORE WIDGETS section, you see all of the app widgets available. Tap Add (+) to add a widget to the Widget Center. The widget moves to the top of the screen, and you can drag it up or down to change the order in which it appears.

When you are finished configuring app widgets, tap Done. You see the new set of app widgets at the bottom of the Widget Center.

Adding Widgets to Home Screens

You can put your favorite widgets on Home screens to make them more accessible.

(1) Open the Widget Center, put it in Edit mode, and swipe the screen so you can see the widget you want to put on a Home screen.

(2) Drag the widget to the right until the Home screen appears.

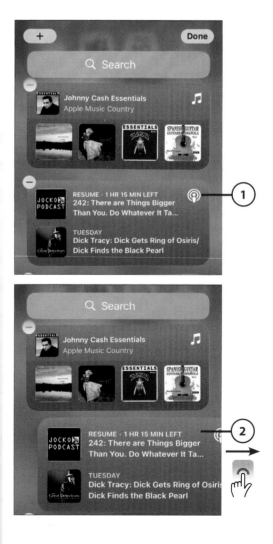

3 Drag the widget around the Home screen or onto a different Home screen until it is where you want it to be placed.

4 Tap Done.

Configuring Widgets on Home Screens

You can configure widgets on Home screens just as you can apps and folders. For more details, see "Customizing Your Home Screens" in Chapter 7.

Notifications keep you aware
of activity

Configure
notifications, the
Control Center,
the Widget Center,
and more

Badges inform
you about new
items

In this chapter, you learn how to turn an iPhone into your iPhone by making it work the way you want it to. Topics include the following:

→ Getting started
→ Configuring and working with notifications
→ Working with and configuring the Control Center
→ Using the Do Not Disturb mode
→ Configuring keyboards
→ Setting accessibility options

Making Your iPhone Work for You

Your iPhone has many tools and options you can use to make it work for you the way you want it to.

Getting Started

In this chapter, you'll learn how to employ a number of very useful elements of your iPhone, including the following:

- **Notifications**—There's a lot of activity going on with your iPhone. Visual, auditory, and vibratory notifications enable you to be aware of that activity. At times, these notifications can be overwhelming. Fortunately, you can determine exactly how and when you receive notifications so that you're informed about what is important to you and not bothered with a lot of stuff that isn't important.

- **Do Not Disturb**—Sometimes, we all want a little peace and quiet. When you put your iPhone in Do Not Disturb mode, it won't bother you with notifications. You can even set Do Not Disturb to activate automatically, such as at night.

- **Control Center**—The Control Center provides quick access to a number of areas of your iPhone so that you can make changes with just a few taps. You can determine the controls that are on the Control Center so that it works even better for you.

- **Keyboards**—Typing is a fundamental task for emails, messages, notes, and much more. In Chapter 3, "Using Your iPhone's Core Features," you learn how to use keyboards to input text and emojis. In this chapter, you learn how to configure the keyboards available to you for all your typing needs.

- **Accessibility Options**—Not everyone sees, hears, or interacts with the world, or an iPhone, in the same way. Accessibility Options enable you to adjust the way your iPhone works to adapt better to how you see, hear, and interact with it.

Configuring and Working with Notifications

Your iPhone uses notifications to keep you informed of important (and at times, not-so-important) information. There are three basic types of notifications:

Banner notifications let you know something has happened

Badges indicate how much new activity there is, such as new messages

- **Visual**—Visual notifications appear on the iPhone's screen. There are two types of visual notifications: Badges and Alerts.

  Badges, which appear as red circles on app icons, indicate the number of new events in an app, such as the number of new emails.

Alerts are onscreen messages about specific activity. You can choose alerts to appear on the Lock screen, Notification Center, or as Banners (you can use any combination of these). Banners can be temporary, which means they appear briefly on the screen and then disappear, or persistent, meaning that you have to take action to clear the notification.

When an alert appears, you can read its information and take action on it, such as responding to a text message.

- **Sounds**—Different kinds of sounds can indicate events that are happening. Obvious examples are the ringtone your iPhone plays when someone calls you, but you can configure auditory notifications for many other events including new emails or FaceTime requests. You can even have different sounds for events of the same kind, such as a different ringtone for specific people.

- **Vibrations (Haptic)**—You can feel this kind of notification when the iPhone vibrates. Like the other types of notifications, you can configure the specific vibrations you feel for certain events.

As you learned at the beginning of the chapter, you can configure the notifications that your phone uses to tailor them to meet your preferences. This allows you to ensure your phone keeps you informed without being overly distracting or annoying.

Working with Visual Notifications

Badge showing
six new emails

Badges appear on an app's or a folder's icon to let you know something has changed, such as new email, messages, or invitations.

Badges are purely informational, meaning you can't take any action on them directly. They inform you about events so that you can take action, such as downloading and installing an update to your iPhone's iOS software or reading new text messages.

Badge showing
activity for all the
apps in the folder

When apps in folders have badges enabled, the badge you see on a folder is a total count of the badges on the apps within that folder. The only way to know which apps in a folder have badges is to open the folder so you can see the individual icons and badges.

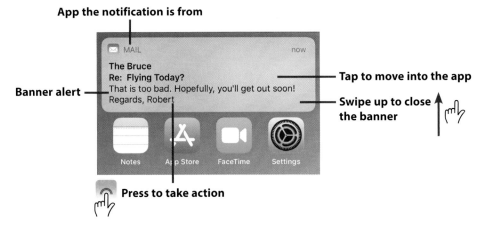

App the notification is from

Banner alert

Tap to move into the app

Swipe up to close
the banner

Press to take action

Alerts appear when activity happens that you might want to know about, such as receiving email or a calendar invitation.

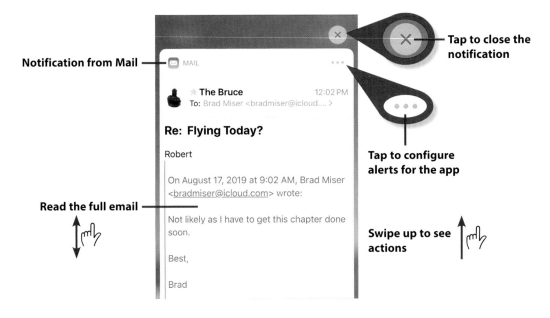

Notification from Mail

Tap to close the
notification

Tap to configure
alerts for the app

Read the full email

Swipe up to see
actions

In addition to providing information for you, alerts enable you to take action related to the activity that generated the notification. For example, when you press on an email notification, you can read the entire message. If you swipe up, you see actions you can take, such as deleting the message.

When your iPhone is unlocked, banner alerts appear at the top of the screen. They provide a summary of the app and the activity that has taken place, such as a new email or text message. When a banner appears, you can view its information; if it is a temporary banner, it rotates off the screen after displaying for a few seconds; if it is a persistent banner, you need to do something to cause it to disappear. You can tap it to move into the app to take some action; for example, to read an email. You can swipe up from the bottom of the banner to close it. For some apps, such as Mail, you can press on the notification to open a menu of commands.

Tap to close the alert

New message

Alert on the Lock screen

Press to take action

Tap to configure alerts for the app

Reply to the message

Alerts can also appear on the Lock screen, which is convenient because you can read and take action on them directly from that screen. If your phone is asleep, the alerts appear briefly on the screen and then it goes dark again (unless the phone is in Do Not Disturb mode, in which case this doesn't happen); you can press the Side button or the Touch ID/Home button or raise your phone to see your alerts without unlocking the iPhone. You can swipe up or down the screen to browse the alerts.

To respond to an alert or take other action on it, press it to open it and then take an action, such as replying to a message. In some cases, you might need to unlock your phone to complete an action associated with an alert. In those cases, you're prompted to use Touch ID, Face ID, or your passcode to proceed.

Working with the Notification Center

Swipe down to open the Notification Center

Swipe up to open the Notification Center

Alerts and other notifications appear in the Notification Center, which you can open by swiping down from the top of the screen when your phone is unlocked or by swiping up on the Lock screen. The Notification Center opens and displays notifications grouped by day and the app from which they come. Notifications from the same app are "stacked" on top of each other so you can see more notifications on the screen.

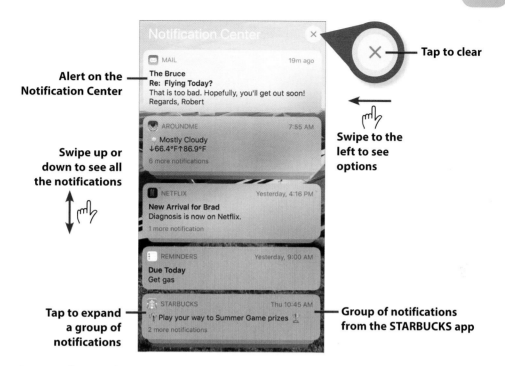

Tap to clear

Alert on the
Notification Center

Swipe to the
left to see
options

Swipe up or
down to see all
the notifications

Tap to expand
a group of
notifications

Group of notifications
from the STARBUCKS app

Using the Notification Center can be efficient because you can see a lot of notifications at the same time without having to be bothered by a banner for each one.

You can read the notifications by swiping up and down the screen. You can work with the notifications on the Notification Center just as you work with individual notifications. For example, press on a notification to pop it open to read more of it or to take action on it.

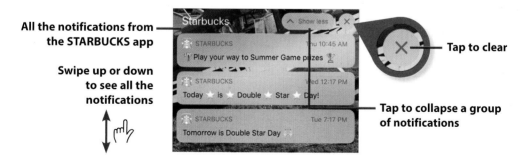

All the notifications from
the STARBUCKS app

Swipe up or down
to see all the
notifications

Tap to clear

Tap to collapse a group
of notifications

When you tap a stack of notifications, they expand so that you can see each notification in the stack. You can work with the individual notifications in the stack or you can collapse the stack again by tapping Show less.

Tap to clear

To clear notifications from a group, tap clear (x). Then tap Clear. The notifications are deleted from the Notification Center.

You can remove all notifications from the Notification Center by tapping clear (x) at the top of the screen and then tapping Clear. All the notifications are deleted and the Notification Center starts collecting new notifications as they come in.

Tap to view the notification

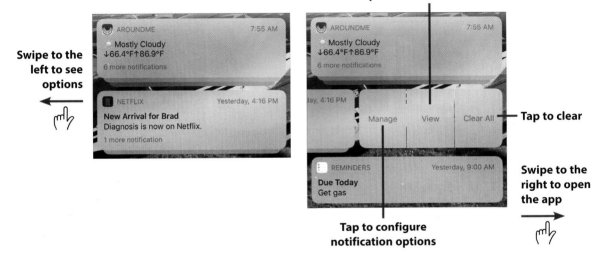

Swipe to the left to see options

Tap to clear

Swipe to the right to open the app

Tap to configure notification options

When you swipe to the left on a notification or group of notifications, you can tap Clear or Clear All to delete the notifications. Tap View to open a notification. Tap Manage to configure notifications for the associated app (this is explained in "Configuring Notifications" later in this chapter).

If you swipe to the right on a notification, you open the app that generated it (if your iPhone is locked, you need to unlock it to move into the app).

To close the Notification Center when your iPhone is unlocked, move back to the Home screen by swiping up from the bottom of the screen (iPhones without a Home button) or pressing the Touch ID/Home button (iPhones with a Home button). When the iPhone is locked, just press the Side button to put the phone to sleep to hide the Notification Center. When you view it on the Lock screen, the Notification Center closes automatically after a few seconds of inactivity.

Working with Sounds and Vibrations

Sounds are audible notifications that something has happened. For example, you can be alerted to a new email message by a specific sound. Auditory notifications can get your attention and provide some level of information. For example, if you have a specific ringtone for a person, you can tell he is calling you just by hearing his ringtone. You can't take any action directly from auditory notifications, but they certainly can get your attention.

Vibrations are a physical indicator that something has happened. Your phone can vibrate using different patterns to indicate what has happened. For example, one pattern might indicate a new email while another indicates an incoming phone call. Vibrations can be useful when you don't want to disturb others with sound, but you can't see your iPhone's screen.

You learn to configure the sounds and vibrations your iPhone makes in "Configuring Sounds and Vibrations."

Configuring Notifications for Specific Apps

You can configure how apps provide notifications and, if you allow notifications, which type. You also can configure other aspects of notifications, such as whether an app's notifications appear in the Notification Center or if they appear on the Lock screen. Apps can support different notification options; some apps, such as Mail, support notification configuration by account (for example, you can set a different alert sound for new mail in each account). You can follow the same general steps to configure notifications for each app; you should explore the options for the apps you use most often to ensure they work the best for you.

When you configure notifications for apps that support multiple accounts (such as Mail), you configure notifications for each account separately. For example, you might want a different sound for new email sent to your iCloud account than the sound you hear for new email sent to your Google Gmail account.

When you configure notifications for an app that doesn't support different accounts, you configure all notifications for the app at the same time.

The steps in the following task show you how to configure Mail's notifications, which is a good example because it can include notifications for multiple accounts and supports a lot of notification features; the notification settings for other apps might have fewer features or might be organized slightly differently.

When you configure notifications for an app that doesn't support accounts, you set all the options from one screen as opposed to using a different screen for each account used by that app. But configuring the notifications for any app follows a similar pattern as exemplified by the steps for Mail's notification settings.

To configure notifications from the Mail app, perform the following steps:

(1) Tap Settings on the Home screen.

(2) Tap Notifications.

(3) Tap Show Previews. As you saw earlier, alerts from an app can contain a preview of the information related to the alert, such as part of an email message.

(4) To have alerts always show the preview, tap Always; tap When Unlocked to show previews only when your iPhone is unlocked; or tap Never to hide previews.

(5) Tap Back (<).

(6) Tap Siri Suggestions. Siri can suggest shortcuts to accomplish tasks using apps on your phone. On the Siri Suggestions screen, you see the apps installed on your iPhone. The switch next to each app determines whether these suggestions can be made on the Lock screen.

No Suggestions on the Lock Screen

To prevent Siri from displaying any suggestions on the Lock screen, set the Suggestions on the Lock Screen switch to off (white). To access Siri Suggestions, you need to unlock your phone.

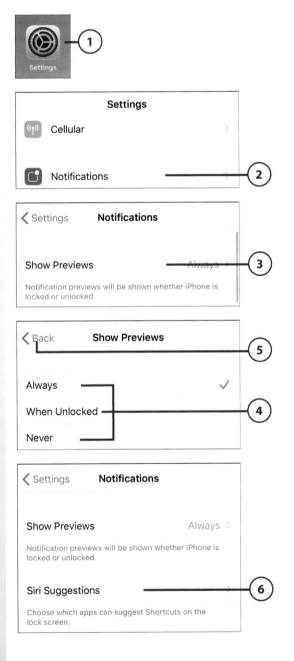

7 Set an app's switch to on (green) if you want to see these suggestions on the Lock screen.

8 Set the switches to off for apps for which you don't want to see shortcut suggestions on the Lock screen. (If you set an app's switch to off, Siri still creates suggestions for the app; you just won't see them on the Lock screen.)

9 Tap Back (<). In the NOTIFICATION STYLE section, you see all the apps installed on your phone that can provide notifications. Along with the app name and icon, you see the current status of its notifications.

10 Swipe up or down the screen to locate the app whose notifications you want to configure. (The apps are listed in alphabetical order.)

11 Tap the app whose notifications you want to configure.

12 If you want the app to provide notifications, set the Allow Notifications switch to on (green) and move to step 13. If you don't want notifications from the app, set the Allow Notifications switch to off (white) and skip to step 35.

13 Tap the account for which you want to configure notifications; if the app doesn't support accounts, skip to step 15.

Notification options enabled

14 If you don't want any notifications for the account, set the Allow Notifications switch to off (white) and skip to step 30. If you do want notifications for the account, set the switch to on (green).

15 Tap Lock Screen so it has a check mark if you want notifications to be visible on the Lock screen; if you don't want to receive notifications on the Lock screen, tap the Lock Screen icon so it doesn't have a check mark.

16 Tap Notification Center so it has a check mark if you want notifications to appear in the Notification Center.

17 Tap Banners so it has a check mark if you want to see banner notifications (which appear at the top of the screen).

18 If you enabled Banners, tap Banner Style; if not, skip to step 22.

19 Tap Temporary if you want the alert banners to appear on the screen, remain there for a few seconds, and then disappear. Temporary banners keep you informed but don't interrupt what you are doing.

20 Tap Persistent if you want the alert banners to remain on screen until you take action on them. For example, if the alert is for a calendar event, you might want it to be persistent so that it really gets your attention.

21 Tap back (<). The specific label you tap to move back to the previous screen depends on where you came from. For example, if you're configuring notifications for an iCloud account, the label is iCloud.

22 Tap Sounds.

23 Use the resulting Sounds screen to choose the alert sound and vibration for new email messages to the account (see "Configuring Sounds and Vibrations" later in this chapter for the details about configuring sounds and vibrations).

24 Tap back (<).

25 To display the app's badge (which shows the number of new items in that app or account), set the Badges switch to on (green). (If you set this to off [white] for an account, new items sent to that account won't be included in the count of new items shown on the app's badge.)

26 Tap Show Previews.

< iCloud   **Banner Style**   **21**

Temporary   ✓

Persistent

Banner Style   Temporary >

Sounds   None >   **22**

24

< iCloud   **Sounds**

Vibration   Default >

STORE

Tone Store   **23**

Download All Purchased Tones

This will download all ringtones and alerts purchased using the "bradmacosx@mac.com" account.

ALERT TONES

None (Default)

Aurora

✓ Bamboo

Chord

Circles

Complete

Hello

Sounds   Bamboo >

Badges   **25**

OPTIONS

Show Previews   Always (Default)   **26**

27 Tap Always (Default) if you always want previews to appear in notifications; When Unlocked if you want them to appear only when your iPhone is unlocked; or Never if you don't want previews to be displayed at any time.

28 Tap back (<).

29 Tap back (<).

30 Configure notifications for the other accounts used in the app.

31 Configure notifications for VIP email, threads, or other special types of messages or locations. These notifications override the account notifications. For example, if you've disabled notifications for your iCloud email but have notifications for VIP messages, you receive notifications for messages from VIPs sent to your iCloud email account.

32 Tap Notification Grouping.

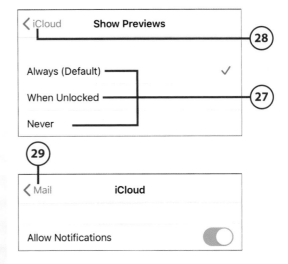

33 Tap Automatic if you want notifications grouped in the Notification Center automatically; tap By App if you want them to be grouped by the app they come from, or Off if you don't want them grouped at all (they appear individually in the Notification Center, assuming you've enabled them to appear there in step 16).

34 Tap back (<).

35 Tap Notifications (<).

36 Repeat steps 11 through 35 for each app shown on the Notifications screen. Certain apps might not have all the options shown in these steps, whereas others might have more options, but the process to configure their notifications is similar.

37 Swipe up until you reach the bottom of the screen.

38 Configure any special notifications you see. What you see here depends on the country or region your phone is associated with. For example, where I live in the United States, the GOVERNMENT ALERTS section includes three notifications. AMBER Alerts are issued when a child is missing and presumed abducted, whereas Emergency Alerts are issued for things such as national crises, local weather, and so on. Public Safety Alerts can be used to warn about events that might be a threat to life or property. You can use the switches to enable (green) or prevent (white) these types of alerts, but you can't configure them.

>>>Go Further

NOTIFY THIS

Here are some other hopefully useful notification tidbits for your consideration:

- **App settings override general settings**—When you configure an app's notifications, such as setting the sound for new email for an account, it overrides the general sounds and vibrations that you set using the Sounds & Haptics settings (covered in "Configuring Sounds and Vibrations").

- **Special sounds and vibrations for contacts**—You can override some app's sounds and vibration notification settings for individuals in your Contacts app. For example, you can configure a specific ringtone, new text tone, and vibrations for calls or texts from a contact. You do this using the contact information screen as explained in Chapter 8, "Managing Contacts." These override both specific app notification settings as well as any set using other Settings tools.

- **Installed app not shown**—You must have opened an app at least once for it to appear on the Notifications screen.

- **Initial notification prompt**—The first time you open many apps, you are prompted to allow that app to send you notifications. If you allow this, the app is able to send notifications about its activity. If you deny this, the app isn't able to send notifications. You can always configure the app's notifications using the steps in this task regardless of your initial decision.

- **Lots of apps**—If you have a lot of apps or activity on your iPhone, notifications can become disruptive. It can take some time to set each app's notifications, but making sure you receive only the notifications that are important to you prevents your iPhone from bothering you unnecessarily. For less important apps, have their notifications appear grouped only in the Notification Center so you can review them at your leisure. For more important apps, use banner alerts and sounds to make sure you're aware of the activity for that app.

- **Important alerts**—When the alerts in the GOVERNMENT ALERTS section are enabled, they activate even if Do Not Disturb is on.

>>>*Go Further*

CONFIGURING NOTIFICATIONS FROM NOTIFICATIONS

As you receive notifications, you can configure the notifications for the associated app directly from the notification. This is convenient because you can configure notifications for apps as they happen, which sometimes makes it easier to decide which options you want to set.

Swipe to the left on an alert and tap Manage. You see the following options:

- **Deliver Quietly**—Tap this to continue to receive alerts in the Notification Center, but not on the Lock screen or via banners. Sounds and vibrations are also disabled.

  If you set the notifications for an app to Deliver Quietly, you can restore them to their prior settings by opening the Manage notifications dialog and tapping Deliver Prominently.

- **Turn Off**—Tap Turn Off and then tap Turn Off All Notifications to disable all notifications for the app. (You need to use the Notifications screen in the Settings app to see them again.)

- **Settings**—Tapping this takes you to the app's notification settings screen in the Settings app.

Configuring General Sounds and Vibrations

Earlier, you learned how to configure the notifications (visual, sounds, and vibrations) that apps use to communicate to you. You can also configure general sounds and vibrations that your iPhone uses to get your attention (when an app-specific notification doesn't override the general setting).

To configure your iPhone's general sounds, do the following:

(1) On the Settings screen, tap Sounds & Haptics.

Settings

Notifications

Sounds & Haptics ————————— (1)

(2) Set the Vibrate on Ring switch to on (green) if you want your iPhone to also vibrate when it rings.

(3) Set the Vibrate on Silent switch to on (green) if you want your iPhone to vibrate when you have it muted.

(4) To have your iPhone automatically reduce loud sound when using headphones, tap Reduce Loud Sounds.

(5) Set the Reduce Loud Sounds switch to on (green).

(6) Drag the volume slider to the left to lower the sound level at which your iPhone reduces sound levels or to the right to increase it. As you make changes, you see examples of the sound level, such as 80 decibels, which is represented by a noisy restaurant. The lower you make this setting, the more the iPhone will limit louder sounds.

(7) Tap Back (<).

(8) Set the volume of the ringer and alert tones by dragging the slider to the left (quieter) or right (louder).

(9) Set the Change with Buttons switch to on (green) if you want to also be able to change the ringer and alert volume using the Volume buttons on the side of the phone.

(10) Tap Ringtone. On the resulting screen, you can set the sound and vibration your iPhone uses when a call comes in.

< Settings   **Sounds & Haptics**

VIBRATE

Vibrate on Ring   — (2)

Vibrate on Silent   — (3)

HEADPHONE AUDIO

Reduce Loud Sounds   Off — (4)

— (7)

< Back   **Reduce Loud Sounds**

Reduce Loud Sounds   — (5)

80 decibels
As loud as a noisy restaurant

— (6)

Your iPhone can analyze headphone audio and reduce any sound that is over a set decibel level. Learn More...

< Settings   **Sounds & Haptics**

HEADPHONE AUDIO

Reduce Loud Sounds   80 decibels >

RINGER AND ALERTS

— (8)

Change with Buttons   — (9)

The volume of the ringer and alerts can be adjusted using the volume buttons.

SOUNDS AND VIBRATION PATTERNS

Ringtone   Ring — (10)

11 Swipe up and down the screen to see all the ringtones available to you. There are two sections of sounds on this screen: RINGTONES and ALERT TONES. These work in the same way; alert tones tend to be shorter sounds. At the top of the RINGTONES section, you see any custom ringtones you have configured on your phone; a dark line separates those from the default ringtones that are below the custom ones.

Individual Ringtones and Vibrations

The ringtone and vibration you set in steps 10–19 are the default or general settings. These are used for all callers except for people in your Contacts app for whom you've set specific ringtones or vibrations. In that case, the contact's specific ringtone and vibration are used instead of the defaults.

12 Tap a sound, and it plays; tap it again to stop it.

13 Repeat steps 11 and 12 until you have selected the sound you want to have as your general ringtone.

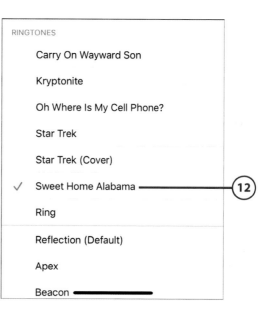

(14) If necessary, swipe down the screen so you see the Vibration section at the top.

(15) Tap Vibration. A list of Standard and Custom vibrations is displayed.

(16) Swipe up and down the screen to see all the vibrations available. The STANDARD section contains the default vibrations, and in the CUSTOM section you can tap Create New Vibration to create your own vibration patterns, as discussed in the "Sounding Off" sidebar at the end of this section.

(17) Tap a vibration. It "plays" so you can feel it; tap it again to stop it.

(18) Repeat steps 16 and 17 until you've selected the general vibration you want to use; you can tap None at the bottom of the Vibration screen below the CUSTOM section if you don't want to have a general vibration.

(19) Tap Ringtone.

(20) Tap Back (<). The ringtone you selected is shown on the Sounds and Haptics (or Sounds) screen next to the Ringtone label.

(21) Tap Text Tone.

< Back **Ringtone**

Vibration Heartbeat > ──(15)

STORE

Tone Store

Download All Purchased Tones ──(14)

This will download all ringtones and alerts purchased using the _____ account.

RINGTONES

Carry On Wayward Son

< Ringtone **Vibration** Edit ──(19)

STANDARD

Accent (Default)

Alert

Heartbeat ✓──(17)

Quick

Rapid

S.O.S.

Staccato

Symphony

CUSTOM ──(16)

My Vibe

Create New Vibration >

None

< Back **Ringtone** ──(20)

SOUNDS AND VIBRATION PATTERNS

Ringtone Sweet Home Alabama >

Text Tone Aurora ──(21)

22 Use steps 11–18 with the Text Tone screen to set the sound and vibration used when you receive a new text. The process works the same as for ringtones, though the screens look a bit different. For example, the ALERT TONES section is at the top of the screen because you are more likely to want a short sound for new texts.

23 When you're done setting the text tone, tap Back (<).

24 Using the same process as you did for the ringtone and text tone, set the sound and vibrations for the rest of the events on the list.

25 If you don't like the audible feedback when you tap keys on the iPhone's virtual keyboard, slide the Keyboard Clicks switch to off (white) to disable that sound. The keyboard is silent as you type on it.

26 If you don't want your iPhone to make a sound when you lock it, slide the Lock Sound switch to off (white). Your iPhone no longer makes this sound when you press the Side button to put it to sleep and lock it.

27 Set the System Haptics switch to off (white) if you prefer not to experience vibratory feedback when you make changes to settings, such as when you tap Start when creating a new event to set its start time.

‹ Back   **Text Tone**

STORE

Tone Store

Download All Purchased Tones

This will download all ringtones and alerts purchased using the ~~account~~ account.

ALERT TONES

None

Note (Default)

Aurora

✓ Bamboo

Chord

Circles

‹ Settings   **Sounds & Haptics**

The volume of the ringer and alerts can be adjusted using the volume buttons.

SOUNDS AND VIBRATION PATTERNS

Ringtone   Sweet Home Alabama ›

Text Tone   Bamboo ›

New Voicemail   Tri-tone ›

New Mail   None ›

Sent Mail   Swoosh ›

Calendar Alerts   None ›

Reminder Alerts   Chord ›

AirDrop   Pulse ›

Keyboard Clicks   ◯

Lock Sound   ◯

System Haptics   ◯

Play haptics for system controls and interactions.

>>>Go Further

SOUNDING OFF

Following are three more sound- and vibration-related pointers:

- System haptics are much more subtle than the general vibrations you set (for example, when your phone rings). These system vibrations are very short and occur in response to something you do, such as changing a setting. You can't change the way system haptics work; they're either on or off.

- You can tap Tone Store on the Ringtone, Text Tone, and other screens to move to the iTunes Store, where you can purchase and download ringtones and other sounds to your iPhone. You can then select these sounds for various events.

- If you've previously downloaded tones from the iTunes Store, tap Download All Purchased Tones to make sure you have all of them available on your iPhone. After you've done that, the Download All Purchased Tones option disappears and you see only Tone Store.

Working with and Configuring the Control Center

The Control Center provides quick access to a number of very useful controls. It includes a number of tools by default, but you can also add, remove, or re-organize it so that it works even better for you.

Working with the Control Center

Swipe down from the upper-right corner of the screen to open the Control Center (iPhones without a Home button)

Swipe up from the bottom of the screen to open the Control Center (iPhones with a Home button)

Open the Control Center by swiping down from the upper-right corner of the screen (iPhones without a Home button) or swiping up from the bottom of the screen (iPhones with a Home button). You can open the Control Center while you're on any Home screen, on an app screen, or on the Lock screen. (If your iPhone is asleep, you need to wake it.)

Control Center Tip

Some apps have their own Dock at the bottom of the screen. When you're using such an app on a model with a Home button, make sure you don't touch an icon on the Dock when you're trying to open the Control Center because you'll do whatever the icon is for instead. Just swipe up on an empty area of the app's Dock and the Control Center opens. On an iPhone that doesn't have a Home button, this isn't an issue because you swipe down from the upper-right corner of the screen instead.

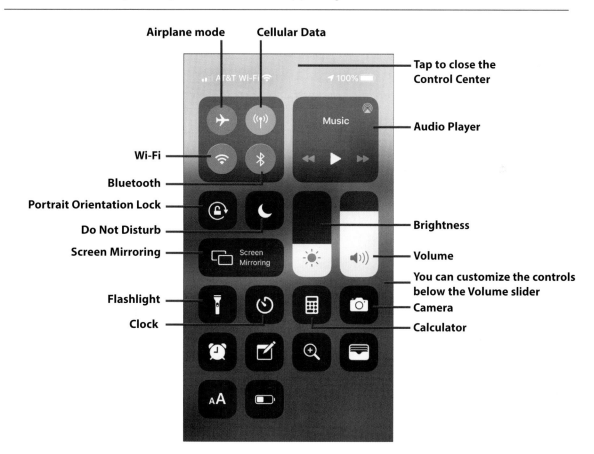

In the top-left quadrant of the Control Center are icons you can use to turn on or turn off and configure important functions, which are Airplane mode, Cellular Data, Wi-Fi, Bluetooth, Portrait Orientation Lock, and Do Not Disturb mode. To enable or disable one of these functions, tap its icon, which changes color to show its status. When the function is enabled, the buttons have color, such as blue, orange, or green. To disable a function, tap the icon so that it becomes gray to show you it's inactive. For example, to lock the orientation of the iPhone's screen in the portrait orientation (vertical), tap Portrait Orientation Lock so it becomes white with a red icon. Your iPhone screen's orientation no longer changes when you rotate the phone from vertical to horizontal. To make the orientation change when you rotate the phone again, tap Portrait Orientation Lock to turn it off again.

In the upper-right quadrant, you see the Audio Player. You can use this to control music, podcasts, and other types of audio that are playing in their respective apps.

Just below the Audio Player are the Brightness and Volume sliders. You can swipe up or down on these to increase or decrease the screen's brightness or the volume of whatever you're hearing on your phone. It's handy to be able to get to either of these controls quickly. For example, when someone is scanning your phone's screen, such as when you are boarding a plane with a boarding pass on your phone, you might be asked to make the screen brighter so it can be scanned more easily. Just open the Control Center and swipe up on the Brightness slider and then tap outside the Control Center to close it. (Remember to lower the brightness again because having the screen very bright increases the rate at which battery power is used.)

Screen Mirroring enables you to broadcast your iPhone's screen onto an Apple TV or a Mac computer.

The controls above the first row of four icons are always on the Control Center; you can't change them in any way. However, below those is a section of controls you can change. By default, you see the Flashlight, Clock, Calculator, Camera, and others in this area. Like the icons toward the top of the screen, tap these icons to perform the associated action, such as using the iPhone's flash as a flashlight, or opening an app—the Calculator app, for example. You can configure the controls that are in this area by adding, removing, and organizing them; you learn how in the next section.

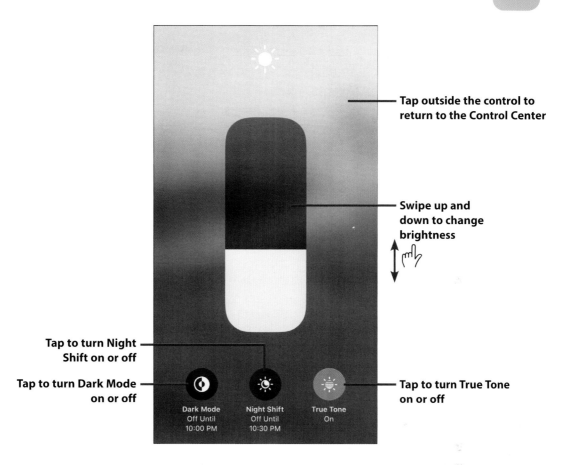

Tap outside the control to return to the Control Center

Swipe up and down to change brightness

Tap to turn Night Shift on or off

Tap to turn Dark Mode on or off

Dark Mode
Off Until
10:00 PM

Night Shift
Off Until
10:30 PM

True Tone
On

Tap to turn True Tone on or off

When you press on some of the controls, you see additional options. You should try pressing the controls to see what options are presented to you; you might find some of these very useful. For example, when you press the Brightness slider, you see a larger slider and have access to the Dark Mode, Night Shift, and True Tone icons. You can swipe on the slider to change the brightness level or tap one of the icons to enable or disable the associated function.

When you're done using the Control Center, tap anywhere on the screen except on one of its icons to close it.

Configuring the Control Center

You can configure the controls toward the bottom of the Control Center by performing the following steps:

(1) Open the Settings app and tap Control Center.

(2) To be able to access the Control Center while you're using apps, set the Access Within Apps switch to on (green). If you set this to off, you need to move back to the Home or Lock screen to use the Control Center.

(3) If you use the Home app to automate and remotely control specific functions, such as lighting, set the Show Home Controls switch to on (green). A Home section appears on the Control Center below the Screen Mirroring, Brightness, and Volume controls. You can add controls to this section to make accessing devices controlled by the Home app easier. If you don't use the Home app, leave this switch off (white).

(4) Swipe up the screen to further customize the Control Center. You see two sections: INCLUDED CONTROLS shows the controls currently installed in your Control Center, whereas MORE CONTROLS shows controls that are not installed in the Control Center, but are available for you to add.

(5) To remove a control from the Control Center, tap Unlock (–).

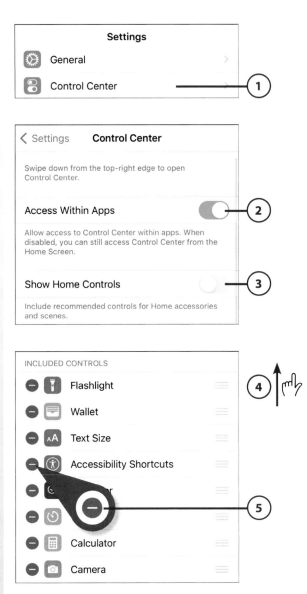

6 Tap Remove. The control is moved from the INCLUDED CONTROLS list to the MORE CONTROLS list. It no longer appears on the Control Center.

7 Swipe up the screen to see the MORE CONTROLS list.

8 To add a control to the Control Center, tap Add (+). The control moves to the bottom of the INCLUDED CONTROLS list and is added to the Control Center.

9 Move a control higher on the Control Center by dragging its Order icon (three lines) up the INCLUDED CONTROLS list or move it lower by dragging its Order icon down the list. The top four controls on the list appear first in the customizable part of the Control Center; the next four are below those, and so on.

10 Repeat steps 5 through 9 until you have all the controls you want on the Control Center in the order you want them. The next time you open the Control Center, it reflects the changes you made.

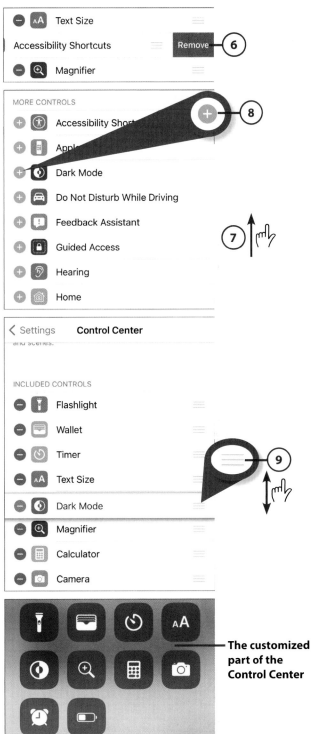

The customized part of the Control Center

Using the Do Not Disturb Mode

All the notifications your iPhone uses to communicate with you are useful, but at times, they can also be annoying or distracting. When you put your iPhone in Do Not Disturb mode, its visual, audible, and vibration notifications are disabled so that they don't activate. For example, the phone won't ring if someone calls you (unless you specify certain contacts whose calls you do want to receive while your phone is in this mode).

You can activate Do Not Disturb manually at any time. You can also configure your iPhone so that it goes into Do Not Disturb mode automatically at specific times.

Activating Do Not Disturb Manually

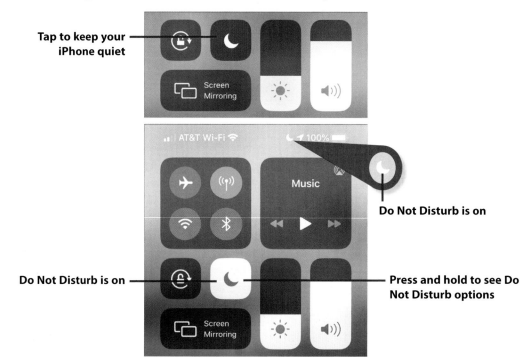

Tap to keep your iPhone quiet

Do Not Disturb is on

Do Not Disturb is on

Press and hold to see Do Not Disturb options

To put your iPhone in Do Not Disturb mode, open the Control Center and tap Do Not Disturb. It becomes purple and the Do Not Disturb: On status briefly appears at the top of the Control Center. Your iPhone stops notifications. The Do Not Disturb status icon appears at the top of the screen so you know your iPhone is silent.

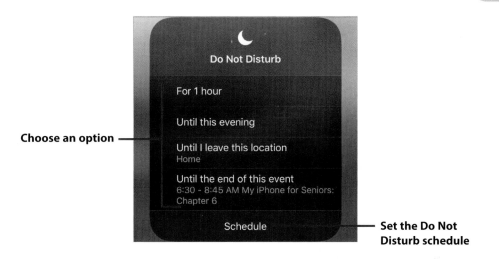

Choose an option ——

Set the Do Not Disturb schedule

If you press Do Not Disturb, you see options, such as For 1 hour, Until this evening, Until I leave this location, or Until the end of this event. Tap an option to activate it. You can tap Schedule to set a Do Not Disturb schedule.

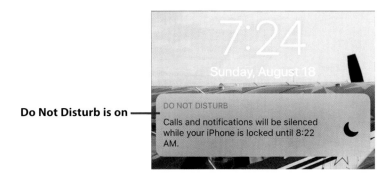

Do Not Disturb is on ——

When Do Not Disturb is active, you see its current status on the Lock screen. On iPhones with a Home button, you also see the Do Not Disturb icon on the status bar at the top of the screen.

To make your notifications active again, open the Control Center and tap Do Not Disturb so it turns gray; your iPhone resumes trying to get your attention when it's needed, and the Do Not Disturb status message disappears from the Lock screen. Of course, Do Not Disturb turns off automatically if it's set to do so, such as at a specific time or when an event ends.

Activating Do Not Disturb Automatically

You can use the following steps to configure quiet times during which notifications are automatically silenced:

(1) Open the Settings app and tap Do Not Disturb.

(2) To activate Do Not Disturb manually, set the Manual switch to on (green). (This is the same thing as tapping Do Not Disturb in the Control Center.)

(3) To configure Do Not Disturb to activate automatically on a schedule, set the Scheduled switch to on (green).

(4) Tap the From box.

(5) Set the time you want Do Not Disturb to start by using the keyboard to enter the time and tapping AM or PM.

(6) Tap To.

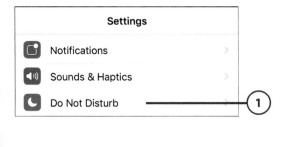

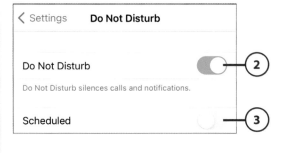

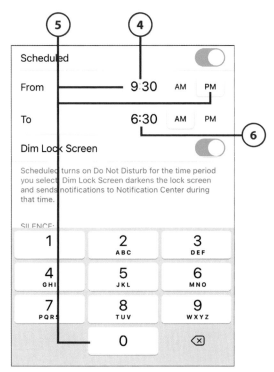

⑦ Set the time you want Do Not Disturb to end by using the keyboard to enter the time and tapping AM or PM.

⑧ Swipe up the screen to close the keyboard.

⑨ If the Do Not Disturb period is at night or at another time during which you really don't want to be disturbed, set the Dim Lock Screen switch to on (green). In addition to calls and notifications being silenced, the Lock screen is also dimmed.

⑩ If you want notifications to be silenced during the Do Not Disturb period only when your phone is locked, tap While iPhone is Locked. This setting presumes that if your iPhone is unlocked, you won't mind taking calls or having notifications even if it is within the Do Not Disturb period because you are probably using the phone. Tap Always if you want notifications to be silenced regardless of the Lock status.

⑪ Tap Allow Calls From.

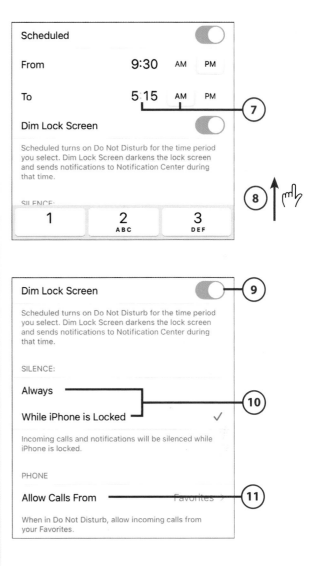

(12) Tap the option for the group of people whose calls should be allowed during the Do Not Disturb period. The options are Everyone, which allows all calls to come in; No One, which sends all calls directly to voicemail; Favorites, which allows calls from people on your Favorites lists to come through but calls from all others go to voicemail; or one of your contact groups, which allows calls from anyone in the selected group to come through while all others go to voicemail.

(13) Tap Back (<).

(14) Set the Repeated Calls switch to on (green) if you want a second call from the same person within three minutes to be allowed through. This feature is based on the assumption that if a call is really important, the person calling you will try again immediately.

(15) Tap Activate in the DO NOT DISTURB WHILE DRIVING section.

(16) To have Do Not Disturb activate automatically when you're driving, tap Automatically to have this based on your iPhone's motion (once the iPhone's accelerometer detects that the phone has reached a particular speed) or When Connected to Car Bluetooth to have Do Not Disturb active whenever your iPhone is connected to your car's Bluetooth system; or to prevent this type of automatic activation, tap Manually.

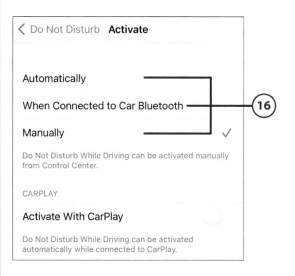

(13)

< Back   **Allow Calls From**

Everyone

No One

Favorites   ✓   **(12)**

GROUPS

All Contacts

Book Group

Allow Calls From   Favorites >

When in Do Not Disturb, allow incoming calls from your Favorites.

Repeated Calls   ◯   **(14)**

A second call from the same person within three minutes will not be silenced.

DO NOT DISTURB WHILE DRIVING

Activate   Manually   **(15)**

< Do Not Disturb   **Activate**

Automatically

When Connected to Car Bluetooth   **(16)**

Manually   ✓

Do Not Disturb While Driving can be activated manually from Control Center.

CARPLAY

Activate With CarPlay   ◯

Do Not Disturb While Driving can be activated automatically while connected to CarPlay.

17 If you use your iPhone with CarPlay, set the Activate With CarPlay switch to on (green) if you want Do Not Disturb to be on whenever CarPlay is active. (CarPlay enables some of your iPhone's controls and information to appear on your car's audio system so you can control your iPhone using that system's controls.)

18 Tap Do Not Disturb (<).

19 Tap Auto-Reply To.

20 Configure to whom you want automatic replies to be sent when Do Not Disturb is on (regardless of how it was activated) by tapping No One to prevent automatic replies; Recents to send replies to people on your recent lists (such as calls you have recently received); Favorites to send replies to your favorites; or All Contacts to automatically reply to anyone on your Contacts lists.

21 Tap Back (<).

22 Tap Auto-Reply.

23 Type the message you want to be automatically sent.

24 Tap Back (<). During the Do Not Disturb period or based on the DO NOT DISTURB WHILE DRIVING setting, your iPhone is silent, except for any exceptions you configured. Automatic replies are sent according to your configuration. When the scheduled Do Not Disturb period ends, your iPhone resumes its normal notification activity.

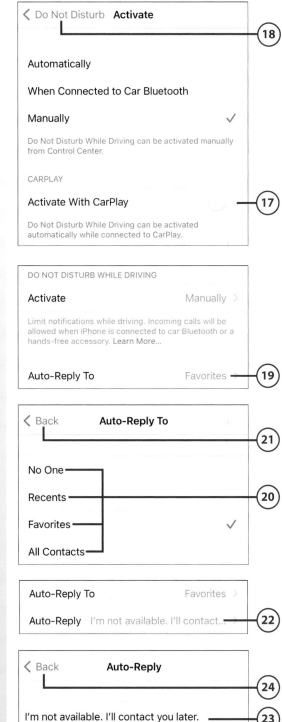

Configuring Keyboards

As explained in Chapter 3, you use the iPhone's keyboard to input text in many apps, including Mail, Messages, and so on. To make typing better, you can configure the keyboards available to you along with a number of preferences that determine how those keyboards work.

By default, you have access to one keyboard for the language you selected when you started your iPhone for the first time and the Emoji keyboard that you can use to insert images into your text. To change your keyboard configuration, perform the following steps:

1. On the Settings screen, tap General.
2. Swipe up the screen.
3. Tap Keyboard.
4. Tap Keyboards. At the top of the screen, you see at least two keyboards: One is based on the language you selected when you first turned your iPhone on. The other is the Emoji keyboard. You can activate more keyboards so that you can choose a specific language's keyboard when you're entering text.
5. Tap Add New Keyboard.

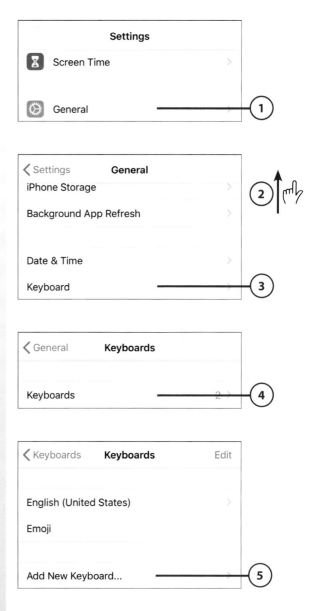

6. Swipe up and down the screen to browse the available keyboards. These are organized into two sections. THIRD-PARTY KEYBOARDS shows keyboards you have added to access their additional functionality. OTHER IPHONE KEYBOARDS are the default keyboards provided with the iPhone.

7. Tap the keyboard you want to add.

8. Tap the keyboard you added in step 7.

9. Tap the keyboard layout you want to use. (Not all keyboards support options; if the one you're configuring doesn't, skip this step.)

10. Tap Back (<).

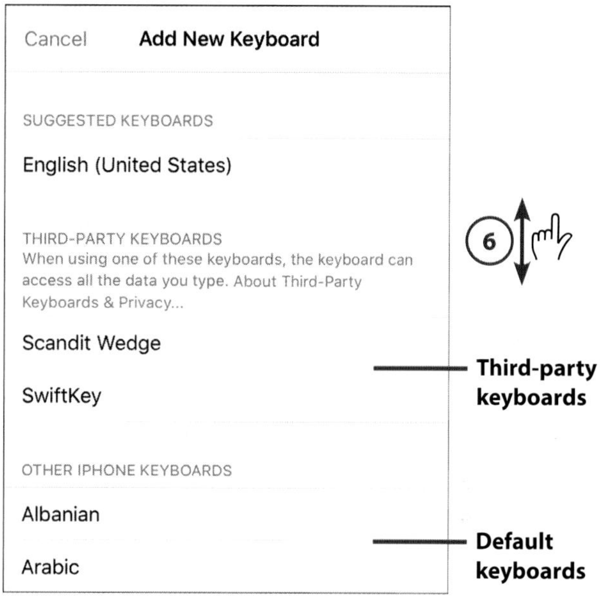

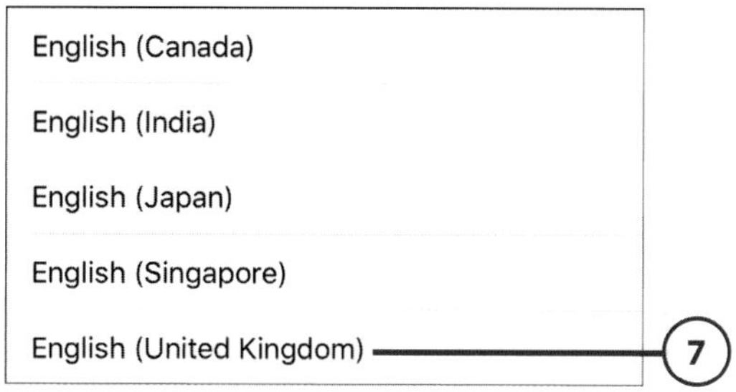

Cancel   **Add New Keyboard**

SUGGESTED KEYBOARDS

English (United States)

THIRD-PARTY KEYBOARDS
When using one of these keyboards, the keyboard can access all the data you type. About Third-Party Keyboards & Privacy...

Scandit Wedge

SwiftKey ———————————— **Third-party keyboards**

OTHER IPHONE KEYBOARDS

Albanian

Arabic ———————————— **Default keyboards**

English (Canada)

English (India)

English (Japan)

English (Singapore)

English (United Kingdom) ———————— 7

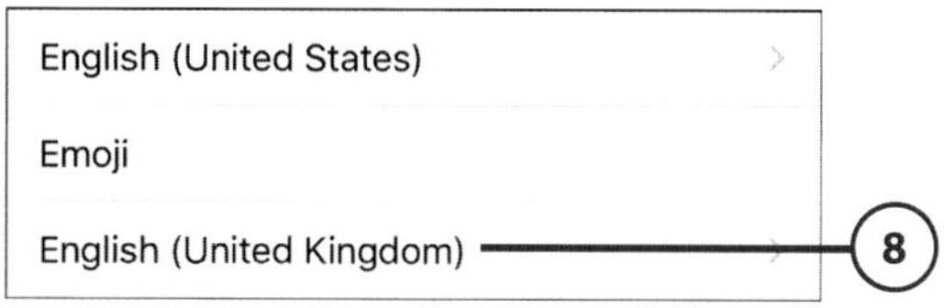

English (United States)   >

Emoji

English (United Kingdom)   > ———————— 8

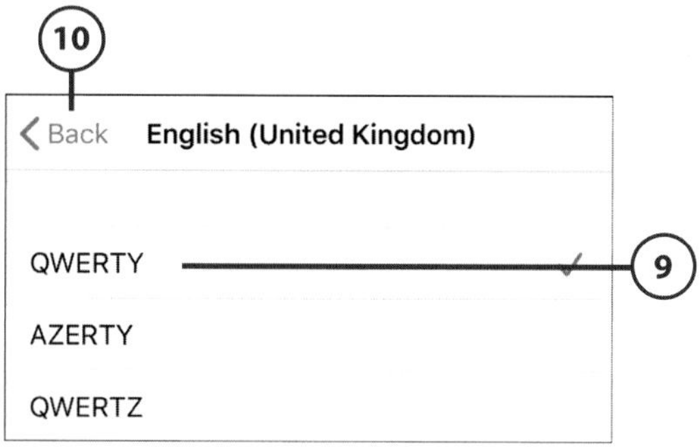

10

< Back   **English (United Kingdom)**

QWERTY ———————————— ✓ 9

AZERTY

QWERTZ

(11) Repeat steps 5–10 to add and configure additional keyboards.

(12) Tap Keyboards (<).

(13) Tap One Handed Keyboard.

(14) To be able to use the one-handed keyboard (which squishes all the keys toward one side of the screen), tap Left to place it on the left side or Right to put it on the right side of the screen; tap Off if you don't want to use the one-handed keyboard.

(15) Tap Back (<).

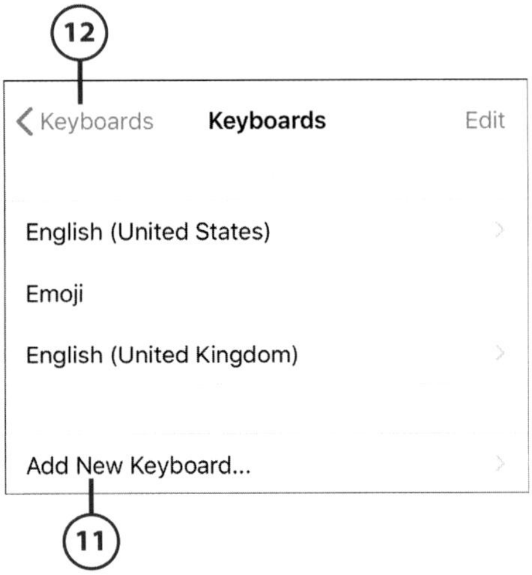

(12)

‹ Keyboards	**Keyboards**	Edit

English (United States)

Emoji

English (United Kingdom)

Add New Keyboard...

(11)

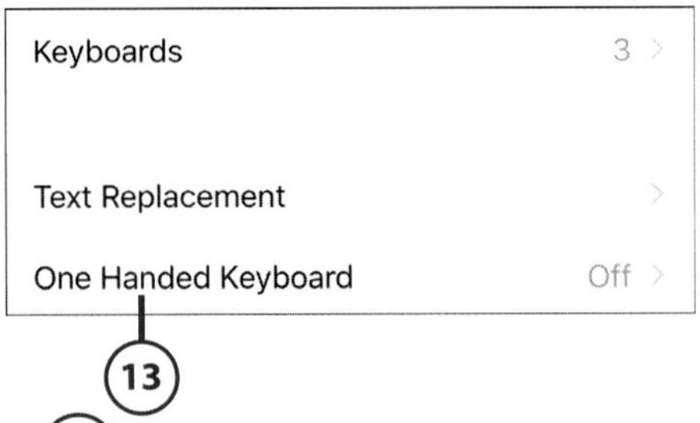

Keyboards	3 ›
Text Replacement	›
One Handed Keyboard	Off ›

(13)

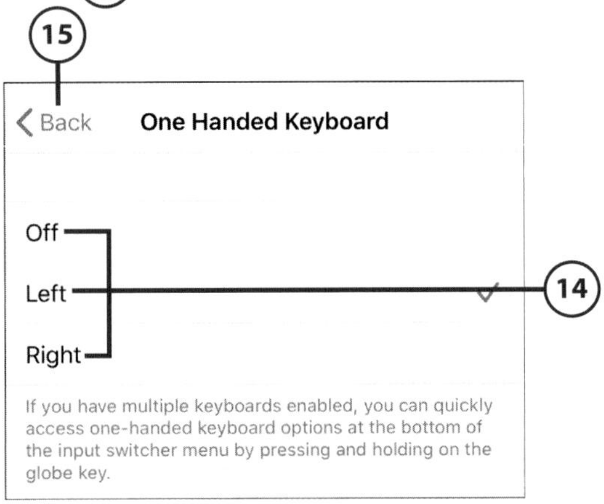

(15)

‹ Back	**One Handed Keyboard**

Off

Left ✓ (14)

Right

If you have multiple keyboards enabled, you can quickly access one-handed keyboard options at the bottom of the input switcher menu by pressing and holding on the globe key.

(16) To prevent your iPhone from automatically capitalizing as you type, set Auto-Capitalization to off (white).

(17) To disable the automatic spell checking/correction, set Auto-Correction to off (white).

(18) To disable the iPhone's Spell Checker, set the Check Spelling switch to off (white).

(19) To disable the Caps Lock function, set the Enable Caps Lock to off (white).

(20) To disable the iPhone's Predictive Text feature, set the Predictive switch to off (white).

(21) To prevent the iPhone from automatically trying to correct your punctuation, set the Smart Punctuation switch to off (white).

(22) If you don't want to be able to slide your finger on the keyboard to type, set Slide to Type to off (white). You have to touch each character individually to type.

(23) If Slide to Type is on (green) but you don't want to be able to delete words by swiping the delete key, set Delete Slide-to-Type by Word to off (white). You have to tap the delete key to delete text.

(24) To prevent the character you type from being shown in a magnified pop-up as you type it, set the Character Preview switch to off (white).

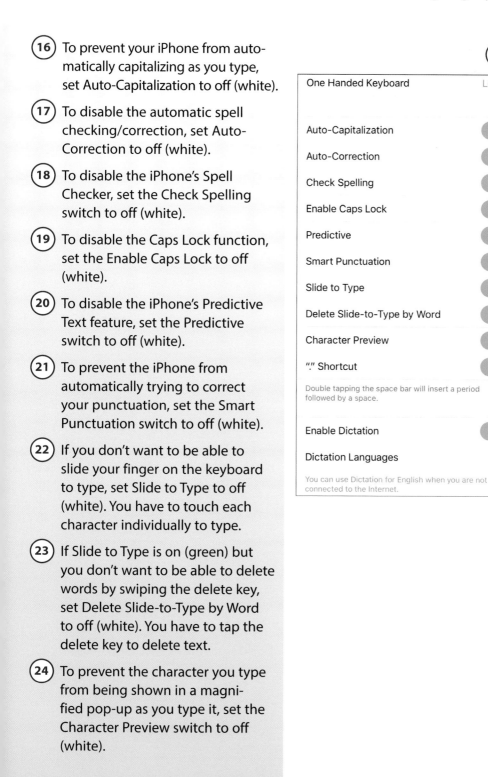

(25) To disable the shortcut that types a period followed by a space when you tap the space bar twice, set the "." Shortcut switch to off (white). You must tap a period and the spacebar to type these characters when you end a sentence.

(26) To disable the iPhone's dictation feature, set the Enable Dictation switch to off (white). The microphone key won't appear on the keyboard and you won't be able to dictate text.

(27) If you have enabled keyboards for more than one language and have dictation enabled, tap Dictation Languages.

(28) Tap the languages in which you want to be able to dictate so they have a check mark.

(29) Tap the languages in which you don't want to be able to dictate so they don't have a check mark.

(30) Tap Back (<).

(31) If you want to be able to send Memoji and Animoji stickers using the Emoji keyboard, set the Memoji Stickers switch to on (green). In the apps that support these objects, such as Messages, you can select Memoji and Animoji from the Emoji keyboard.

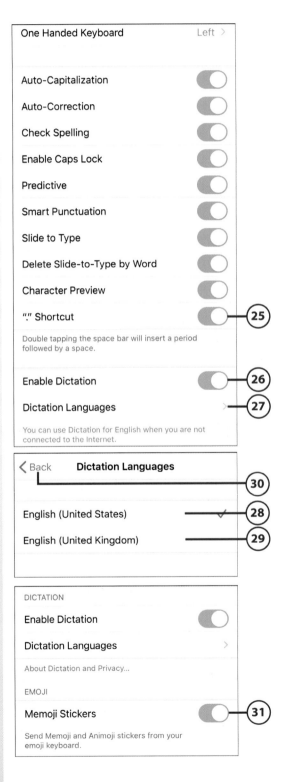

>>>Go Further
MORE ON KEYBOARDS AND RELATED TOPICS

You use your iPhone's keyboards constantly as you work with your iPhone. Following are a few more keyboard pointers to keep in mind:

- **Changing Keyboards**—To delete a keyboard, move to the Keyboards Settings screen and swipe to the left on the keyboard you want to remove. Tap Delete. The keyboard is removed from the list of activated keyboards and is no longer available to you when you type. (You can always activate it again later.) To change the order in which keyboards appear, move to the Keyboards screen, tap Edit, and drag the keyboards up and down the screen. When you've finished, tap Done. This changes the order in which you cycle through keyboards.

- **Text Replacements**—You can use the Text Replacement option in the Keyboards screen to create your own text shortcuts. For example, suppose you type "Leaving to fly in Plainfield" regularly. You can create a replacement such that you only have to type "lp," which gets replaced by the phrase associated with it. Tap Text Replacement and tap Add (+). Type the phrase you want to enter, the shortcut you want to use to enter it, and then Save. From that point forward, when you type the shortcut, the phrase replaces the shortcut's letters.

- **Third-party Keyboards**—You can install and use keyboards from third parties (meaning not Apple) on your iPhone. To do this, open the App Store app and search for "keyboards for iPhone," or you can search for a specific keyboard by name if you know of one you want to try. After you've downloaded the keyboard you want to use, use steps 1–5 to move back to the Keyboards Settings screen. When you open the Add New Keyboard screen, you see the additional keyboards you have installed. Tap a keyboard in this section to activate it as you do with the default keyboards. When you move back to the Keyboards screen, you see the keyboard you just activated. Tap it to configure its additional options. Then you can use the new keyboard just like the others you've activated. Make sure you check out the documentation for any keyboards you download so you take advantage of all of their features.

- **Language and Region Preferences**—The iPhone automatically applies formatting to various items, such as addresses and temperature, based on the language you selected and your location. You can make changes to these by opening the Settings app, tapping General, and then tapping Language & Region. On the Language & Region screen, you can change the language your iPhone uses, the preferred order of languages you use with your iPhone, the region you want to use for things such as address formats, the calendar type, and temperature unit. Samples of the current formats are shown at the bottom of the screen.

Setting Accessibility Options

The iPhone has many features designed to help people who have hearing impairments, visual impairments, or other physical challenges to be able to use it effectively.

You can enable and configure the Accessibility features on the Accessibility Settings screen.

(1) Open the Settings app.

(2) Swipe up the screen until you see Accessibility.

(3) Tap Accessibility. The Accessibility screen is organized into different sections for different kinds of limitations. The first section is VISION, which includes options to assist people who are visually impaired.

(4) Use the controls in the VISION section to change how the iPhone's screens appear. Some of the options include the following:

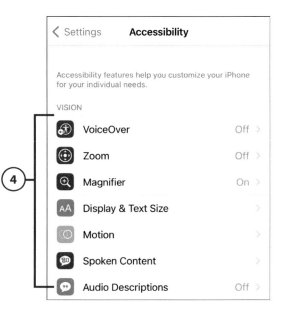

- **VoiceOver**—The iPhone guides you through screens by speaking their contents. To configure this, tap VoiceOver and set the VoiceOver switch to on (green) to turn it on. The rest of the settings configure how VoiceOver works. For example, you can set the rate at which the voice speaks, what kind of feedback you get, and many more options.

- **Zoom**—This setting magnifies the entire screen. Tap Zoom and then turn Zoom on. Use the other settings to change how zoom works, such as whether it follows where you're focused on the screen or remains fixed.

- **Magnifier**—This feature enables you to use your iPhone's camera like a magnifying glass. When you enable this, you can triple-press the Side or Touch ID/Home button to activate it. You also can add its control to the Control Center using the steps provided earlier in the chapter.

- **Display & Text Size**—These options change how your iPhone presents text and other information on the screen. For text, you can configure bold text, make text larger, change button shapes, and turn labels off or on. You also can increase the contrast on the screen, remove the reliance on color to differentiate objects, and use the two Invert options function to reverse the color on the screen so that what is light becomes dark and vice versa. The Color Filters tool enables you to customize how colors appear on the screen. The Reduce White Point switch, when enabled, reduces the intensity of bright colors. The Auto-Brightness switch controls whether the iPhone's screen automatically dims.

- **Motion**—These controls enable you to reduce the perceived change in appearance of icons when you hold the phone in different orientations. You can also enable or disable effects in the Messages app and to enable or prevent video previews from playing automatically. Effects in messages and video previews play automatically.

- **Spoken Content**—Under the Spoken Content option, Speak Selection causes a Speak button to appear when you select text, and Speak Screen provides the option to have the screen's content spoken. You can determine whether you hear feedback while you type, and you can configure the voices used, the rate of speech, and pronunciations.

- **Audio Descriptions**—This causes an audio description of media to be played when available.

5) Swipe up to see the PHYSICAL AND MOTOR section.

6) Use the controls in this section to adjust how you can interact with the iPhone. The controls here include the following:

- **Touch**—These controls help improve interaction when physical dexterity is limited.

  AssistiveTouch makes an iPhone easier to manipulate; if you enable this, a white button appears on the screen at all times. You can tap this to access the Home screen, Notification Center, and other areas. You also can create new gestures to control other functions on the iPhone. There are many other options you can use to change how you can physically interact with the iPhone to control it.

  When you enable Reachability, you can jump to the top of the screen by swiping down on the bottom of the screen.

  The Haptic Touch option adjusts how long you have to touch something to perform a press. The options are Fast or Slow. You can test each option on the Haptic Touch screen.

  You can use the Touch Accommodations options to make it easier for you to use the touch screen. For example, you can change the amount of time you must touch the screen before it is recognized as a touch.

  Turn off Tap to Wake Set if you don't want be able to wake the phone by tapping the screen.

The Shake to Undo setting enables you to turn off the shake motion to undo a recent action.

The Vibration setting enables you to enable or disable vibrations. It overrides the vibration settings in other areas, such as notifications.

Use Call Audio Routing to configure where audio is heard during a phone call or FaceTime session. If you select Automatic, the iPhone chooses the routing based on how it's configured. You can select Bluetooth Headset or Speaker to always use one of those options first instead. You also can have the phone automatically answer calls.

The Back Tap tool enables you to perform specific actions by double- or triple-tapping on the back of your phone. You can choose which action results for each of these. For example, you can enable the App Switcher to be opened by choosing App Switcher on the Double Tap screen.

- **Face ID & Attention**—These controls can be set to emphasize someone looking at the phone before related action is taken. You can also enable haptic feedback when Face ID is used successfully.

- **Switch Control**—The controls on this screen enable you to configure an iPhone to work with an adaptive device so that you can control the iPhone with that device.

- **Voice Control**—This area enables you to configure the iPhone to be controlled by voice commands. This is similar to Siri, except instead of performing tasks, you can actually control the phone, such as navigating on it. For example, you can scroll on the screen or move around within text using your voice to speak commands.

- **Side Button**—Use these controls to set the rate at which you press the Side button to register as a double- or triple-press. On iPhones with a Home button, this is the Home button, but it performs the same purpose, which is to configure how pushes on the Touch ID/Home button are registered. You can also determine if holding the Side button activates Siri, Classic Voice Control, or is turned off. You can also determine how Side button confirmation is handled, either via Switch Control or through AssistiveTouch.

- **Apple TV Remote**—Use this area to enable directional buttons on the iPhone Apple TV Remote instead of swipe gestures.

- **Keyboards**—Using these options, you can show or hide lowercase letters and change how the keys react to your touches.

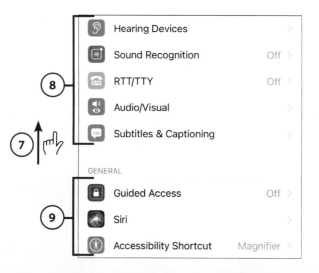

(7) Swipe up the Accessibility screen to see the HEARING section.

(8) Use the controls in this section to configure sounds and to configure the iPhone to work with hearing-impaired people. The controls in this section include the following:

- **Hearing Devices**—When you activate this setting, you can pair an iPhone to work with MFi hearing aids and other devices. (You pair other types of hearing aids using Bluetooth.) You can also set the iPhone for maximum hearing aid compatibility. When your iPhone is paired with a hearing aid, you can configure the hearing aid, such as bass level, using the phone.

- **Sound Recognition**—This feature enables your phone to listen for certain sounds and notifies when those sounds are recognized. When you enable this feature, tap Sounds and select the sounds you want the phone to try to recognize. For example, if you set the Cat switch to on (green), when the iPhone recognizes a sound that might be a cat, it displays a visual alert.

- **RTT/TTY**—These controls enable you to use your iPhone with an RTT or TTY device.

- **Audio/Visual**—These control various audio and visual aspects of your iPhone.

  Headphone Accommodations enables you to adjust how certain Apple or Beats earbuds or headphones sound.

  Mono Audio causes the sound output to be in mono instead of stereo.

  Phone Noise Cancellation turns noise cancellation on and off. Noise cancellation reduces ambient noise when you're using the Phone app.

Use the Balance slider to change the balance of stereo sound between left and right.

When you set the LED Flash for Alerts switch to on (green), the flash flashes whenever an alert plays on the phone.

- **Subtitles & Captioning**—Use these controls to enable subtitles and captions for video and choose the style of those elements on the screen.

(9) Use the controls in the GENERAL section to configure the following:

- **Guided Access**—Use the Guided Access setting if you want to limit the iPhone to using a single app at a time.

- **Siri**—Using these controls, you can enable or disable Type to Siri and choose when Siri provides voice feedback. You can also prevent the phone from listening for "Hey Siri" when it is facing down or is covered.

- **Accessibility Shortcut**—Use the Accessibility Shortcut control to determine what happens when you press the Side or Touch ID/Home button three times. For example, if you choose Magnifier, you can quickly magnify something by pressing the Side or Touch ID/Home button three times (Magnifier must be enabled in the VISION section for this to work).

Put widgets on your
Home screens

Customize the layout of the icons on
your Home screens by placing icons
where you want them

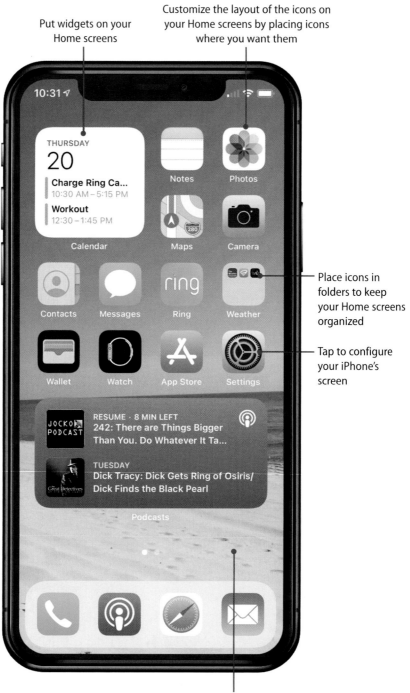

Place icons in
folders to keep
your Home screens
organized

Tap to configure
your iPhone's
screen

Choose the image you want as wallpaper

In this chapter, you learn how to make an iPhone look the way you want it to. Topics include the following:

→ Getting started

→ Customizing your Home screens

→ Setting the screen's appearance, brightness, lock/wake, text, view, and wallpaper options

Customizing How Your iPhone Looks

There are lots of ways that you can customize the way your iPhone looks. You can design and organize your Home screens; choose the Dark or Light mode; and set the screen's brightness, text size, and wallpaper.

Getting Started

In Chapter 6, "Making Your iPhone Work for You," you learned how to change many aspects of how your iPhone works, the most significant of which were configuring notifications and choosing ringtones. This chapter focuses on how you can change the way your iPhone looks and is organized. You learn how to customize your iPhone in two major groups of tasks:

- **Home screens**—The iPhone's Home screens are the starting point for most everything you do because these screens contain the icons that you tap to access the apps that you want to use. You see and use the Home screens constantly, so it's a good idea to customize them to your preferences. You can place icons and widgets on specific screens, hide Home screens you don't need to see, and you can use folders to make your Home screens work better for you.

- **Appearance, screen brightness, Auto-Lock, Raise to Wake, text, view, and wallpaper options**—There are a number of ways you can change how your iPhone's screen looks and works. For example, you can set its brightness level and text size.

Customizing Your Home Screens

You see and use the Home screens constantly, so it's a good idea to customize them to your preferences.

In the background of the Lock screen and every Home screen is the wallpaper image. In the section called "Setting the Wallpaper on the Home and Lock Screens," you learn how to configure your iPhone's wallpaper in both locations.

As you know, you access apps on your Home screens by tapping them; the Home screens come configured with icons in default locations. You can change the location of these icons to be more convenient for you.

As you install more apps on your iPhone, it's a good idea to organize your Home screens so that you can quickly get to the apps you use most frequently. You can move icons around the same screen, move icons between the pages of the Home screen, and organize icons within folders. You can even change the icons that appear on the Home screen Dock. You can hide Home screens that contain apps you don't use often, which makes navigating the screens you do use easier and faster. Using the App Library, you can still use any apps stored on hidden pages.

You can also install widgets on Home screens; you can move and organize widgets just like you manage the apps on your phone.

Moving Icons Around Your Home Screens

You can move icons around on a Home screen, and you can move icons among screens to change where they are located.

(1) If you aren't already on one, move to a Home screen.

Going Home
If your iPhone doesn't have a Touch ID/Home button, you move to a Home screen by swiping up from the bottom of the screen. If your iPhone has a Touch ID/Home button, press it once to move to a Home screen.

(2) Swipe to the left or right across the Home screen until the page containing an icon you want to move appears.

(3) Touch (don't tap because if you do, the app opens instead) and hold on any icon. After a moment, the Quick Actions menu appears.

(4) Tap Edit Home Screen. The icons begin jiggling, which indicates that you can move icons on the Home screens. You also see the Delete icons (x) in the upper-left corner of some icons, which indicates that you can delete both the icon and app.

No Icon Required
If you touch and hold on the screen where there isn't an icon, the iPhone moves directly into Edit mode.

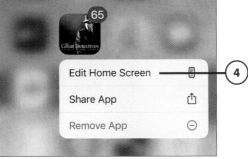

(5) Touch and hold an icon you want to move; it becomes darker to show that you've selected it.

(6) Drag the icon to a new location on the current screen; as you move the icon around the page, other icons separate and are reorganized to enable you to place the icon in its new location.

(7) When the icon is in the location you want, lift your finger. The icon is set in that place. (You don't have to be precise; the icon automatically snaps into the closest position when you lift your finger off the screen.)

(8) Tap and hold on an icon you want to move to a different page.

(9) Drag the icon to the left edge of the screen to move it to a previous page or to the right edge of the screen to move it to a later page. As you reach the edge of the screen, you move to the previous or next page.

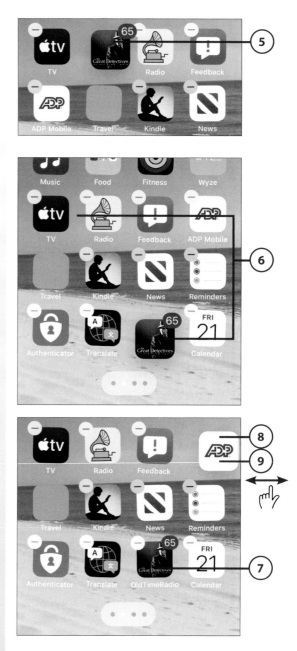

10 Drag the icon to its location on the new Home screen and lift your finger off the icon.

11 Continue moving icons until you've placed them in the locations you want; then tap Done (iPhones without a Home button) or press the Touch ID/ Home button (iPhones with a Home button). The icons are locked in their current positions, they stop jiggling, and the Delete icons disappear.

Done Is Done

Yet another way to end the process is to tap on the screen where there isn't an icon.

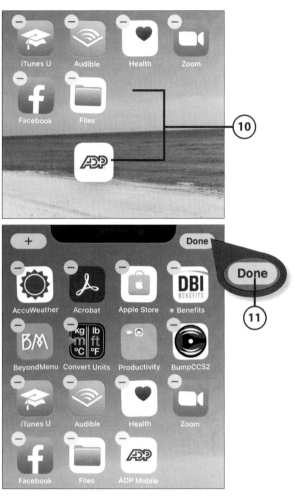

Creating and Naming Folders to Organize Apps on Your Home Screens

You can place icons into folders and name these folders to keep them organized and to make more icons available on the same page. To create a folder, do the following:

1 Move to the Home screen containing icons you want to place in a folder.

2 Touch and hold an icon until the Quick Action menu appears.

3 Tap Edit Home Screen. The icons start jiggling; the Delete icons (x) appear on most icons.

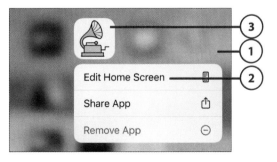

(4) Drag one icon on top of another one that you want to be in the new folder together.

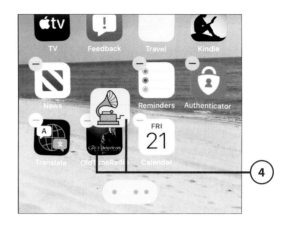

(5) When the first icon is on top of the second and a folder appears, lift your finger. The two icons are placed into the new folder, which is named based on the type of icons you place within it. The folder opens and you see its default name.

(6) To delete the current name so you can create a completely new name, tap Delete (x); if you want to keep all or part of the current name, skip to step 9.

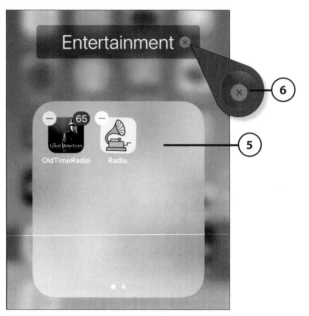

(7) Change the default name for the folder or type a completely new name if you deleted the previous one.

(8) Tap done.

(9) Tap outside the folder to close it.

(10) If you're done organizing the Home screens, tap Done (iPhones without a Home button) or press the Touch ID/Home button (iPhones with a Home button). The icons are locked into place and stop jiggling.

Locating Folders

You can move a folder to a new location in the same way you can move any icon. Touch and hold (don't tap, or it opens instead) the folder's icon until the Quick Action menu appears. Tap Edit Home Screen. The icons start jiggling. Drag the folder icon to where you want it to be.

The new folder

Placing Icons in Existing Folders

You can add icons to an existing folder like so:

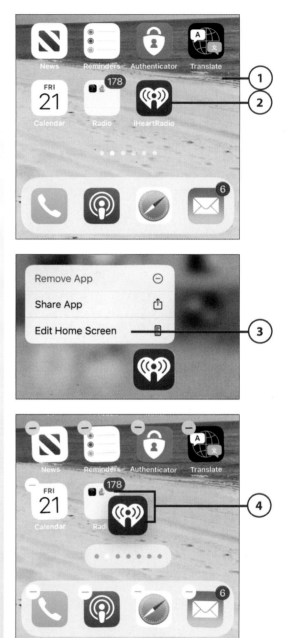

(1) Move to the Home screen containing an icon you want to place in a folder.

(2) Touch and hold an icon until the Quick Action menu appears. (It doesn't matter which icon you tap and hold because this starts the screen edit mode and you can get to all the icons once you are in this mode.)

(3) Tap Edit Home Screen. The icons start jiggling; Delete icons (x) appear on most apps.

Skippin' Steps

You can perform steps 2 and 3 by touching and holding on an empty spot on a Home screen. After a second or two, the Home screens move into edit mode.

(4) Drag the icon you want to place into a folder on top of the folder's icon; the folder opens. (The icon doesn't have to be on the same Home screen page; you can drag an icon from one page and drop it on a folder on a different page.)

5 When the folder opens, lift your finger from the screen. The icon is placed within the folder.

Adding Apps to Folders Quickly

If you don't want to change the icon's location when you place it in the folder, lift your finger as soon as the folder's icon is highlighted; this places the icon in the folder but doesn't cause the folder to open. This is faster than waiting for the folder to open, but doesn't allow you to position the icon within the folder.

6 Drag the new icon to its location within the folder.

7 Tap outside the folder. The folder closes.

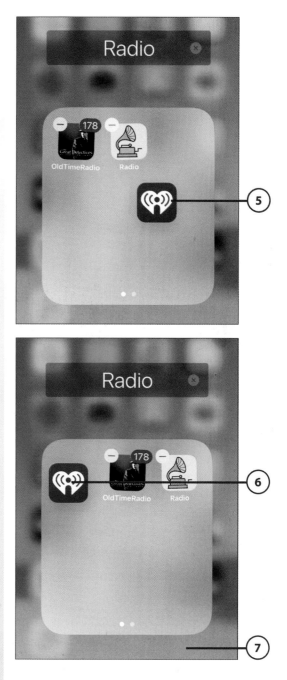

8) When you're done adding icons to folders, tap Done (iPhones without a Home button) or press the Touch ID/Home button (iPhones with a Home button), or tap in an empty spot on the screen.

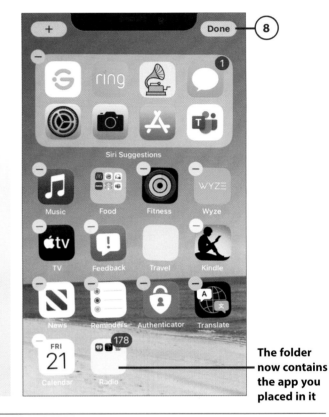

The folder now contains the app you placed in it

Removing Icons from Folders

To remove an icon from a folder, tap the folder from which you want to remove the icon to open it. Touch and hold the icon until the Quick Action menu appears. Tap Edit Home Screen. Drag the icon you want to remove from inside the folder to outside the folder. When you cross the border of the folder, the folder closes and you can place the icon on a Home screen. When you move the last icon out of a folder, the folder is deleted.

Folders and Badges

When you place an icon that has a badge notification (the red circle with a number in it that indicates the number of new items in an app) in a folder, the badge transfers to the folder so that you see it on the folder's icon. When you place more than one app with badge notifications in the same folder, the badge on the folder becomes the total number of new items for all the apps in the folder. You need to open a folder to see the badges for the individual apps it contains.

Configuring the Home Screen Dock

The Dock on the bottom of the Home screen appears on every page, making accessing those icons quick and easy. You can place any four icons on the Dock that you want, including folder icons.

(1) Move to the Home screen containing an icon you want to place on the Dock.

(2) Touch and hold an icon until the Quick Action menu appears.

(3) Tap Edit Home Screen. The icons start jiggling; the Delete icons (x) appear on most apps.

(4) Drag an icon that is currently on the Dock from the Dock onto the Home screen to create an empty space on the Dock.

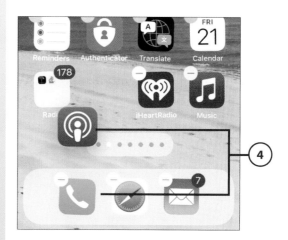

5 Drag an icon or folder from the Home screen onto the Dock.

6 Drag the icons on the Dock around so they're in the order you want them to be.

7 If you're done organizing the Home screen, tap Done (iPhones without a Home button) or press the Touch ID/Home button (iPhones with a Home button) or tap on an empty space on the screen. The icons are locked into place and stop jiggling.

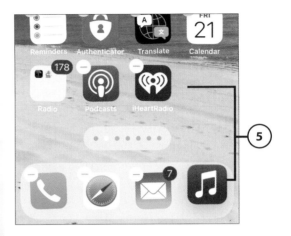

Hiding Home Screens

As you accumulate apps, you can also accumulate Home screens. Having too many Home screens is cumbersome and makes your phone harder to use than it needs to be. You can hide the Home screens you don't use.

1 Move to a Home screen.

2 Touch and hold on an empty space on the screen until the Home screens switch into Edit mode.

3 Tap the screen position indicator at the bottom of the screen. You see the Edit Pages screen.

4 Hide a page by tapping its circle so that it doesn't have a check mark.

5 Show a page by tapping its circle so it has a check mark.

6 Tap Done. When you move back to the Home screens, you no longer see the hidden pages. You can access any apps contained on hidden pages using the App Library (see Chapter 5, "Customizing Your iPhone with Apps and Widgets"), or you can unhide the pages again.

Gone But Not Forgotten?

You can delete apps and widgets that you no longer need. Refer to "Deleting Apps" in Chapter 5 for the details.

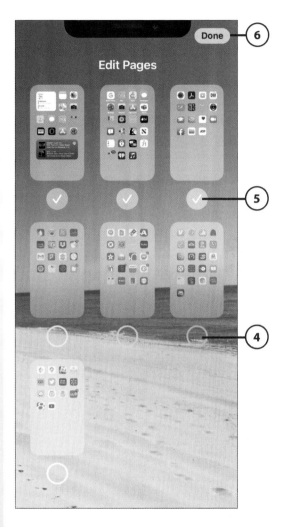

Adding Widgets to Home Screens

Widgets are mini-apps that can provide useful information. You can place widgets on Home screens to make them fast and easy to access.

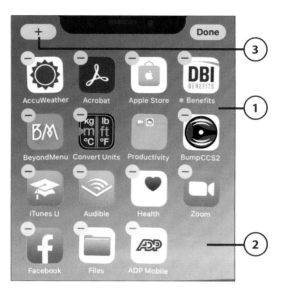

Widgets on Home Screens

In Chapter 5, you learn about widgets and how you can configure the widgets you want to use on the Widget Center. You should understand what widgets are and how they work before following the steps in this section.

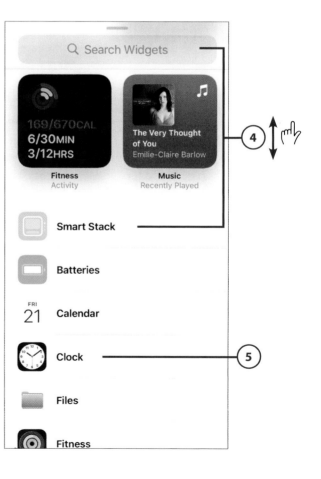

1. If you aren't already on one, move to a Home screen to which you want to add a widget (this doesn't matter too much because you can move widgets around Home screens just as you can move other icons).

2. Touch and hold on an empty space on the screen to put the Home screens into Edit mode.

3. Tap Add (+). You see the Widgets sheet.

4. Search or browse (swipe up or down on the sheet) for the widget (or smart stack) you want to add.

5. Tap the widget (or smart stack) you want to add.

6 Swipe to the left or right until the version of the widget you want to add is shown.

7 Tap Add Widget. The widget is added to the upper-left corner of the page.

8 Drag the widget to where you want it located.

(9) Tap an empty area of the screen
or tap Done. The widget is locked
into place and the Home screens
exit Edit mode.

>>>Go Further
MORE ON ORGANIZING HOME SCREENS

Organizing your Home screens can make the use of your iPhone more efficient. Here are a few
more things to keep in mind:

- You can place many icons in the same folder. When you add more than nine, any
  additional icons are placed on new pages within the folder. As you keep adding icons,
  pages keep being added to the folder to accommodate the icons you add to it. You can
  swipe to the left or right within a folder to move among its pages, just as you can to move
  among your Home screens. You also can drag icons between pages in a folder, just like
  you drag them between Home screens.

- To delete a folder, remove all the icons from it. The folder is deleted as soon as you remove
  the last icon from within it.

- To return your Home screens to their original state when you first started using your iPhone, open the Settings app, tap General, Reset (you might need to swipe up the screen to see this), Reset Home Screen Layout, and Reset Home Screen. The Home screens return to their default configurations. Icons you've added are moved onto the later pages.

Setting the Screen's Appearance, Brightness, Lock/Wake, Text, View, and Wallpaper Options

There are a number of settings you can configure to suit your viewing preferences and how your iPhone locks/wakes:

- **Appearance**—Your iPhone has a Light or Dark mode. In the Light mode, the screen's background is bright, and the text is dark. The Dark mode "inverts" the screen so the background is dark and the text is light. You can switch modes manually and set a schedule so your phone changes modes automatically.

- **Brightness**—Because you continually look at your iPhone's screen, it should be the right brightness level for your eyes. However, the screen is also a large user of battery power, so the dimmer an iPhone's screen is, the longer its battery lasts. You should find a good balance between viewing comfort and battery life.

- **True Tone**—This feature attempts to adjust the display so that colors appear the same despite changes to the ambient light around the phone.

The Models Don't Remain the Same

Different models of iPhones have different screen capabilities, so you might or might not see all of or the same options on your phone that are described in this section. If your iPhone doesn't have a specific capability, such as True Tone, you can just ignore information related to that topic.

- **Night Shift**—This feature changes the color profile of the screen after dark. It's supposed to make the light produced by the iPhone more suitable to darker conditions. You can set the color temperature to your preferences and can set a schedule if you want Night Shift to be activated automatically.

- **Auto-Lock**—The Auto-Lock setting causes your iPhone to lock and go to sleep after a specific amount of inactivity. This is good for security as it is less likely someone can pick up and use your phone if you let it sit for a while. It also extends battery life because it puts the iPhone to sleep when you aren't using it.

- **Raise to Wake**—This setting enables you to wake up the iPhone by lifting it. This is useful because you don't even need to touch the screen or press a button, just lift the phone and you see the Lock screen, giving you quick access to the current time, the Audio Player, notifications, and widgets. However, some people find this feature more annoying than helpful so if the phone waking when you lift it bothers you, disable this setting on your phone.

- **Text Size/Bold**—As you use your iPhone, you'll be constantly working with text, so it's important to configure the text size to meet your preferences. You can use the Bold setting to bold text to make it easier to read.

- **View**—Some iPhone models offer two views. The Standard view maximizes screen space and the Zoomed view makes things on the screen larger, making them easier to see, but less content fits on the screen. You can choose the view that works best for you.

- **Wallpaper**—Wallpaper is the image you see "behind" the icons on your Home screens. Because you see this image so often, you might as well have an image that you want to see or that you believe makes using the Home screens easier and faster. You can use the iPhone's default wallpaper images, or you can use any photo available on your iPhone. You also can set the wallpaper you see on the iPhone's Lock screen (you can use the same image as on the Home screens or a different one). Although it doesn't affect productivity or usability of the iPhone very much, choosing your own wallpaper to see in the background of the Home and Lock screens makes your iPhone more personal to you and is just plain fun.

Configuring Light and Dark Modes Using the Settings App

To configure the Appearance mode your iPhone uses, perform the following steps:

(1) In the Settings app, tap Display & Brightness.

(2) Tap Light to use the Light mode or Dark to use the Dark mode.

(3) To have your iPhone automatically change modes, set Automatic to on (green). The default setting is to have your iPhone in Light mode until sunset at which time it changes to Dark mode. At sunrise, it changes back to Light mode. If you don't want this change to happen automatically, set Automatic to off (white) and skip the rest of these steps.

(4) To change the schedule for the automatic mode change, tap Options.

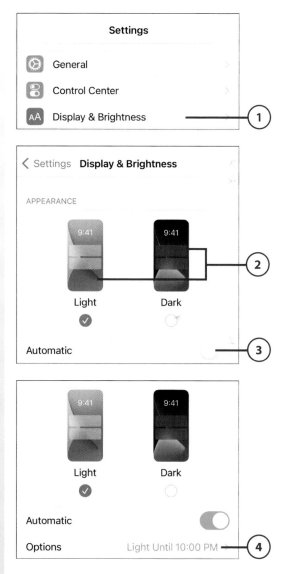

5 Tap Custom Schedule.

6 Use the Light time tool to set the time at which your iPhone switches to Light mode.

7 Use the Dark time tool to set the time at which your iPhone switches to Dark mode.

8 Tap Sunset to Sunrise to return the automatic mode to its default.

How Does It Know?

Your iPhone gets time and date information through your cellular or Wi-Fi network. The iPhone gathers your location through its GPS system. Based on your location, the time, and the date, the iPhone can determine the time for sunrise and sunset. So, the actual times at which the Appearance mode changes vary with the change in sunrise and sunset times.

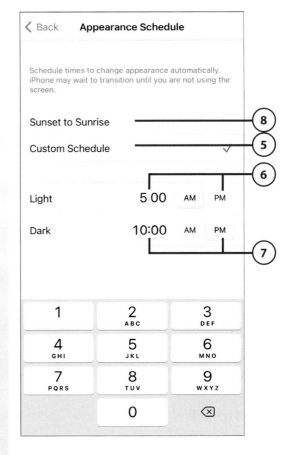

Setting Screen Brightness, True Tone, and Night Shift Using the Settings App

To set the screen brightness and Night Shift, perform the following steps:

1 In the Settings app, tap Display & Brightness.

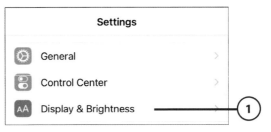

(2) Drag the slider to the right to raise the brightness or to the left to lower it. A brighter screen uses more battery power but is easier to see.

(3) If you don't want the True Tone feature to be active, set the True Tone switch to off (white).

(4) To use True Tone, set the True Tone switch to on (green).

(5) Tap Night Shift.

(6) To have Night Shift activate automatically, set the Scheduled switch to on (green); if you don't want it to activate automatically, skip to step 13.

(7) Tap the From/To setting.

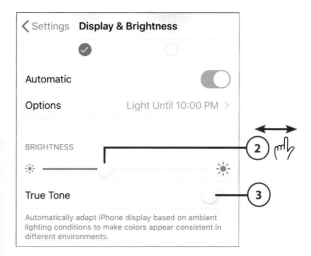

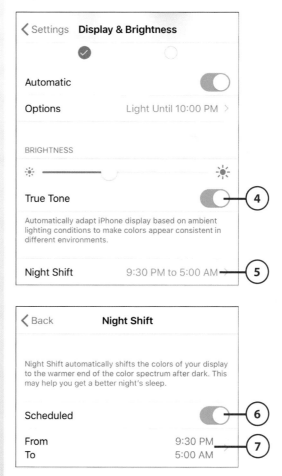

8 To have Night Shift on between sunset and sunrise, tap Sunset to Sunrise and skip to step 12.

9 To set a custom schedule for Night Shift, tap Custom Schedule.

10 Use the Turn On tool to set the time when you want Night Shift to activate.

11 Use the Turn Off tool to set when you want Night Shift to turn off.

12 Tap Night Shift.

13 To manually turn Night Shift on at any time, set the Manually Enable Until Tomorrow switch to on (green). Night Shift activates and remains on until sunrise when it shuts off automatically. (You can manually turn off Night Shift by setting the Manually Enable Until Tomorrow switch to off [white]).

14 Drag the COLOR TEMPERATURE slider to the right to make the Night Shift effect more pronounced or to the left to make it less pronounced. If Night Shift isn't active when you drag the slider, it goes into effect as you move the slider so you can see the effect of the temperature you select.

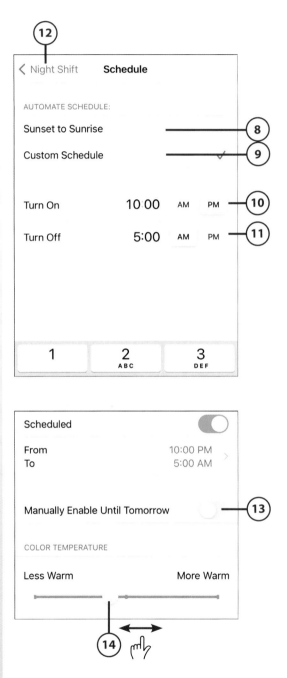

Setting the Appearance, Screen Brightness, and Night Shift Using the Control Center

You can use the Control Center to quickly configure your iPhone's screen as follows:

1. Open the Control Center by swiping down from the upper-right corner of the screen (iPhones without a Home button) or swiping up from the bottom of the screen (iPhones with a Home button).

2. Swipe up or down on the Brightness slider to increase or decrease the brightness, respectively.

3. To access more controls, touch and hold on the Brightness slider.

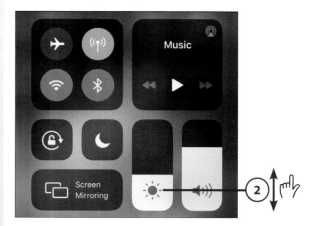

4 Use the Brightness slider to change the screen's brightness (this does the same thing as using the slider on the Control Center, but it's larger here, so it's a bit easier to use).

5 To enable or disable True Tone, tap True Tone. The current status is indicated under the icon so you know when it's on or off. The icon is blue when True Tone is on and black when it's off.

6 Tap Night Shift to turn Night Shift on. When on, the Night Shift icon is orange.

7 Tap Night Shift to turn Night Shift off. When off, the Night Shift icon is black.

8 Tap Appearance to change modes. You see the current status of the Dark mode under the icon and the icon itself reflects the mode. When Dark mode is off, the icon is mostly dark; when Dark mode is on, the icon is mostly light (which seems backward to me).

9 Tap outside the tools to return to the Control Center.

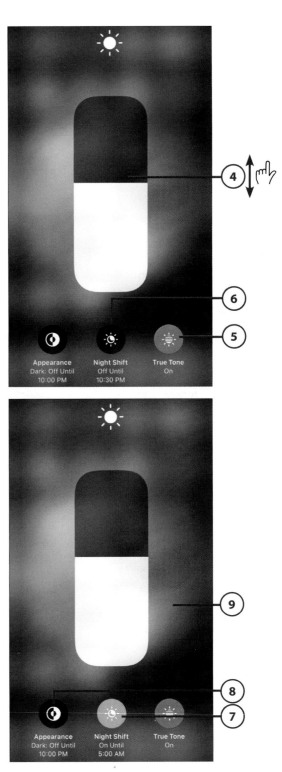

10 Tap on any area on the background of the Control Center to close it.

Setting Auto-Lock and Raise to Wake

To configure Auto-Lock or Raise to Wake, perform the following steps:

1 Open the Display & Brightness settings screen.

2 Tap Auto-Lock.

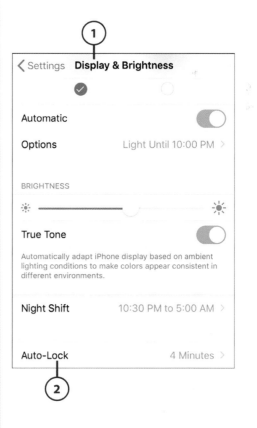

(3) Tap the amount of idle time you want
to pass before the iPhone automatically locks and goes to sleep. You can
choose from 30 seconds or 1 to 5 minutes; choose Never if you want to only
manually lock your iPhone. I recommend that you keep Auto-Lock set to a
relatively small value to conserve your
iPhone's battery and to make it more
secure. Of course, the shorter you set
this time to be, the more frequently
you have to unlock your iPhone.

You Looking at Me?

iPhones without a Home button can detect
when you're looking at the screen. This is cool
because the iPhone won't lock while you're
looking at it but not touching the screen,
such as when you are reading a book. This
means you can set a very short Auto-Lock time
and not have the iPhone sleep while you are
looking at it.

(4) Tap Back (<).

(5) If you want to be able to wake your
phone by lifting it, set the Raise to
Wake switch to on (green); to disable
this feature, set the switch to off
(white).

Setting Text Size and Bold

To change the text size or make all text bold,
perform the following steps:

(1) Open the Display & Brightness settings
screen.

(2) Tap Text Size. This control changes the
size of text in all the apps that support
the iPhone's Dynamic Type feature.

< Back   **Auto-Lock**

30 Seconds

1 Minute

2 Minutes

3 Minutes

4 Minutes

5 Minutes ✓

Never

Attention is detected when you are looking at the
screen. When attention is detected, iPhone does not dim
the display.

Auto-Lock   5 Minutes >

Raise to Wake

< Settings   **Display & Brightness**

Automatic

Options   Light Until 10:00 PM >

BRIGHTNESS

True Tone

Automatically adapt iPhone display based on ambient
lighting conditions to make colors appear consistent in
different environments.

Night Shift   10:30 PM to 5:00 AM >

Auto-Lock   5 Minutes >

Raise to Wake

Text Size

(3) Drag the slider to the right to increase the size of text or to the left to decrease it. As you move the slider, the text at the top of the screen resizes so you can see the effect of the change you are making.

(4) When you're happy with the size of the text, tap Back (<).

(5) If you want to make all of the text on your iPhone bold, set the Bold Text switch to on (green).

(6) To remove the bold on text, set the Bold Text switch to off (white).

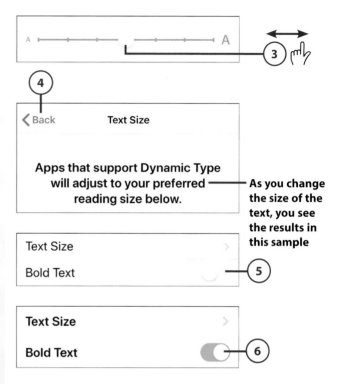

As you change the size of the text, you see the results in this sample

Setting Text Size Using the Control Center

To change the text size with the Control Center, perform the following steps:

(1) Open the Control Center by swiping down from the upper-right corner of the screen (iPhones without a Home button) or swiping up from the bottom of the screen (iPhones with a Home button).

(2) Tap the Text Size icon. (If you don't see this icon, you need to add it to the Control Center, as explained in Chapter 6.)

3 Tap above the shaded area to increase the text size.

4 Tap below the shaded area to decrease the text size.

5 Tap outside the tool to return to the Control Center.

6 Tap on any area on the background of the Control Center to close it.

As you change the size of the text, you see the results in this sample

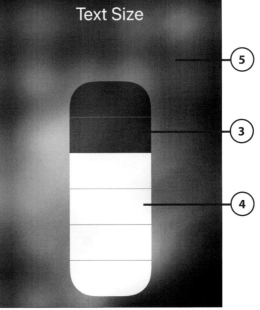

Choosing a View

Some iPhone models enable you to choose between a Standard or Zoomed view. To configure the view you use, perform the following steps:

(1) In the Settings app, tap Display & Brightness to open the Display & Brightness settings screen.

(2) Tap View; if you don't see this option, your iPhone doesn't support it and you can skip the rest of these steps.

(3) Compare the two screens; the images on the screens change so you see examples of how various types of information will appear in each view. The images are representations and don't show any detail so it can be hard to see the difference between the two views. The best way to decide is to try each view and see which one you prefer.

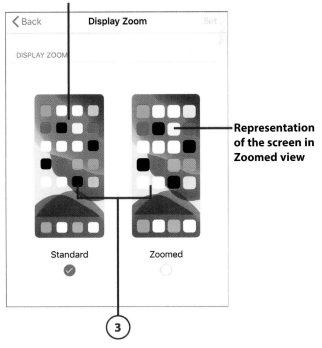

Representation of the screen in Standard view

Representation of the screen in Zoomed view

4. Tap the View you want to use.

5. Tap Set. (If Set is in gray, you're already using that view and you can skip the rest of these steps.)

6. Tap Use Zoomed or Use Standard. The iPhone restarts. When it starts up, you see your iPhone in the selected view.

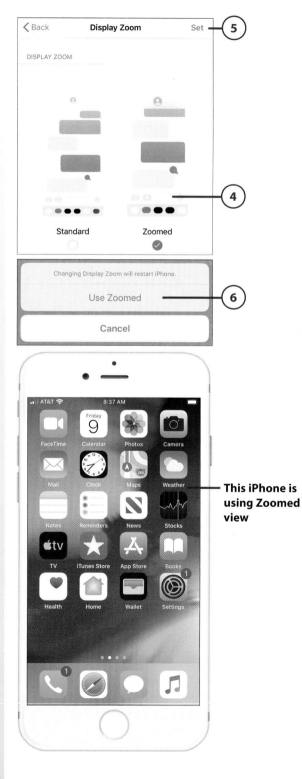

This iPhone is using Zoomed view

Setting the Wallpaper on the Home and Lock Screens

To configure your wallpaper, perform the following steps:

(**1**) In the Settings app, tap Wallpaper. You see the current wallpaper set for the Lock and Home screens.

(**2**) If you want your wallpaper to darken when your iPhone is in Dark mode, set the Dark Appearance Dims Wallpaper to on (green).

(**3**) Tap Choose a New Wallpaper. The Choose screen has two sections. The top section enables you to choose one of the default wallpaper images (Dynamic, Stills, or Live), whereas the lower section shows you the photos available on your iPhone. If you don't have any photos stored on your iPhone, you can choose only from the default images. To choose a default image, continue with step 4; to use one of your photos as wallpaper, skip to step 9.

(**4**) Tap Dynamic if you want to use dynamic wallpaper, Stills if you want to use a static image, or Live if you want to use a Live Photo. These steps show selecting the Stills option, but using a dynamic image or Live Photo is similar.

Current wallpaper on your Lock screen

Current wallpaper on your Home screens

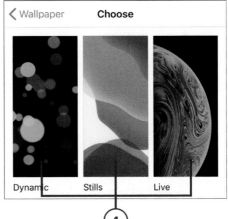

5 Swipe up and down the screen to browse the images available to you.

6 Tap the image you want to use as wallpaper.

7 Tap Perspective to use the Perspective view of the wallpaper; when the Perspective option is disabled, there is a slash through its icon.

8 Tap Set and move to step 16.

Keep It in Perspective

With the Perspective option, the images on "top" of the wallpaper shift slightly when you move the phone. For example, when you tilt the phone with the Lock screen displayed, the time and date move relative to the wallpaper's image. When you disable this, there is no motion. This is a fairly subtle effect so you might have to try it on and off a few times to see if you can tell the difference.

Dynamic Wallpaper

Dynamic wallpaper has motion (kind of like a screen saver on a computer). This type of wallpaper doesn't have any options. You can only choose to set it.

Live Wallpaper

Live wallpaper uses a Live Photo (see Chapter 15, "Taking Photos and Video with Your iPhone" for a description of Live Photos). When you select a Live Photo as wallpaper, you can enable or disable the Live aspect of that Photo. You can also set perspective. Since you have to press the screen to see a Live Photo's motion, I don't really get the point of this one, but it's an option you might want to check out.

9 To use a photo as wallpaper, swipe up the screen to browse the sources of photos available to you; these include All Photos, Favorites, Selfies, Albums, and so on. (To learn how to work with the photos on your iPhone, see Chapter 16, "Viewing and Editing Photos and Video with the Photos App.")

10 Tap the source containing the photo you want to use.

11 Swipe up and down the selected source to browse its photos.

12 Tap the photo you want to use. (If the photo is not currently stored on your iPhone, it downloads, which can take a few seconds.) The photo appears and you can resize and move the image around the screen.

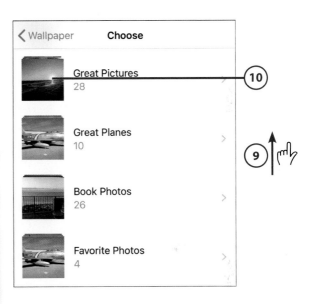

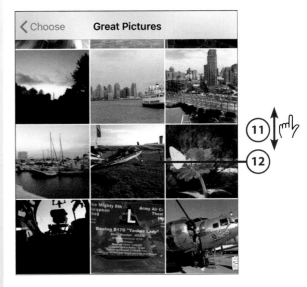

(13) Use your fingers to unpinch to zoom in or pinch to zoom out, and hold down and drag the photo around the screen until it appears how you want the wallpaper to look.

(14) Tap Perspective to use the Perspective view of the wallpaper; if you selected a Live Photo, tap Live Photo to enable or disable its motion.

(15) Tap Set.

(16) Tap Set Lock Screen or Set Home Screen to apply the wallpaper to only one of those screens; tap Set Both to apply the same wallpaper in both locations. The next time you move to the screen you selected, you see the wallpaper you chose.

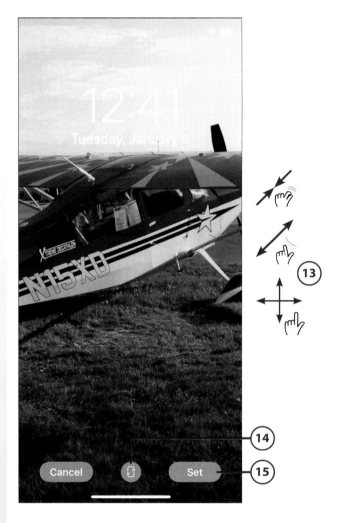

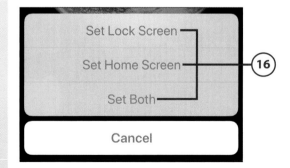

(17) If you set the wallpaper in only one location, tap Choose (not shown on a figure) to move back to the Choose screen and repeat steps 3–16 to set the wallpaper for the other location.

Updating Current Wallpaper

To update the settings for the current wallpaper without changing the image, move to the Wallpaper screen and tap the wallpaper (tap the Lock or Home screen) you want to change. The Wallpaper screen opens and you can change the options for the type of wallpaper you're using. For example, tap Perspective to enable or disable perspective for the photo. Or, if you're working with a photo you've taken, you can resize and move it. To save your changes, tap Set; to leave it as it is, tap Cancel.

New wallpaper on the Home screen

New wallpaper on the Lock screen

Use Settings to configure how contacts are displayed

Tap here to work with your contact information

Use your contact information in many apps

In this chapter, you learn how to ensure that your iPhone has the contact information you need when you need it. Topics include the following:

→ Getting started
→ Creating contacts on your iPhone
→ Working with contacts on your iPhone
→ Managing your contacts on your iPhone

Managing Contacts

You'll be using your iPhone to make calls, get directions, send emails, and do many other tasks that require contact information, including names, phone numbers, email addresses, and physical addresses. It would be time-consuming and a nuisance to have to remember and retype this information each time you use it. Fortunately, you don't have to do either because the Contacts app puts all your contact information at your fingertips (literally).

Getting Started

The Contacts app makes using your contact information extremely easy. This information is readily available on your phone in all the apps in which you need it, including Mail, Messages, and Phone. And you don't need to remember or type the information because you can enter it by choosing someone's name, a business's name, or other information that you know about the contact. You also can access your contact information directly in the Contacts app and take action on it (such as placing a call).

You can store information in the Contacts app in several ways. When you configure an online account on your iPhone—such as iCloud or Google—to include contact information, the contact information stored in that account is immediately available on your phone.

You should store your contacts in an online account (for example, iCloud or Google) so that the information is accessible on many devices, and is also backed up on the Internet cloud. If you don't have an online account, contact information is stored only on your iPhone, which isn't good because, if something happens to your phone, you can lose all of your contacts.

If you haven't already configured an online account to store your contact information, move to Chapter 4, "Setting Up and Using an Apple ID, iCloud, and Other Online Accounts," and do so before continuing here.

You can manually add more contact information to the Contacts app by capturing that information when you perform tasks (such as reading email). You also can enter new contact information directly in the Contacts app.

The Contacts app makes it easy to keep your contact information current by doing such things as adding more information, updating existing contacts, or removing contacts you no longer need.

Settings for Contacts

Like almost all apps on your iPhone, there are settings that determine how your contacts look or work. If your contacts look a bit different than those in the figures in this chapter, it's likely your settings are not the same as mine. Refer to the Go Further sidebar "Contacts Settings," at the end of this chapter to learn about the more important contacts settings and how to change them.

Creating Contacts on Your iPhone

You can create new contacts on an iPhone in a number of ways. You can start with some information, such as the email address on a message you receive, and create a contact from it, or you can create a contact by manually filling in the contact information. In this section, you learn how to create contacts in both ways.

Creating New Contacts from Email

When you receive an email, you can easily create a contact to capture email addresses associated with a message. (To learn how to work with the Mail app, see Chapter 10, "Sending, Receiving, and Managing Email.")

(1) On the Home screen, tap Mail.

(2) Use the Mail app to read an email message. If your iPhone can associate a name with an email address, you see the name. If not, you see the email address instead.

(3) Tap the name or email address for which you want to create a new contact. The names or addresses turn blue to indicate you can work with them. If the names or addresses are already blue, skip this step.

(4) Tap the name or email address that you want to capture as a contact. The contact's Info screen appears. You see as much information as could be gleaned from the email address, which is typically the sender's name and email address.

(5) Tap Create New Contact. The New Contact screen appears. The name, email address, and any other information that can be identified are added to the new contact. The email address is labeled with iPhone's best guess, such as other or home.

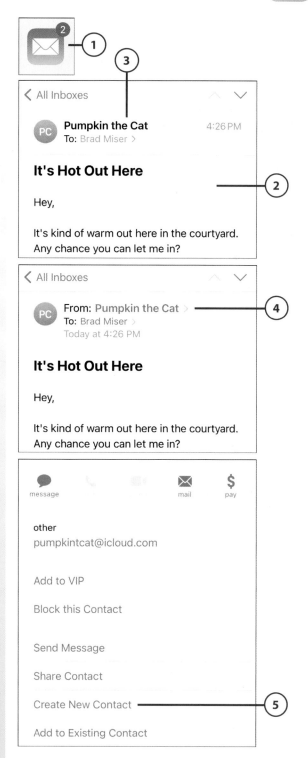

6. Use the New Contact screen to enter more contact information or update the information that was added (such as the label applied to the email address) and save the new contact by tapping Done. This works just like when you create a new contact manually, except that you already have some information—most likely, a name and an email address. For details on adding or changing information for the contact, see the next task, "Creating Contacts Manually."

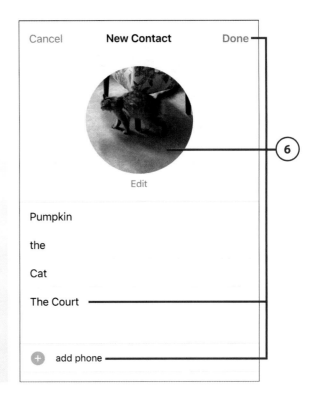

Cancel   **New Contact**   Done

Edit

Pumpkin

the

Cat

The Court

add phone

>>>Go Further
MORE ON CREATING CONTACTS FROM APPS

It's useful to be able to create contact information by starting with some information in an app. Keep these points in mind:

- Mail is only one of the apps from which you can create contacts. You can start a contact in just about any app you use to communicate, such as Messages, or get information, for example, Maps. The steps to start a contact in these apps are similar to those for Mail. Tap Info (i) for the person or address for which you want to create a contact, and then tap Create New Contact. The Contacts app fills in as much of the information as it can, and you can complete the rest yourself.

- You can also add more contact information to an existing contact from an app you're currently using. You can do this by viewing the contact information (such as a phone number) and tapping Add to Existing Contact (instead of Create New Contact). You then search for and select the contact to which you want to add the additional information. After it's saved, the additional information is associated with the contact you selected. For example, suppose you've created a contact for a company, but all you have is its phone number. You can quickly add the address to the contact by using the Maps app to look it up and then add the address to the company's existing contact information by tapping the location, tapping Add to Existing Contact, and selecting that company in your contacts.

Creating Contacts Manually

Most of the time, you'll want to get some basic information for a new contact from an app, as the previous task demonstrated, or through an online account, such as contacts stored in your Google account. If these aren't available, you also can start a contact from scratch and manually add all the information you need to it. Also, you use the same steps to add information to or change information for an existing contact that you do to create a new one, so even if you don't start from scratch often, you do need to know how to make changes to existing contacts.

The Contacts app leads you through creating each type of information you might want to capture. You can choose to enter some or all of the default information on the New Contact screen, or add additional fields as needed.

The following steps show creating a new contact that contains the most common contact information you're likely to need:

(1) On the Home screen, tap Contacts. (If you don't see the Contacts app on the Home screen, tap the Extras folder to open it and you should see the app's icon. You might want to move the Contacts icon from this folder to a more convenient location on your Home screen; see Chapter 7, "Customizing How Your iPhone Looks," for the steps to do this.) The Contacts screen displays.

If you see the Groups screen instead, tap Done to move to the Contacts screen. (Groups are covered later.)

② Tap Add (+). The New Contact screen appears with default fields. (You can add more fields as needed using the add field command.)

③ Tap Add Photo to associate an image with the contact. There are many options for the images you can use for contacts. You can choose or create a Memoji, take a photo, use an emoji, add text, or use an existing photo stored on your phone. These steps show using an existing photo. See the "Taking Photos" note for the steps to take a new photo.

④ Tap Photos.

Options, Options

After you work with contact images, you see Suggestions. If one of the images in the Suggestions section is what you want to use, tap it and then tap Assign to Contact. You can also create an Animoji for the contact by tapping it and then following the onscreen directions to pose the Animoji and assign it to the contact.

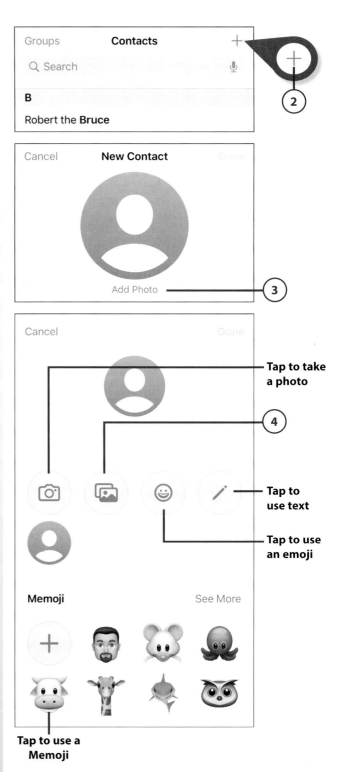

Tap to take a photo

Tap to use text

Tap to use an emoji

Tap to use a Memoji

5 Use the Photos app to move to, select, and configure the photo you want to associate with the contact (see Chapter 16, "Viewing and Editing Photos and Video with the Photos App," for help with the Photos app). You can configure the photo by pinching or unpinching on it and dragging it around so that the part you want to use is shown within the circle.

6 Tap Choose.

7 Tap the filter you want to apply to the image.

8 Tap Done. You return to the image selection screen.

Taking Photos

To take a new photo for the contact, tap the Camera icon in step 4. The Camera app's screen appears. Use the camera to capture the photo you want to use (taking photos is covered in Chapter 15, "Taking Photos and Video with Your iPhone"). Use the Move and Scale screen to adjust the photo and then tap Use Photo.

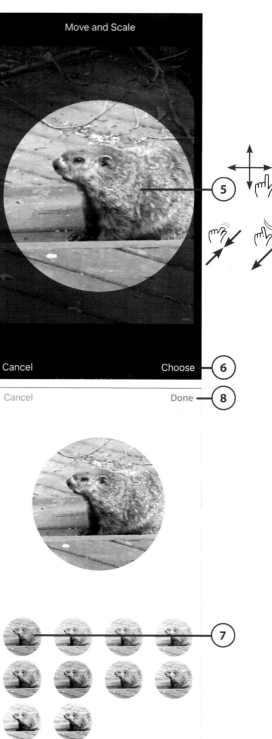

9 Tap Done. You return to the New Contact screen where the photo you selected is displayed.

10 Tap in the First name field and enter the contact's first name; if you are creating a contact for an organization only, leave both name fields empty and skip to step 12.

11 Tap in the Last name field and enter the contact's last name.

12 Enter the organization, such as a company, with which you want to associate the contact, if any.

13 Tap add phone to add a phone number (you might have to swipe up on the screen to see this). A new phone field appears along with the numeric keypad.

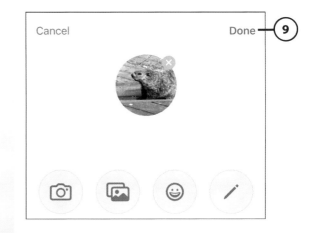

(14) Use the numeric keypad to enter the contact's phone number. Include any prefixes you need to dial it, such as area code and country code. The Contacts app formats the number for you as you enter it.

(15) Tap the label for the phone number; for example home, to change it to another label. The Label screen appears.

(16) Swipe up and down the Label screen to see all the options available.

(17) Tap the label you want to apply to the number; for example, iPhone. That label is applied and you move back to the New Contact screen.

(18) Repeat steps 13–17 to add more phone numbers to the contact.

(19) Swipe up the screen until you see add email.

(20) Tap add email. The keyboard appears.

A Rose by Any Other Name Isn't the Same

The labels you apply to contact information, such as phone numbers, become important when a contact has more than one type of that information. For example, a person might have several phone numbers, such as for home, an iPhone, and work. Applying a label to each of these numbers helps you know which number is for which location. This is especially useful when you use Siri because you can tell Siri which number to use to place a call, such as "Call Sir William Wallace home" to call the number labeled as home on William's contact card.

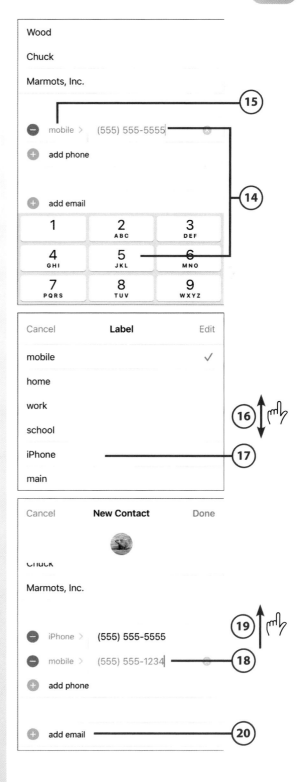

21. Type the contact's email address.

22. Tap the label for the email address to change it.

23. Tap the label you want to apply to the email address.

24. Repeat steps 20–23 to add more email addresses.

25. If necessary, swipe up the screen until you see Ringtone.

26. Tap Ringtone. The list of ringtones and alert tones available on your iPhone appears.

27. Set the Emergency Bypass switch to on (green) if you want sounds and vibrations for phone calls or new messages associated with the contact you are creating to play even when Do Not Disturb is on.

28. Swipe up and down the list to see all of the tones available.

29. Tap the ringtone you want to play when the contact calls you. When you tap a ringtone, it plays so you can experiment to find the one that best relates to the contact. Setting a specific ringtone helps you identify a caller without looking at your phone.

30. Tap Vibration and use the resulting screen if you want to set a specific vibration for the contact. When you are finished, tap Back (<) located in the upper-left corner of the screen.

31. Tap Done. You return to the New Contact screen.

add phone

| home | > | wchuck123@icloud.com |

Cancel **Label** Edit

home ✓

work

school

iCloud

work > wchuck123@icloud.com

add email

Ringtone Default

Cancel **Ringtone** Done

Emergency Bypass

Emergency Bypass allows sounds and vibrations from this person even when the ring switch is set to silent, or when Do Not Disturb is on.

Vibration Default

STORE

Tone Store

Download All Purchased Tones

This will download all ringtones and alerts purchased using the "bradmacosx@mac.com" account.

DEFAULT

Ring

RINGTONES

✓ Carry On Wayward Son

Kryptonite

32 Using the pattern you have learned in the previous steps, move to the next item you want to set and tap it. For example, tap add address to enter a physical address or add url to add a website.

33 Use the resulting screens to enter the information you want to configure for the contact. After you've done a couple of the fields, it's easy to do the rest because the same pattern is used throughout.

34 When you've added all the information you want to capture, tap Done. The New Contact screen closes and the new contact is created and ready for you to use in Contacts and other apps. It's also moved onto other devices with which your contact information is synced. See the Go Further sidebar, "Creating Contacts Expanded," for additional information on syncing and maintaining contacts.

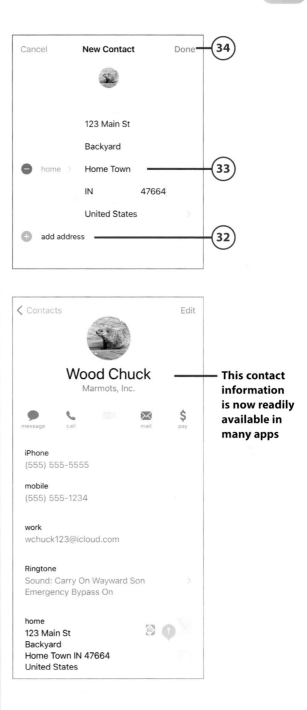

This contact information is now readily available in many apps

>>>*Go Further*

CREATING CONTACTS EXPANDED

Contacts are useful in many ways, so you should make sure you have all the contact information you need. Here are a few points to ponder:

- You can (and should) back up your contacts and sync them on multiple devices (computers and other iOS devices) by using iCloud, Gmail, or other similar online accounts. Refer to Chapter 4 for the details of setting up online accounts.

- Syncing your contacts works in both directions. Any new contacts you create or any changes you make to existing contact information on your iPhone move back to your other devices through the sync process.

- To remove a field in which you've entered information, tap Edit, tap the red circle with a dash in it next to the field, and then tap Delete. If you haven't entered information into a field, just ignore it because empty fields don't appear on a contact's screen.

- The address format on the screens in the Contacts app is determined by the country you associate with the address. If the current country isn't the one you want, tap it and select the country in which the address is located before you enter any information. The fields appropriate for that country's addresses appear on the screen.

- If you want to add a type of information that doesn't appear on the New Contact screen, swipe up the screen and tap add field to see a list of fields you can add. Tap a field to add it; for example, tap Nickname to add a nickname for the contact. Then, enter the information for that new field.

- When you add more fields to contact information, those fields appear in the appropriate context on the Info screen. For example, if you add a nickname, it's placed at the top of the screen with the other "name" information. If you add an address, it appears with the other address information.

- As you learn in Chapter 9, "Communicating with the Phone and FaceTime Apps," you can configure the Phone app to announce the name of callers when you receive calls. This can be even more useful than setting a specific ringtone for your important contacts.

Working with Contacts on Your iPhone

There are many ways to use contact information. The first step is always finding the contact information you need, typically by using the Contacts app. Whether you access it directly or through another app (such as Mail), it works the same way. Then, you select the information you want to use or the action you want to perform.

Using the Contacts App

You can access your contact information directly in the Contacts app. For example, you can search or browse for a contact and then view the detailed information for the contact in which you are interested.

(1) On the Home screen, tap Contacts. (If the Groups screen appears, tap Done. You move back to the Contacts screen.)

You can find a contact to view by browsing (step 2), using the index (step 3), or searching (step 4). You can use combinations of these, too, such as first using the index to get to the right area and then browsing to find the contact in which you are interested.

(2) Swipe up or down to scroll the screen to browse for contact information; swipe up or down on the alphabetical index to browse rapidly. When you see the contact you want to view, move to step 5.

(3) Tap the index to jump to contact information organized by the first letter of the format selected in the Contact Preferences (last name or first name). When you see the contact you want to view, move to step 5.

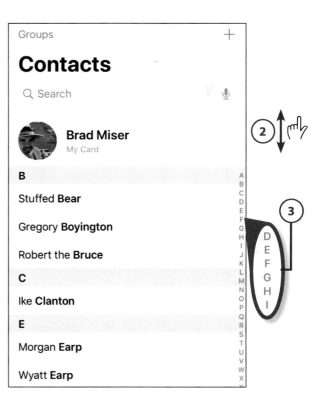

(4) Use the Search tool to search for a specific contact; tap in the tool, type the name (you can type last, first, company, or nickname) of the contact you want to find.

(5) Tap a contact to view that contact's information.

(6) Swipe up and down the screen to view all the contact's information.

(7) Tap the data or icons on the screen to perform actions, including the following:

- **Phone numbers**—Tap a phone number to call it. You can also tap call just under the contact's image to call that contact. If the contact has only one phone number, tap call to dial it. If there is more than one number, you choose the number you want to dial.

- **Email addresses**—Tap an email address or the mail icon or the icon with the email address label (such as other for the email address labeled as other) on it to create a new email message to that address.

- **URLs**—Tap a URL to open Safari and move to the associated website.

- **Addresses**—Tap a physical address to show it in the Maps app.

- **FaceTime**—Tap video or FaceTime to start a FaceTime call with the contact. (If an icon, such as video, is disabled, you don't have that type of information stored for the contact.)

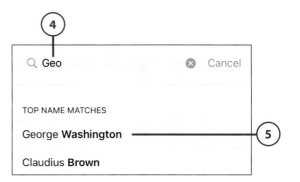

- **Text**—Tap the message icon or tap Send Message and choose the phone number or email address to which you want to send a text message.

- **Pay**—Tap pay to use Apple Pay to send money to the contact.

- **Share Contact**—Tap Share Contact. The Share menu appears. To share the contact via email, tap Mail; to share it via a text, tap Message; or to share it using AirDrop, tap AirDrop. You can also share via Twitter or Facebook. Then, use the associated app to complete the task.

- **Favorites**—Tap Add to Favorites and choose the phone number or email address you want to designate as a favorite. You can use this in the associated app to do something faster. For example, if it's the Phone app, you can tap the Favorites tab to see your favorite contacts and quickly dial one by tapping it. You can add multiple items (such as cell and work phone numbers) as favorites for the same contact.

- **Emergency Contacts**—Tap Add to Emergency Contacts if this person is someone you want to be contacted if you use the emergency features of the phone. When you activate that feature, your contacts receive notifications about the situation.

⑧ To return to the prior screen without performing an action, tap Search (<) located in the upper-left corner of the screen.

>>>Go Further

MAKE CONTACT

When working with your contacts, keep the following points in mind:

- **Last known contact**—The Contacts app remembers where you last were and takes you back there whenever you move into the Contacts app. For example, if you view a contact's details and then switch to a different app to send a message, and then back to Contacts, you return to the screen you were last viewing. To move back to the main Contacts screen, tap Back (it is labeled with the previous screen's name, such as Search or Contacts) in the upper-left corner of the screen.

- **Groups**—In a contact app on a computer, such as Contacts on a Mac, contacts can be organized into groups, which in turn can be stored in an online account, such as iCloud. (If you've never created contact groups elsewhere, you can skip any information related to contact groups.) When you sync, the groups of contacts move onto the iPhone. You can limit the contacts you browse or search. To access your groups, tap Groups on the Contacts screen.

  The Groups screen displays the accounts (such as iCloud or Google) with which you are syncing contact information; under each account are the groups of contacts stored in that account. If a group has a check mark next to it, its contacts are displayed on the Contacts screen. To hide a group's contacts, tap it so that the check mark disappears. To hide or show all of a group's contacts, tap All *account*, where *account* is the name of the account in which those contacts are stored. To make browsing contacts easier, tap Hide All Contacts to hide all the groups and contacts; then, tap each group whose contacts you want to show on the Contacts screen.

  Tap Done to move back to the Contacts screen.

- **Speaking of contacts**—You can use Siri to speak commands to work with contacts, too. You can get information about contacts by asking for it, such as "What is William Wallace's work phone number?" If you want to see all of a contact's information, you can say "Show me William Wallace." When Siri displays contact information, you can tap it to take action, such as tapping a phone number to call it. (See Chapter 13, "Working with Siri," for more on using Siri.)

It's Not All Good

Managing Contact Groups

When you create a new contact, it is associated with the account you designate as the default in the Contacts settings and is stored at the account level (not in any of your groups). You can't create groups in the Contacts app, nor can you change the group with which contacts are associated. You have to use a contacts app on a computer to manage groups and then sync your iPhone (which happens automatically when you use an online account, such as iCloud) to see the changes you make to your contact groups.

Accessing Contacts from Other Apps

You can also access contact information while you're using a different app. For example, you can use a contact's email address when you create an email message. When you perform such actions, you use the Contacts app to find and select the information you want to use. The following example shows using contact information to send an email message; using your contact information in other apps (such as Phone or Messages) is similar.

(**1**) Open the app from which you want to access contact information (this example uses Mail).

(**2**) Tap Compose.

(**3**) Tap Add (+) in the To field.

___

Another Way to Address

Rather than performing steps 3 through 5, you can type the contact's name directly in the field and then choose the address you want to use from the list that's presented.

___

(4) Search, browse, or use the index to find the contact whose information you want to use. You see the results of your search under the Search bar. The text that matches your search is shown in bold. For example, if you search for Wi, you see people named William, Wil, and so on, and "Wi" is shown in bold in each result. The search looks within the contact too. For example, when you search for an email address within Mail, the email addresses are searched in addition to the name information.

(5) Tap the contact whose information you want to use. (If a contact doesn't have relevant information; for example, if no email address is configured when you are using the Mail app, that contact is grayed out and can't be selected.) You move back to the previous screen and the appropriate information is entered.

Multiple Contacts

If the contact has multiple entries of the type you're trying to use, the contact's information screen appears. Tap the specific information you want to use—in this scenario, the email address. The information is entered in the appropriate location on the new email screen.

(6) Complete the task you're doing, such as sending an email message.

Choose a contact to mail

Q Wi ✖ Cancel

TOP NAME MATCHES

Sir William **Wallace**

OTHER RESULTS

Gregory **Boyington**
dont_mess_**wi**th_pappy@me.com

SIRI FOUND IN APPS

Cancel

New Message

To: Sir William Wallace

Cc/Bcc, From: bradmiser@icloud.com

Subject:

Best,

Brad

Siri Makes It Even Easier

Rather than performing steps 1 through 5, you can activate Siri and say "Send email to Wood Chuck." Siri creates a new email message addressed to the lovable rodent. (If you have more than one email address for the contact, Siri prompts you to select the address you want to use.) Siri then asks you to speak the subject followed by the message. When that's done, you can say "Send" to send the email.

Managing Your Contacts on Your iPhone

When you sync contacts with an iCloud, Google, or other account, the changes go both ways. For example, when you change a contact on the iPhone, the synced contact manager application, such as Outlook, makes the changes for those contacts on your computer. Likewise, when you change contact information in a contact manager on your computer, those changes move to the iPhone. If you add a new contact in a contact manager, it moves to the iPhone, and vice versa. You can also change contacts manually in the Contacts app on your iPhone.

Updating Contact Information

You can change any information for an existing contact, such as adding new email addresses, deleting outdated information, and changing existing information.

(1) Use the Contacts app to find and view the contact whose information you want to change.

(2) Tap Edit. The contact screen moves into Edit mode, and you see Unlock icons.

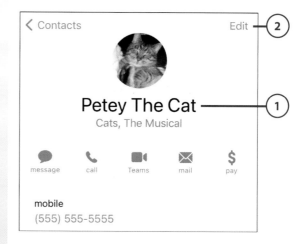

3 Tap current information to change it; you can change a field's label by tapping it, or you can change the data for the field by tapping the information you want to change. Use the resulting tools, such as the phone number entry keypad, to make changes to the information. These tools work just like when you create a new contact (refer to "Creating Contacts Manually," earlier in this chapter).

4 To add more fields, tap add (+) in the related section, such as add email in the email address section; then, select a label for the new field and complete its information. This also works just like adding a new field to a contact you created manually.

5 Tap a field's Unlock (–) icon to remove that field from the contact.

6 Tap Delete. The information is removed from the contact.

7 To change the contact's image, tap the current photo, or the word Edit under the current photo, and use the resulting menu and tools to choose a new photo, take a new photo, delete the existing photo, or edit the existing photo.

8 When you finish making changes, tap Done. Your changes are saved, and you move out of Edit mode.

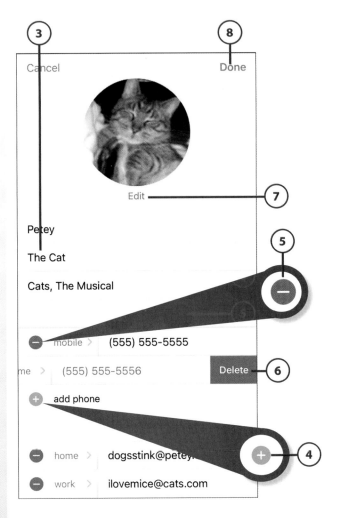

Adding Information to an Existing Contact While Using Your iPhone

As you use your iPhone, it is likely that you'll encounter information related to a contact that isn't currently part of that contact's information. For example, a contact might send an email to you from a different email address than the one you have stored for her. When this happens, you can easily add the new information to the existing contact. Just tap the information to view it (such as an email address), and then tap Add to Existing Contact. Next, select the existing contact to which you want to add the new information. The new information is added to the contact. Depending on your iPhone model, tap Update or Done to save the new information or Cancel, if you decide not to add the new information.

Deleting Contacts

To get rid of contacts, you can delete them from the Contacts app.

1. Find and view the contact you want to delete.

2. Tap Edit.

3. Swipe up to get to the bottom of the screen.

4. Tap Delete Contact.

5. Tap Delete Contact to confirm the deletion. The app deletes the contact, and you return to the Contacts screen.

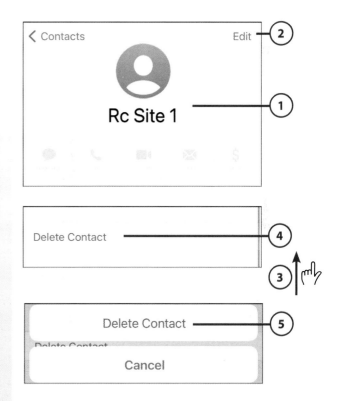

>>>Go Further

CONTACTS SETTINGS

Using Contacts settings, you can change certain aspects of how contacts work for you. You can probably work with contacts just fine without changing any of these settings. However, if you want to make adjustments, open the Settings app, scroll up the screen, and tap Contacts. Following are some of the more useful settings:

- **Sort Order**—Tap First, Last to have contacts sorted by *first name* and then *last name* or tap Last, First to have contacts sorted by *last name* and then *first name*.

- **Display Order**—To show contacts in the format *first name, last name,* tap First, Last. To show contacts in the format *last name, first name,* tap Last, First.

- **Short Name**—You can choose whether short names are used and, if they are, what form they take. Short names are useful because more contact information can be displayed in a smaller area, and they look "friendlier." To use short names, move the Short Name switch to the on position (green). Tap the format of the short name you want to use. You can choose from a combination of *initial* and *name* or just *first* or *last name*. If you want nicknames for contacts used for the short name when available, set the Prefer Nicknames switch to on (green).

- **My Info**—Use this setting to find and tap your contact information in the Contacts app, which it can insert for you in various places and which Siri can use to call you by name; your current contact information is indicated by the label "me" next to the alphabetical index.

- **Default Account**—Tap the account in which you want new contacts to be created by default (which is then marked with a check mark). If you have only one account configured for contacts or if you only store contacts on your iPhone, you don't have this option.

Tap to configure
Phone and FaceTime
settings

Tap to hear and
see the people
you want to talk to

Tap to make calls,
listen to voicemail,
and more

In this chapter, you explore all the cell phone and FaceTime functionality that your iPhone has to offer. Topics include the following:

→ Getting started
→ Making voice calls
→ Managing in-process voice calls
→ Receiving voice calls
→ Managing voice calls
→ Using visual voicemail
→ Communicating with FaceTime

Communicating with the Phone and FaceTime Apps

Although it's also a lot of other great things, such as a music player, web browser, and email tool, there's a reason the word *phone* is in iPhone. It's a feature-rich cell phone that includes some amazing functionality, including visual voicemail and FaceTime. Other useful capabilities include a speakerphone, conference calling, and easy-to-use onscreen controls. Plus, your iPhone's phone functions are integrated with its other features. For example, when using the Maps application, you might find a location, such as a business, that you're interested in contacting. You can call that location just by tapping the number you want to call directly on the Maps screen.

Getting Started

Here are some of the key concepts you'll learn about in this chapter:

- **Phone app**—The iPhone's cell phone functionality is provided by the Phone app. You use this app whenever you want to make calls, answer calls, or listen to voicemail.

- **Visual Voicemail**—The Phone app shows you information about your voice-mails, such as the person who left each message, a time and date stamp, and the length of the message. The Phone app provides a lot of control over your messages, too; for example, you can easily fast forward to specific parts of a message that you want to hear. (This is particularly helpful for capturing information, such as phone numbers.) And if that wasn't enough, you also can read transcripts of voicemails so you don't have to listen to them at all.

- **FaceTime**—The FaceTime app enables you to have videoconferences with other people using iPhones, iPads, or Mac computers so that you can both see and hear each other. Using FaceTime is intuitive so you won't find it any more difficult than making a phone call.

- **FaceTime Audio**—You can make FaceTime calls using only audio; this is similar to making a phone call.

Making Voice Calls

There are a number of ways to make calls with your iPhone; after a call is in progress, you can manage it in the same way no matter how you started it.

All in One Place

If you use an iPhone that has a Home button, you see the signal strength, provider, and Wi-Fi Calling status in the top-left corner of the Home screen and on the Lock screen.

Signal strength (more filled-in bars indicate stronger signal)

The strength of the signal your phone is receiving—indicated by the number of shaded bars in the Signal Strength icon at the top of the Home or Lock screens—tells you whether you can make or receive calls. As long as you see at least one bar shaded in, you should be able to place and receive calls via the cellular network. More shaded bars mean you have a stronger signal, so the call quality will be better.

Provider name

Signal strength (more filled-in bars indicate strong signal)

To see who your provider is, move to the Lock screen (iPhones without a Home button). You also see the Signal Strength icon here. On iPhones with a Home button, you see the signal strength and provider information in the upper-left corner of the Home and Lock screens.

Provider name

Wi-Fi calling is available

Signal strength (more filled-in bars indicate strong signal)

If the Wi-Fi calling feature is enabled and your phone is connected to a Wi-Fi network, you see the Wi-Fi calling icon for your provider at the top of the Lock screen (iPhones without a Home button) or the Home or Lock screens (iPhones with a Home button).

With a reasonably strong cellular signal or connection to a Wi-Fi network with Wi-Fi calling enabled, you're ready to make calls.

Which Network?

When you leave the coverage area for your provider and move into an area that is covered by another provider that supports roaming, your iPhone automatically connects to the other provider's network. For example, if AT&T is your provider and you travel to Toronto, Canada, the provider might become Rogers instead of AT&T. In most cases, your provider sends a text message to you explaining the change in networks, including information about roaming charges. Although the change to a roaming network is automatic, you need to be aware of roaming charges, which can be significant depending on where you use your iPhone and your default network. Before you travel outside of your default network's coverage, check with your provider to determine the roaming rates that apply to where you are going. Also, see if there is a discounted roaming plan for that location. If you don't do this before you leave and have roaming turned on, you might get a nasty surprise when the bill arrives because roaming charges can be substantial.

On some iPhone models with specific providers, you can disable roaming by opening the Cellular screen in the Settings app, tapping Cellular Data Options, and then disabling all roaming. To prevent roaming for voice calls on any iPhone, put your iPhone in Airplane mode (your phone can't make or receive calls or use any cellular functions when it operates in this mode). Make sure Wi-Fi and Bluetooth are on if you want to use those services. When you return to the area your provider covers, turn Airplane mode off.

Dialing with the Keypad

The most obvious way to make a call is to dial the number.

 On the Home screen, tap Phone. The Phone app opens.

Number of new missed calls and voicemails

(2) If you don't see the keypad on the screen, tap Keypad.

(3) Tap numbers on the keypad to dial the number you want to call. If you dial a number associated with one or more contacts, you see the contact's name and the type of number you've dialed just under the number. (If you make a mistake in the number you're dialing, tap Delete [X] to the right of the receiver icon to delete the most recent digit you entered.)

(4) Tap the green receiver icon. The app dials the number, and the Call screen appears.

(5) If you have Bluetooth headphones connected to your iPhone, tap the option you want to use for the call; for example, tap the Bluetooth headphones to communicate through them. If you don't have Bluetooth headphones connected to your phone, skip this step.

(6) Use the Call screen to manage the call (not shown in the figure); see "Managing In-Process Voice Calls" later in this chapter for the details.

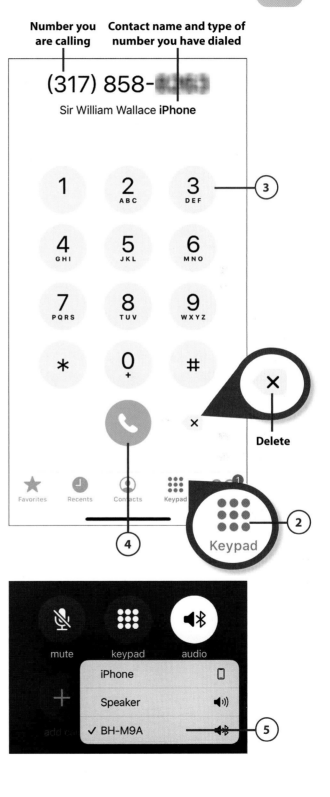

Number you are calling

Contact name and type of number you have dialed

(317) 858-

Sir William Wallace **iPhone**

Delete

Favorites Recents Contacts Keypad

Keypad

mute keypad audio

iPhone

Speaker

✓ BH-M9A

Headphones

Using headphones for phone calls is good because you can typically hear better and you don't have to hold the phone. Further, in many locations, it is illegal to drive while holding your phone so you need to use headphones or your car's audio system.

You can use the headphones that came with your iPhone for this. Simply plug them into the phone and they are used for calls automatically.

You can also connect your iPhone to Bluetooth headphones so you don't have to bother with wires. See "Using Bluetooth to Connect to Other Devices" in Chapter 17, "Working with Other Useful iPhone Apps and Features," for the details of working with Bluetooth devices.

Dialing with Contacts

As you saw in Chapter 8, "Managing Contacts," the Contacts app is a complete contact manager so you can store various kinds of phone numbers for people and organizations. To make a call using a contact, follow these steps.

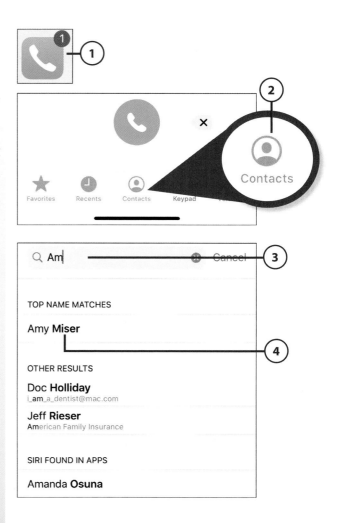

1. On the Home screen, tap Phone.

2. Tap Contacts.

3. Browse the list, search it, or use the index to find the contact you want to call. (Refer to Chapter 8 for information about using the Contacts app.)

4. Tap the contact you want to call.

(5) Tap the number you want to dial. The app dials the number, and the Call screen appears.

(6) If you have Bluetooth headphones connected to your iPhone, tap the option you want to use for the call; for example, tap the Bluetooth headphones to communicate through them. If you don't have Bluetooth headphones connected to your phone, skip this step.

(7) Use the Call screen to manage the call (not shown in the figure); see "Managing In-Process Voice Calls" later in this chapter for the details.

Make the Call

At the top of a contact's screen, you see the call icon. When you tap call, you see all of the person's numbers and can tap the number you want to dial it. Or, if FaceTime is available, tap FaceTime to place a FaceTime call.

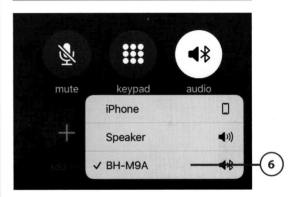

Dialing with Favorites

You can save contacts as favorites to make calling or videoconferencing them even simpler. (You learn how to save favorites in various locations later in this chapter. You learn how to make a contact into a favorite in Chapter 8.)

(1) On the Home screen, tap Phone.

(**2**) Tap Favorites.

(**3**) Browse the list until you see the favorite you want to call. Under the contact's name, you see the type of favorite, such as a phone number (identified by the label in the Contacts app; for example, iPhone or mobile) or FaceTime.

(**4**) Tap the favorite you want to call; to place a voice call, tap a phone number (if you tap a FaceTime contact, a FaceTime call is placed instead). The app dials the number, and the Call screen appears.

(**5**) If you have Bluetooth headphones connected to your iPhone, tap the option you want to use for the call; for example, tap the Bluetooth headphones to communicate through them. If you don't have Bluetooth headphones connected to your phone, skip this step.

(**6**) Use the Call screen to manage the call (not shown in the figure); see "Managing In-Process Voice Calls" later in this chapter for the details.

Nobody's Perfect

If your iPhone can't complete the call for some reason, such as not having a strong enough signal, the Call Failed screen appears. Tap Call Back to try again and maybe try moving to another location that might have a stronger signal or tap Done to give up. When you tap Done, you return to the previous screen.

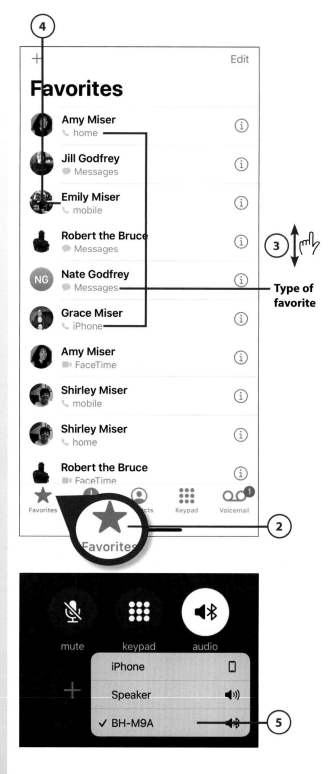

Dialing with Recents

As you make, receive, or miss calls, your iPhone tracks all the numbers on the Recents list. You can use the Recents list to make calls.

1. On the Home screen, tap Phone.

2. Tap Recents. You see a list of recent calls you've made or received. Calls in red are missed calls that you didn't answer for one reason or another. Completed calls are in black.

3. Tap All to see all calls.

4. Tap Missed to see only calls you missed.

5. If necessary, browse the list of calls.

6. To call the number associated with a recent call, tap the title of the call, such as a person's name, or the number if no contact is associated with it. The app dials the number, and the Call screen appears. Skip to step 10.

7. To get more information about a recent call, for example, to see exactly what time yesterday the call came in or how long the call was, tap Info (i). The Info screen appears.

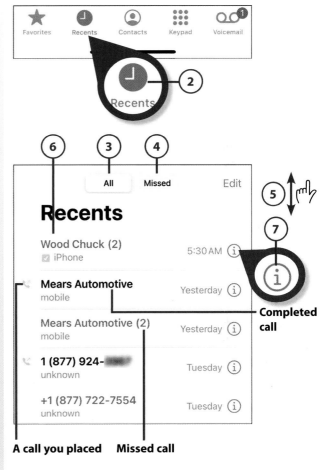

Info on the Recents Screen

If you have a contact on your iPhone associated with a phone number, you see the person's name and the label for the number (such as mobile). If you don't have a contact for a number, you see the number itself. If a contact or number has more than one call associated with it, you see the number of recent calls in parentheses next to the name or number. If you initiated a call, you see the phone icon next to the contact's name and label.

8 Read the information about the call or calls. For example, if the call is related to someone in your Contacts list, you see detailed information for that contact. If there are multiple recent calls, you see information for each call, such as its status (Missed Call, Canceled Call, or Outgoing Call, for example) and time.

9 Tap a number on the Info screen. The app dials the number, and the Call screen appears.

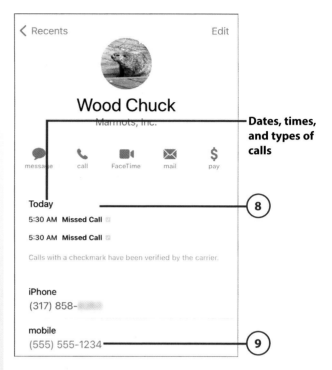

Dates, times, and types of calls

Going Back

To return to the Recents screen without making a call, tap Recents (<) in the upper-left corner of the screen.

---

10 If you have Bluetooth headphones connected to your iPhone, tap the option you want to use for the call; for example, tap the Bluetooth headphones to communicate through them. If you don't have Bluetooth headphones connected to your phone, skip this step.

11 Use the Call screen to manage the call (not shown in the figure); see "Managing In-Process Voice Calls" later in this chapter for the details.

Managing In-Process Voice Calls

When you place a call, there are several ways to manage it. The most obvious is to place your iPhone next to your ear and use your iPhone like any other phone you've ever used. As you place your iPhone next to your ear, the controls on its screen become disabled so you don't accidentally tap onscreen icons with the side of your face or your ear. When you take your iPhone away from your ear, the Call screen appears and the Phone app's controls become active again.

When you are on a call, press the Volume buttons on the left side of the iPhone to increase (top button) or decrease (bottom button) its volume. Some of the other things you can do while on a call might not be so obvious, as you learn in the next few tasks.

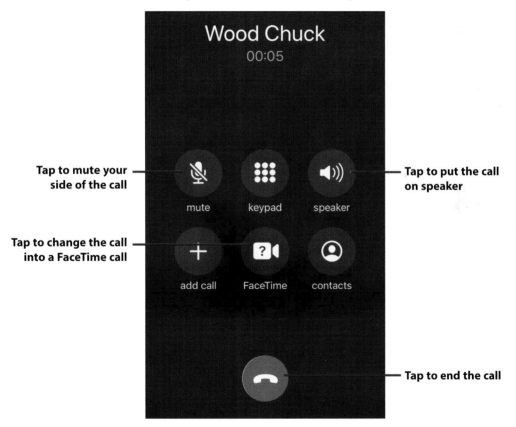

Tap to mute your side of the call

Tap to change the call into a FaceTime call

Wood Chuck
00:05

mute   keypad   speaker

add call   FaceTime   contacts

Tap to put the call on speaker

Tap to end the call

Following are some of the icons on the Call screen that you can use to manage an active call:

- Mute your side of the call by tapping mute. You can hear the person on the other side of the call, but he can't hear anything on your side.

- Tap speaker to use the iPhone's speakers to hear the call. You can speak with the phone held away from your face, too.

- Tap FaceTime to convert the voice call into a FaceTime call (read more on FaceTime later in this chapter).

- When you're done with the call, tap the red receiver icon to end it.

Contact Photos on the Call Screen

If someone in your contacts calls you, or you call her, the photo or other image associated with the contact appears on the screen. Depending on how the image was captured and if you are on the Lock screen or an active screen, it either appears as a small icon at the top of the screen next to the contact's name or fills the entire screen as the background wallpaper. If there isn't an image for the contact, you only see the contact's name and label of the number being used to call, such as Home.

Entering Numbers During a Call

You sometimes need to enter numbers during a call, such as to log in to a voicemail system, access an account, or make a selection from a menu.

(1) Place a call using any of the methods you've learned so far.

(2) Tap keypad.

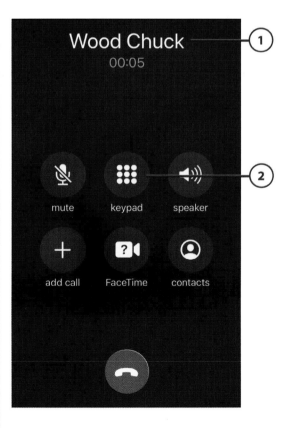

3 Tap the numbers you want to enter.

4 When you're done, tap Hide. You return to the Call screen.

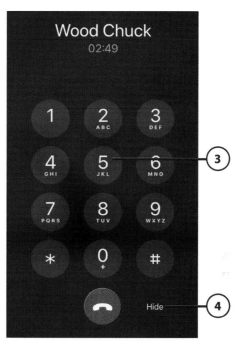

Making Conference Calls

Your iPhone makes it easy to talk to multiple people at the same time. You can have two separate calls going on at any point in time. You can also create conference calls by merging them together. Not all cell providers support two on-going calls or conference calling, though. If yours doesn't, you won't be able to perform the steps in this section.

1 Place a call using any of the methods you've learned so far.

2 Tap add call.

(3) Tap the icon you want to use to place the next call. Tap Favorites to call a favorite, tap Recents to use the Recents list, tap Contacts to place the call using the Contacts app, or tap Keypad to dial the number. These work just as they do when you start a new call.

(4) Place the call using the option you selected in step 3.

Similar but Different

If you tap contacts instead of add call in step 2, you move directly into the Contacts screen. This might save you one screen tap if the person you want to add to the call is in your Contacts app.

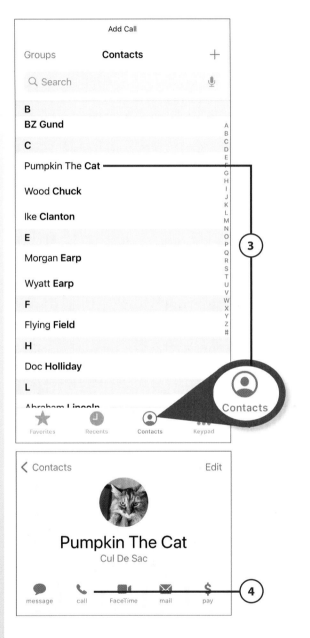

5 If the person has multiple call options, tap the one you want to use.

As you place the second call, the first call is placed on hold and you move back to the Call screen while the Phone app makes the second call. The first call's information appears at the top of the screen, including the word HOLD so you know the first call is on hold. The app displays the second call just below that, and it becomes the active call.

6 Talk to the second person you called; the first remains on hold.

7 To switch to the first call, tap it on the list or tap swap. This places the second call on hold and moves it to the top of the call list, while the first call becomes active again.

8 Tap merge calls to join the calls so all parties can hear you and each other. The iPhone combines the two calls so you now have a three-way call, and you see a single entry at the top of the screen to reflect this.

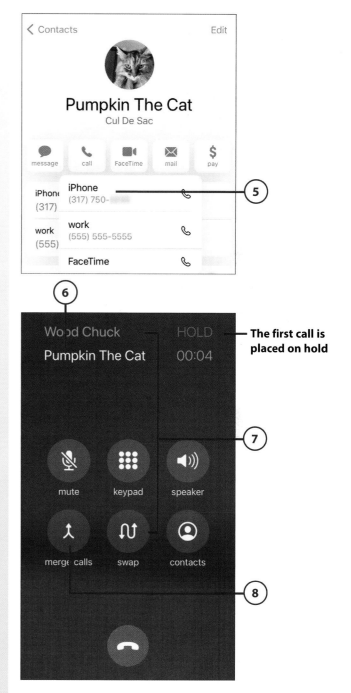

The first call is placed on hold

(9) To add another call, repeat steps 2–8. Each time you merge calls, the second line becomes free so you can add more calls.

(10) To manage a conference call, tap Info (i) at the top of the screen.

(11) To speak with one of the callers privately, tap Private (if the Private icons are disabled, you can't do this with the current calls). Doing so places the conference call on hold and returns you to the Call screen showing information about the active call. You can merge the calls again by tapping merge calls.

(12) Tap End to remove a caller from the call. The app disconnects that caller from the conference call. When you have only one person left on the call, you return to the Call screen and see information about the active call.

(13) Tap Back (<) to move back to the Call screen. You move to the Call screen and can continue working with the call, such as adding more people to it.

(14) To end the call for all callers, tap the receiver icon.

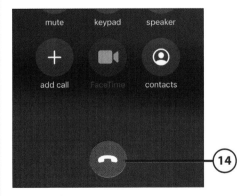

Merging Calls

As you merge calls, your iPhone attempts to display the names of the callers at the top of the Call screen. As the text increases, your iPhone scrolls it so you can read it. Eventually, the iPhone replaces the names with the word "Conference."

Number of Callers

Your provider and the specific technology of the network you use can limit the number of callers you place in a conference call. When you reach the limit, perhaps up to five people, the add call function is disabled.

Using Another App During a Voice Call

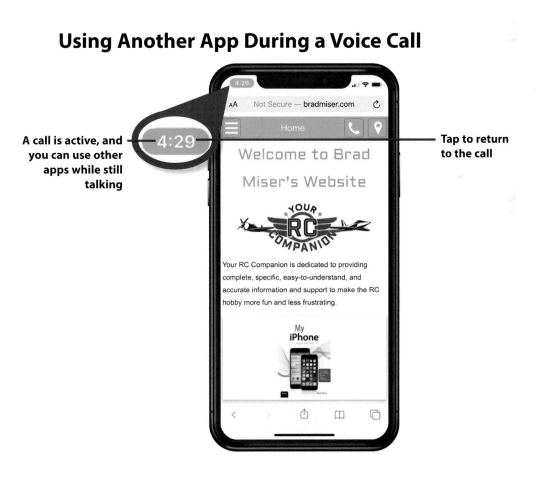

A call is active, and you can use other apps while still talking

Tap to return to the call

If your provider's technology supports it, you can use your iPhone for other tasks while you're on a call. When you're on a call, move to a Home screen by swiping up from the bottom of the screen (iPhones without a Home button) or pressing the Touch ID/Home button (iPhones with a Home button), and then tap a different app (placing the call in speaker mode before you switch to a different app or using the headphones are best for this). Or, you can use the App Switcher to move into a different app. The call remains active and you see the active call information in a green oval in the upper-left corner of the screen (iPhones without a Home button) or green bar across the top of the screen (iPhones with a Home button). You can perform other tasks, such as looking up information, sending emails, or visiting websites. You can continue to talk to the other person just like when the Call screen is showing. To return to the call, tap the green oval or the green bar.

Receiving Voice Calls

Receiving calls on your iPhone enables you to access the same great tools you can use when you make calls, plus a few more for good measure.

>>>*Go Further*

RINGTONES

Of course, we all know that your ringtone is the most important thing about your iPhone, and you'll want to make sure your ringtones are just right. Use the iPhone's Sounds & Haptics settings to configure custom or standard ringtones and other phone-related sounds, including the new voicemail sound. You also can configure the way your phone vibrates when you receive a call. These options are explained in Chapter 6, "Making Your iPhone Work for You."

You can have different ringtones and vibrations for specific people so you can know who is calling just by the sound and feel when a call comes in (configuring contacts is explained in Chapter 8).

Answering Calls

When your iPhone rings, it's time to answer the call—or not. If you configured the ringer to ring, you hear your default ringtone or the one associated with the caller's contact information when a call comes in. If vibrate is turned on, your iPhone vibrates regardless of whether the ringer is on. If you enable the announce feature, the name of the caller, if available, is announced. And if those ways aren't enough, a message appears on your iPhone's screen to show you information about the incoming call. If the number is in your Contacts app, you see the contact with which the number is associated, the label for the number, and the contact's image if there is one. If the number isn't in your contacts, you see the number only.

The details of answering calls are different if your phone is unlocked versus locked or if you are using Bluetooth headphones.

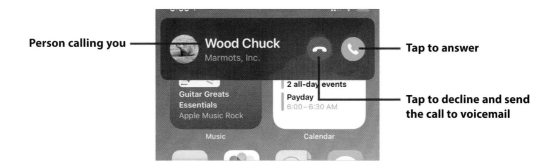

When your phone is unlocked, you see the phone banner that shows who is calling. Tap the red receiver icon to decline the call and send it to voicemail. Tap the green receiver to answer the call.

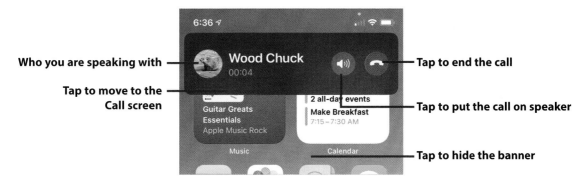

If you accept the call, the banner remains on the screen. Tap the speaker to put the call on speaker, tap the red receiver icon to end the call, or tap the banner to move to the full-screen Call screen (the same one that appears when you place a call). Tap outside the banner to hide it; the call remains active and you can open other apps to work with them.

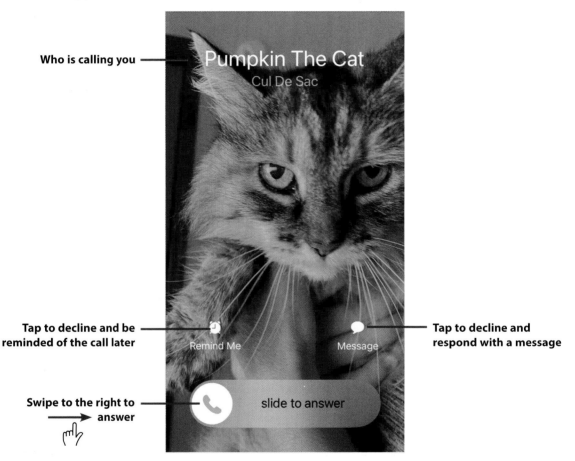

Who is calling you

Tap to decline and be reminded of the call later

Tap to decline and respond with a message

Swipe to the right to answer

If your iPhone is locked when a call comes in, swipe the slider to the right to answer it or use the Remind Me and Message icons.

When you receive a call, you have the following options:

- **Answer**—To accept an incoming call, tap the green receiver icon on the phone banner (the phone is unlocked) or swipe the slider to the right (the iPhone is locked) to take the call; you don't have to unlock the phone to answer a call.

- **Decline**—If you tap the red receiver icon on the Phone banner (when the iPhone is unlocked) or quickly press the Side button twice, the Phone app immediately routes the call to voicemail.

- **Silence the ringer**—To silence the ringer without sending the call directly to voicemail, press the Side button once or press either volume button. The call continues to come in, and you can answer it even though you shut off the ringer.

- **Respond with a message**—When a call comes in while the phone is locked, tap Message to send the call to voicemail and send a message back in response. You can tap one of the default messages, or you can tap Custom to create a unique message (you configure these message using Phone Settings as covered in the Go Further sidebar, "Phone Settings," later in this chapter). Of course, the device the caller is using to make the call must be capable of receiving messages for this to be useful.

- **Decline the call but be reminded later**—When a call comes in while the phone is locked, tap Remind Me and the call is sent to voicemail. Tap When I leave, When I get home, When I get to work, or In 1 hour to set the timeframe in which you want to be reminded. A reminder is created in the Reminders app to call back the person who called you, and it is set to alert you at the time you select.

If your iPhone is connected to wired or Bluetooth headphones when a call comes in, you can answer the call directly from the headphones without even touching your phone (whether your phone is locked or unlocked). How you do this depends on the type of headphones you are using.

If you are using the wired headphones that came with your iPhone, press the center of the button on the right wire to answer a call. If you are using AirPods, tap the outside of one of the AirPods twice. If you are using other types of Bluetooth headphones, there is usually a larger central button that you push to answer calls.

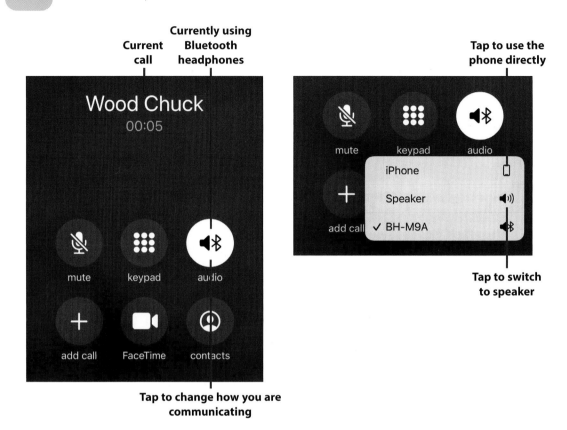

Current call · **Currently using Bluetooth headphones**

Tap to use the phone directly

Tap to switch to speaker

Tap to change how you are communicating

When you've answered a call using headphones, you see the Audio icon on the Phone screen. Tap this to change how you want to communicate; for example, tap Speaker to switch to the speakerphone mode.

Silencio!

To mute your iPhone's ringer, slide the Mute switch located above the Volume switch toward the back so orange appears. The Silent Mode On warning pop ups on the screen for a second or so to let you know you turned off the ringer. To turn it on again, slide the switch forward. The Ringer volume indicator appears on the screen briefly to show you the ringer is active again. To set the ringer's volume, use the Volume buttons when you aren't on a call and aren't listening to an app, such as the Music app.

Blocking Calls

You can block voice and FaceTime calls so that if someone attempts to call you from that number again, the call is prevented from coming to your phone. See "Blocking Unwanted Calls, Messages, or FaceTime Requests," in Chapter 18, "Maintaining and Protecting Your iPhone and Solving Problems." If you change your mind, you can unblock the number. This is also covered in Chapter 18.

Answering Calls During a Call

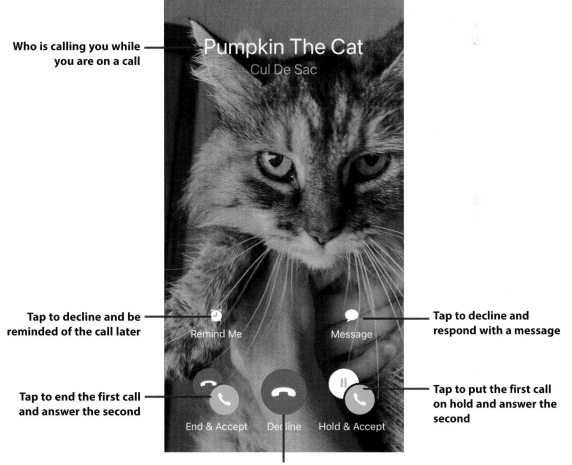

Who is calling you while you are on a call

Tap to decline and be reminded of the call later

Tap to decline and respond with a message

Tap to end the first call and answer the second

Tap to put the first call on hold and answer the second

Tap to send the incoming call to voicemail

As mentioned earlier, your iPhone can manage multiple calls at the same time. If you are on a call and another call comes in, you have a number of ways to respond.

- **Decline incoming call**—Tap Decline to send the incoming call directly to voicemail.

- **Place the first call on hold and answer the incoming call**—Tap Hold & Accept to place the current call on hold and answer the incoming one. After you do this, you can manage the two calls just as when you call two numbers from your iPhone. For example, you can place the second call on hold and move back to the first one, merge the calls, or add more calls.

- **End the first call and answer the incoming call**—Tap End & Accept to terminate the active call and answer the incoming call.

- **Respond with message or get reminded later**—These options work just as they do when you are dealing with any incoming phone call.

Auto-Mute

If you are listening to music, video, audiobooks, or directions from Maps when a call comes in, the app providing the audio, such as the Music app, automatically pauses. When the call ends, that app picks up right where it left off.

Managing Voice Calls

You've already learned most of what you need to know to use your iPhone's cell phone functions. In the following sections, you learn the rest.

Clearing Recent Calls

Previously in this chapter, you learned about the Recents tool that tracks call activity on your iPhone. As you read, this list shows both completed and missed calls; you can view all calls by tapping the All tab or only missed calls by tapping Missed. On either tab, missed calls are always in red, and you see the number of missed calls since you last looked at the list in the badge on the Recents tab. You also learned how you can get more detail about a call, whether it was missed or made.

Over time, you'll build a large Recents list, which you can easily clear.

1. Open the Phone app if you aren't in it already (not shown).

2. Tap Recents.

3. Tap Edit.

4. Tap Clear to clear the entire list.

5. Tap Clear All Recents. The Recents list is reset.

Delete Recents Individually

On the Recents screen, you can delete an individual recent item by swiping all the way to the left on recent call you want to delete.

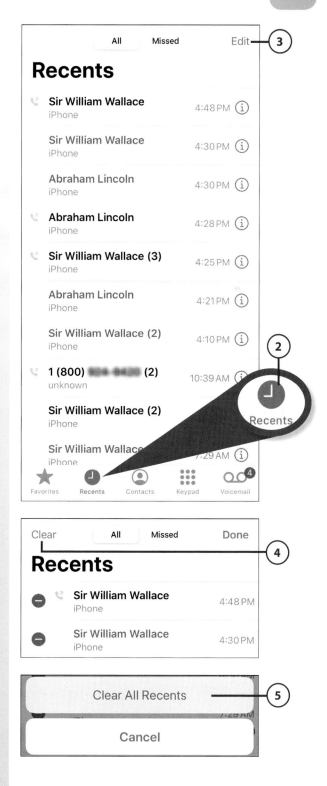

Adding Calling Information to Favorites

Earlier you learned how simple it is to place calls to someone on your Favorites list. There are a number of ways to add people to this list, including adding someone on your Recents list.

1. Move to the Recents list.

2. Tap Info (i) for the person you want to add to your favorites list. The Info screen appears. If the number is associated with a contact, you see that contact's information.

3. Swipe up to move to the bottom of the screen.

4. Tap Add to Favorites. If the person has multiple types of contact information, such as phone numbers and email addresses, you see each type of information available.

5. Tap the type of information you want to add as a favorite, such as Call to make a phone number a favorite.

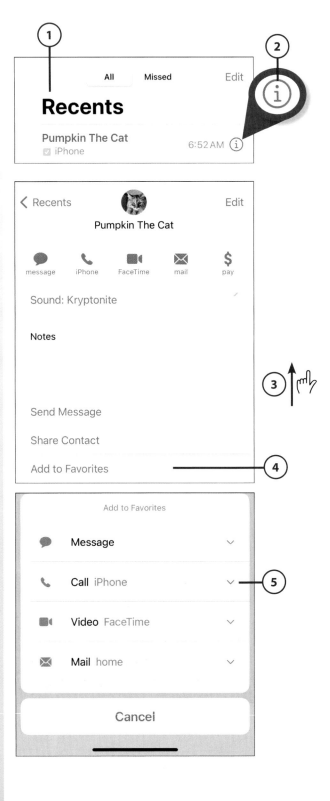

(6) Tap the number or email address you want to add as a favorite.

(7) Repeat steps 5 and 6 if you want to add the contact's other numbers or addresses to the favorites list. (If all the numbers and email addresses are assigned as favorites, Add to Favorites doesn't appear on the contact's screen.)

Add to Favorites

Call iPhone

iPhone
(317) 750-███ ——— (6)

work
(555) 555-5555

FaceTime

Cancel

Make Contact First

To make someone a favorite, he needs to be a contact in the Contacts app. Refer to Chapter 8 to learn how to make someone who has called you into a contact.

Not Your Favorite?

To remove a favorite, open the Favorites tab and swipe to the left on the favorite you want to remove. Tap Delete. The favorite is removed from the list, but the contact is not changed, so you can continue to use its information. You can add the contact as a favorite again if you change your mind.

Using Visual Voicemail

Visual voicemail just might be the best of your iPhone's many great features. No more wading through long, uninteresting voicemails to get to one in which you are interested. You simply jump to the message you want to hear. If that isn't enough for you, you can also jump to any point within a voicemail to hear just that part, such as to repeat a phone number that you want to write down. Even better, the iPhone creates a transcript of voicemails, so you can read them instead of listening to them.

The Phone app can access your voicemails directly so you don't need to log in to hear them.

Recording a New Greeting

The first time you access voicemail, you're prompted to record a voicemail greeting. Follow the onscreen instructions to do so.

You can also record a new greeting at any time.

① Move to the Phone screen and tap Voicemail. (If badge notifications are enabled for the Phone app, you see the number of voicemails you haven't listened to yet in the badge on the Voicemail icon.)

② Tap Greeting.

③ To use a default greeting that provides only the iPhone's phone number, tap Default and skip to step 10.

④ Tap Custom to record a personalized greeting.

⑤ Tap Record. Recording begins.

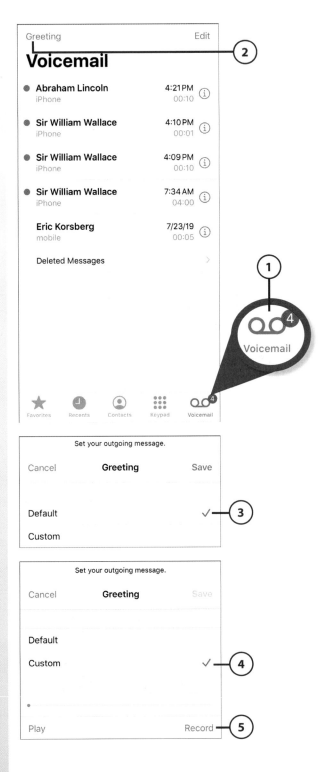

(6) Speak your greeting. As you record your message, the red area of the timeline indicates how long your message is.

(7) When you're done recording, tap Stop.

For the Very First Time

Some providers require that you dial into your voicemail number the first time you use it. You'll know if you tap Voicemail and the phone starts to dial instead of you seeing Visual Voicemail as shown in these figures. You call the provider's voicemail system, and you're prompted to set up your voicemail. When you've completed that process, you can use these steps to record your greeting.

(8) Tap Play to hear your greeting.

(9) If you aren't satisfied, tap Record to replace the recorded message and repeat steps 6–8 to record a new message.

(10) When you're happy with your greeting, tap Save. The Phone app saves the default or custom greeting as the active greeting and returns you to the Voicemail screen.

Set your outgoing message.

Cancel   **Greeting**   Save

Default

Custom   ✓

Play   Stop —(7)

Your message being recorded

Set your outgoing message.

Cancel   **Greeting**   Save —(10)

Default

Custom   ✓

Play   Record

(8)   (9)

No Visual Voicemail?

If your voicemail password isn't stored on your iPhone, when you tap Voicemail your phone dials into your voicemail instead of moving to the Voicemail screen. If that happens, something has gone wrong with your password and you need to reset the voicemail password on your iPhone. If you don't know your current voicemail password, follow your provider's instructions to reset the password. When you have the new password, open the Settings app, swipe up the screen, tap Phone, swipe up the screen, tap Change Voicemail Password, enter the reset password, create a new password, and re-enter your new password. (You need to tap Done after each time you enter a password.)

Change Greeting

To switch between the default and the current custom greeting, move to the Greeting screen, tap the greeting you want to use (which is marked with a check mark), and tap Save. When you choose Custom, you use the custom greeting you most recently saved.

Listening to, Reading, and Managing Voicemails

Unless you turned off the voicemail sound, you hear the sound you selected each time a caller leaves a voicemail for you. The number in the badge on the Phone icon and on the Voicemail icon on the Phone screen increases by 1 (unless you've disabled the badge). Note that the badge number on the Phone icon includes both voicemails left for you and missed calls, whereas the badge number on the Voicemail icon indicates only the number of voicemails left for you. (A new voicemail is one to which you haven't listened, not anything to do with when it was left for you.) If you've configured visual notifications for new voicemails, you see those on the screen as well.

If you receive a voicemail while your iPhone is locked, you see a message on the screen alerting you that your iPhone received a voicemail. (It also indicates a missed call, which is always the case when a call ends up in voicemail.) Press on the notification to jump to the Voicemail screen so that you can work with your messages.

And in yet another scenario, if you're using your iPhone when a message is left, you see a notification (unless you have turned off notifications for the Phone app) that enables you to deal with the new message.

Finding and Listening to Voicemails

Working with voicemails is simple and quick.

(1) Move into the Phone app and tap Voicemail (if you pressed or swiped on a new message notification, you jump directly to the Voicemail screen).

(2) Swipe up and down the screen to browse the list of voicemails. Voicemails you haven't listened to are marked with a blue circle.

(3) To listen to or read a voicemail, tap it. You see the timeline bar and controls and the message plays.

(4) Read the message if you don't want to listen to it.

What is This? Watergate?

The Phone app does its best to transcribe voicemail messages. Sometimes, it can't transcribe words or phrases in a message. When this happens, you see an underscore that shows a gap where the content that couldn't be transcribed occurs. If you can't get the missing bits from context, you have to listen to the message for yourself.

(5) Tap the Pause icon to pause a message.

(6) Tap Speaker to hear the message on your iPhone's speaker.

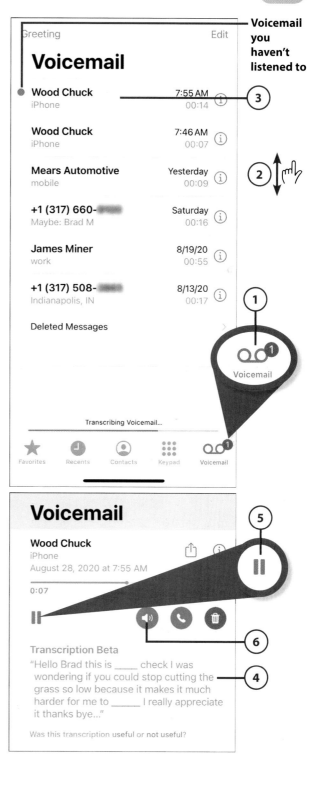

Voicemail you haven't listened to

Greeting   Edit

Voicemail

| Wood Chuck | 7:55 AM |
| iPhone | 00:14 |

| Wood Chuck | 7:46 AM |
| iPhone | 00:07 |

| Mears Automotive | Yesterday |
| mobile | 00:09 |

| +1 (317) 660- | Saturday |
| Maybe: Brad M | 00:16 |

| James Miner | 8/19/20 |
| work | 00:55 |

| +1 (317) 508- | 8/13/20 |
| Indianapolis, IN | 00:17 |

Deleted Messages

Transcribing Voicemail...

Favorites   Recents   Contacts   Keypad   Voicemail

Voicemail

| Wood Chuck |
| iPhone |
| August 28, 2020 at 7:55 AM |

0:07

Transcription Beta

"Hello Brad this is _____ check I was wondering if you could stop cutting the grass so low because it makes it much harder for me to _____ I really appreciate it thanks bye..."

Was this transcription useful or not useful?

(7) To move to a specific point in a message, drag the Playhead to the point at which you want to listen.

(8) Tap the blue receiver icon to call back the person who left the message.

(9) Tap the trash can to delete the message.

(10) Tap Share to share the message, and then tap how you want to share it, such as Message or Mail. For example, when you tap Mail, you send the voicemail to someone else, so he can listen to the message.

(11) Tap Info (i) to get more information about a message. The Info screen appears. If the person who left the message is on your contacts list, you see her contact information. The number associated with the message is highlighted in blue.

(12) Swipe up or down the screen to review the caller's information.

(13) Tap Voicemail.

(14) To listen to a message you have listened to before (one that doesn't have a blue dot), tap the message and then tap the Play icon. It begins to play. You can also read its transcript (if available).

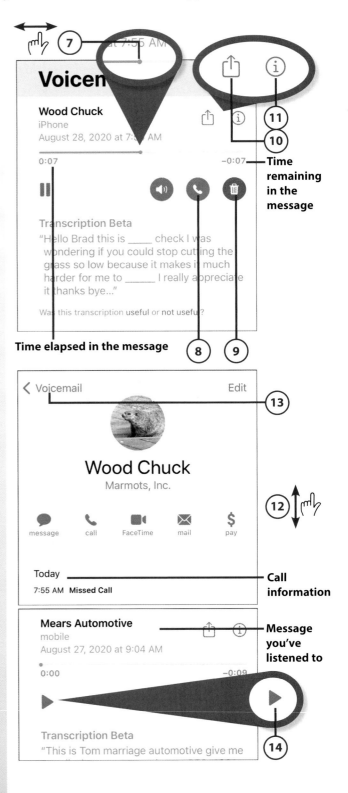

7

Voicen

Wood Chuck
iPhone
August 28, 2020 at 7:5_ AM

0:07   −0:07 — **Time remaining in the message**

11
10

Transcription Beta
"Hello Brad this is _____ check I was wondering if you could stop cutting the grass so low because it makes it much harder for me to _____ I really appreciate it thanks bye..."

Was this transcription useful or not useful?

Time elapsed in the message   8   9

‹ Voicemail   Edit

13

Wood Chuck
Marmots, Inc.

message   call   FaceTime   mail   pay

12

Today — **Call information**
7:55 AM Missed Call

Mears Automotive
mobile
August 27, 2020 at 9:04 AM

— **Message you've listened to**

0:00   −0:09

Transcription Beta
"This is Tom marriage automotive give me

14

>>>Go Further
MORE PHONE TIPS

The Phone app gives you a lot of options to manage your phone calls. Following are some additional things you can do with this great app:

- **Deleting Messages**—To delete a voicemail message that isn't the active message, tap it so it becomes the active message and then tap the trash can. Or swipe quickly all the way to the left on the message to delete it.

- **Listening to and Managing Deleted Voicemails**—When you delete voicemail messages, they move to the Deleted Messages folder. Scroll up the Voicemail screen and tap Deleted Messages to move to the list of deleted messages. On this screen, you can listen to or read messages just as you can on the Voicemail screen. To undelete a message, tap it and then tap the trash can with a slash through it; the message moves back to the Voicemail screen. To get rid of a deleted message permanently, swipe to the left on it. To get rid of all your deleted messages permanently, tap Clear All and then tap Clear All at the prompt. It's a good idea to do this once in a while so you don't have a huge list of deleted voicemail messages.

- **Blocked Messages**—If someone you have blocked leaves a message, you see the Blocked Messages option on the Voicemail screen (just below the Deleted Messages section). You can work with these messages just like the Deleted Messages. When you delete all the Blocked Messages, that section disappears from the screen.

- **What's Missed?**—In case you're wondering, your iPhone considers any call you didn't answer to be a missed call. So if someone calls and leaves a message, that call is included in the counts of both missed calls and new voicemails. If the caller leaves a message, you see a notification informing you that you have a new voicemail and showing who it is from (if available). If you don't answer and the caller doesn't leave a message, it's counted only as a missed call and you see a notification showing a missed call along with the caller's identification (if available).

- **Lost/Forgot Your Password?**—If you have to restore your iPhone or it loses your voicemail password for some other reason and you can't remember it, you need to reset the password to access your voicemail on the iPhone. If you don't know your current voicemail password with your provider, you need to reset it using your provider's support system. For most cell phone providers, this involves calling the customer support number and accessing an automated system that sends a new password to you via a text message. For AT&T, which is one of the iPhone providers in the United States, call 611 on your iPhone and follow the prompts to reset your password (which you receive via a text). No matter which provider you use, it's a good idea to know how to reset your voicemail password because it is likely you will need to do so at some point.

>>>Go Further
PHONE SETTINGS

You can change how the Phone app works by adjusting its settings. The settings for the Phone app depend on the cell phone provider you use. Depending on your provider, you may see more, fewer, or different settings than shown in the following list. It's a good idea to open your Phone settings to see the options available to you.

To access these settings, tap the Settings app on the Home screen, swipe up on the Settings screen, and then tap Phone. Following are some of the more useful settings that you might want to change:

- **Incoming Calls**—Use this option to determine what you see when someone calls you. Tap Banner to have the banner appear as described earlier in the chapter or Full Screen if you prefer the incoming call to take over the screen (similar to what happens when a call comes in while the phone is locked).

- **Announce Calls**—When you enable this setting, the name of the caller (when available) is announced when the phone rings. Tap Always to always have the name announced, Headphones & Car to have the caller announced only when you're using headphones or your car's audio system, Headphones Only to have announcements only when you're using headphones, or Never if you don't want these announcements.

- **Wi-Fi Calling**—When enabled, you can place and receive calls via a Wi-Fi network. This is particularly useful when you're in a location with poor cellular reception, but you have access to a Wi-Fi network. When you set the Wi-Fi Calling on This iPhone switch to on (green), you're prompted to confirm your information. When you do, the service starts and you see the Update Emergency Address option; this is used to record your address so if you place emergency calls via Wi-Fi, your location can still be determined. The Add Wi-Fi Calling For Other Devices switch enables you to take calls on other devices signed into your iCloud account even if they aren't in the same area as your iPhone.

- **Calls on Other Devices**—If you have more than one device configured with the same Apple ID (such as a Mac computer or iPad) and that are connected to the same Wi-Fi network or are connected via Bluetooth, you can take calls made to your iPhone on those devices. You can configure this using the Calls on Other Devices option; if you don't see this option on the Phone settings screen, it is not currently available for you.

When enabled, and your iPhone is on the same network as other devices (such as iPads) or Macs configured with your information, you can take incoming calls and place calls from those devices. This can be useful when you aren't near your phone or simply want to use a different device to have a phone conversation. However, it also can be annoying because when a call comes in, all the devices using this feature start ringing.

When the Allow Calls on Other Devices switch is on (green), you can choose the specific devices calls are allowed on by setting their switches to on (green). If you don't want a device to receive phone calls, set its switch to off (white).

- **Respond with Text**—When calls come in and the phone is locked or you have the Full Screen Incoming Calls option selected, you have the option to respond with text. For example, you might want to say, "Can't talk now, will call later." There are three default text responses, or you can use this setting to create custom text responses.

- **Silence Unknown Calls**—When you enable this setting, calls from unknown sources don't ring and are sent directly to voicemail. Only calls from people in your Contacts app, numbers you've called, or numbers identified by Siri Suggestions cause your phone to ring. Unknown calls go straight to voicemail and appear on the Recents list so you can view their information.

  Although this can be a useful feature, it does mean that you won't receive any calls unless they are recognized. All unknown calls go straight to voicemail. If you are expecting a call from someone who you haven't spoken with before or who isn't in your Contacts app, you might want to disable this feature.

- **Blocked Contacts**—Tap this to see a list of people whom you're currently blocking. You can tap someone on the list to see more information or swipe to the left and tap Unblock to unblock someone. Tap Add New at the bottom of the screen to block someone in your Contacts app.

- **Change Voicemail Password**—Use this option to change your voicemail password. Note that this changes only the password to access voicemail on your iPhone. This password must match the password for your voicemail account with your cell phone provider.

Communicating with FaceTime

FaceTime enables you to see, as well as hear, people with whom you want to communicate. This feature exemplifies what's great about the iPhone; it takes complex technology and makes it simple. FaceTime works great, but two conditions must exist. First, everyone on the FaceTime call must use an Apple device that has the required cameras (this includes iPhone 4s and newer, iPod touches third generation and newer, iPad 2s and newer, and Macs running Snow Leopard and newer) and that has FaceTime enabled (via the settings on an iOS device that are explained on the next page or via the FaceTime application on a Mac). Second, each device has to be able to communicate over a network. An iPhone or cellular iPad can use a cellular data network (if that setting is enabled) or a Wi-Fi network, whereas Macs have to be connected to the Internet through a Wi-Fi or other type of network. When both conditions are true, making and receiving FaceTime calls are simple tasks.

You can have up to 32 people on Group FaceTime conversations.

In addition to making video FaceTime calls, you can make audio-only FaceTime calls. Audio-only FaceTime calls work similarly to making a voice call with the Phone app.

Also, when you use Wi-Fi for a FaceTime call (video or audio), you can place international calls using FaceTime at no additional cost (cellular calls might have fees depending on your plan). If you're making a FaceTime call over the cellular network, the data counts against your data plan, so be careful about this if your plan has limited data.

Configuring FaceTime Settings

FaceTime is a great way to use your iPhone to hear and see someone else. There are a few FaceTime settings you need to configure for FaceTime to work. You can connect with other FaceTime users via your phone number or an email address.

 ① Open the Settings app.

② Swipe up the screen and tap FaceTime.

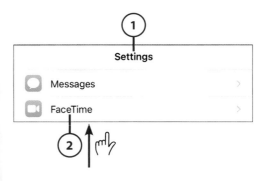

3 If the FaceTime switch is off (white), tap the FaceTime switch to turn it on (green). If the FaceTime switch is on and you see an Apple ID, you're already signed into an account; in this case, you see the current FaceTime settings and can follow along starting with step 7 to change these settings.

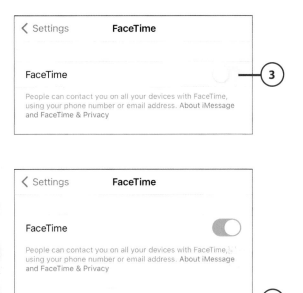

4 To use your Apple ID for FaceTime calls, tap Use your Apple ID for FaceTime (if you haven't signed into an Apple ID account, see Chapter 4, "Setting Up and Using an Apple ID, iCloud, and Other Online Accounts" to do so and then come back here to enable FaceTime). If you don't sign in to an Apple ID, you can still use FaceTime, but it's always via your cellular connection, which isn't ideal because then FaceTime counts under your voice minutes on your calling plan or as data on your data plan.

5 If prompted, tap Sign In to sign into your Apple ID.

Changing Identities

By default, the Apple ID you most recently used for FaceTime appears in the prompt in step 5. If you have more than one Apple ID, you can choose a different Apple ID to use by tapping Use Other Apple ID and signing into that account. This doesn't affect the Apple ID you are using for iCloud.

6. Configure the email addresses you want people to be able to use to contact you for FaceTime sessions by tapping them to enable each address (enabled addresses are marked with a check mark) or to disable addresses (these don't have a check mark). (If you don't have any email addresses configured on your iPhone, you're prompted to enter email addresses.)

7. Tap the phone number or email address by which you will be identified to the other caller during a FaceTime call.

8. Set the Speaking switch to on (green) if you want the video window for the person currently speaking to be larger than the others. This is useful for group FaceTime sessions because you more easily see who is talking.

9. If you want to be able to capture Live Photos during FaceTime calls, set the FaceTime Live Photos switch to on (green).

Make Eye Contact

On some iPhone models, you can set the Eye Contact switch to on (green) to have FaceTime attempt to try to establish more "natural" eye contact. If your phone has this option, I recommend you try it.

Blocking FaceTime

If you tap Blocked Contacts at the bottom of the FaceTime Settings screen, you see the names, phone numbers, and email addresses that are currently blocked from making calls, sending messages, or making FaceTime requests to your iPhone. To block someone else, tap Add New and then tap the contact you want to block. To unblock someone, swipe to the left on that person or number and tap Unblock.

Making FaceTime Calls

FaceTime is a great and usually free way to communicate with people because you can hear and see them (or just hear them if you choose an audio-only FaceTime call). Because iPhones have cameras facing each way, it's also easy to show something to the people you're talking with. You make FaceTime calls starting from the FaceTime, Contacts, or Phone apps and from the FAVORITES widget. No matter which way you start a FaceTime session, you manage it in the same way.

Careful

If your iPhone is connected to a Wi-Fi network, you can make all the FaceTime calls you want (assuming you have unlimited data on that network). However, if you're using the cellular data network, be aware that FaceTime calls might use data under your data plan. If you have a limited plan, it's a good idea to use FaceTime for video primarily when you are connected to a Wi-Fi network.

To start a FaceTime call from the Contacts app, do the following:

1. Use the Contacts app to open the contact with whom you want to chat (refer to Chapter 8 for information about using the Contacts app).

2. To place an audio-only FaceTime call, tap the FaceTime audio icon. (The rest of these steps show a FaceTime video call, but a FaceTime audio-only call is very similar to the voice calls described earlier in this chapter.)

(3) Tap the contact's FaceTime video icon. The iPhone attempts to make a FaceTime connection. You hear the FaceTime "chirping" and see status information on the screen while the call is attempted. When the connection is complete, you hear a different tone and see the other person in the large window and a preview of what she is seeing (whatever your iPhone's front-side camera is pointing at—most likely your face) in the small window. If the person you're trying to FaceTime with isn't available for FaceTime for some reason (perhaps she doesn't have a FaceTime-capable device or isn't connected to the Internet), you see a message saying that the person you're calling is unavailable for FaceTime, and the call terminates. You're given the option to leave a message, cancel, or call back.

(4) After the call is accepted, manage the call as described in the "Managing FaceTime Calls" task later in this chapter.

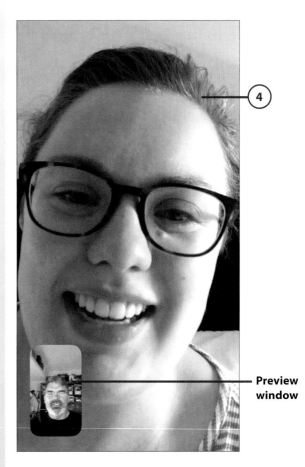

Preview window

>>>Go Further
MORE INFORMATION ABOUT FACETIME CALLS

As you make FaceTime calls, following are some other bits of information for your consideration:

- **Additional Settings**—After you've signed into FaceTime, more settings appear at the top of the FaceTime Settings screen. Use the Incoming Calls to select Banner if you prefer banner notifications when the phone is unlocked or Full Screen if you always want Full Screen requests. Use Announce Calls to determine if and how the caller is announced when the FaceTime request comes in.

- **Failing FaceTime**—If a FaceTime request fails, you can't really tell the reason why. It can be a technical issue, such as none of the contact information you have is FaceTime-enabled, the person is not signed into a device, or the person might have declined the request. If you repeatedly have trouble connecting with someone, contact him to make sure he has a FaceTime-capable device and that you are using the correct FaceTime contact information.

- **Leave a Message**—On the FaceTime Unavailable screen, you can tap Leave a Message to send a text or iMessage message to the person with whom you are trying to FaceTime.

- **Transform a call**—You can transform a voice call into a FaceTime session by tapping FaceTime on the Call screen. The voice call you started from automatically terminates when you make the switch.

- **FaceTime app**—To use the FaceTime app to start a call, tap the FaceTime icon on the Home screen. Tap Add (+) to enter a name, email address, or phone number in the bar at the top of the screen. If the information you enter matches a contact or someone you've communicated with before, tap the person with whom you want to FaceTime. If not, keep entering the information until it's complete to place the call. Then tap the Audio button to place an audio-only FaceTime call or tap Video to place a video FaceTime call. You also can tap a FaceTime call on the Recents list to place another FaceTime call to that person. Once you've connected, you manage the FaceTime session as described in the rest of this chapter.

- **FaceTime with Siri**—You can also place a FaceTime call using Siri by activating Siri and saying "FaceTime *name*" where *name* is the name of the person with whom you want to FaceTime. If there are multiple options for that contact, you must tell Siri which you want to use. After you've made a selection, Siri starts the FaceTime call.

Receiving FaceTime Calls

When someone tries to FaceTime with you, what you see depends on whether your iPhone is currently locked or unlocked.

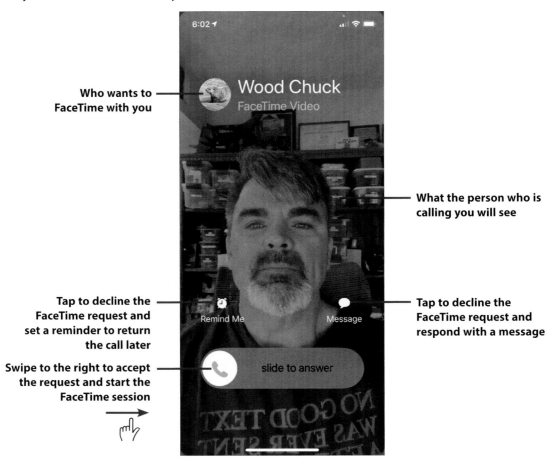

Who wants to FaceTime with you

What the person who is calling you will see

Tap to decline the FaceTime request and set a reminder to return the call later

Tap to decline the FaceTime request and respond with a message

Swipe to the right to accept the request and start the FaceTime session

If your phone is locked. you see the Full Screen FaceTime request screen message showing who is trying to connect with you and the image you're currently broadcasting. Swipe to the right on the slide to answer bar to accept the request and start the FaceTime session. Manage the FaceTime call as described in the "Managing FaceTime Calls" task.

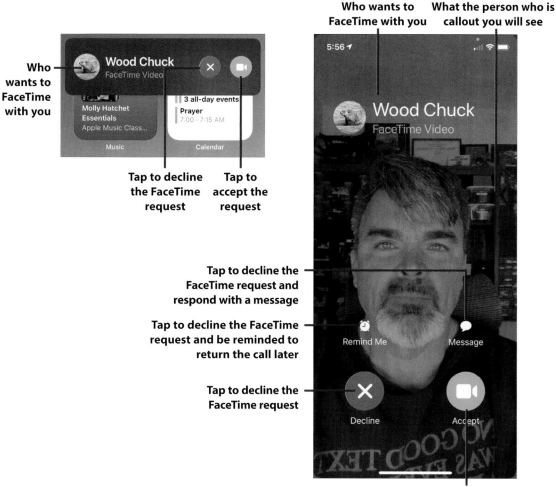

Who wants to FaceTime with you

What the person who is callout you will see

Who wants to FaceTime with you

Tap to decline the FaceTime request

Tap to accept the request

Tap to decline the FaceTime request and respond with a message

Tap to decline the FaceTime request and be reminded to return the call later

Tap to decline the FaceTime request

Tap to accept the request and start the FaceTime session

When your phone is unlocked and the FaceTime Banner setting is enabled, you see a banner with information about the request. Tap the FaceTime icon to accept the call. You move to the Full Screen FaceTime Request screen.

On the Full Screen FaceTime Request screen, you have several options:

- Tap Decline on the FaceTime banner or full screen request screen to decline the request.

- Swipe to the right on the slide to answer bar or tap Accept to start the call.

- Tap Remind Me to decline the FaceTime request and create a reminder. (You can also press the Side button to decline a FaceTime request.)

- Tap Message to decline the request and send a message.

If you decline a FaceTime request, the person trying to call you receives a message that you're not available (and a message if you choose that option). She can't tell whether there is a technical issue or if you simply declined to take the call.

Tracking FaceTime Calls

FaceTime calls are tracked just as voice calls are. Open the FaceTime app and you see the recents list. FaceTime calls are marked with the video camera icon. FaceTime audio-only calls are marked with a telephone receiver icon. FaceTime calls that didn't connect (incoming or outgoing) are in red and are treated as missed calls. You can tap a recent call to place a FaceTime call to the same person.

Managing FaceTime Calls

During a FaceTime call (regardless of who placed the call initially), you can do the following:

Drag to change the location of the preview window

Tap to show the FaceTime controls

- Drag the preview window, which shows the image that the other person is seeing, around the screen to change its location. It "snaps" into place in the closest corner when you lift your finger.

- Move your iPhone and change the angle at which you're holding it to change the images you broadcast to the other person. Use the preview window to see what the other person is seeing.

Tap to hide the controls

Swipe up to open the FaceTime tools palette

Tap to take a photo of the image

Tap to change the camera you are using

Tap to mute or unmute the audio

Tap to apply effects to the call

Tap to end the session

effects   mute   flip   end

- Tap the screen to see the FaceTime controls. After a few seconds of not touching the screen, they disappear or you can tap the screen to hide them immediately. You can show them again at any time by tapping the screen.

- Tap the shutter button to take a photo of the current images; if the Live Photo option is enabled, the result image will be a Live Photo.

- Tap effects to apply effects to the session (see the Go Further sidebar, "Facetime Fun with Animojis and Effects," at the end of the chapter for more information).

- Tap mute to mute your side of the conversation. Your audio is muted and you see the mute icon in the preview window. Video continues to be broadcast so the other person can still see you.

- Tap flip to change the camera you are using. If you're currently using the front-facing camera and tap to flip, the other person now sees the images from the back-facing camera.

- Tap end to terminate the FaceTime session.

- Swipe up from the bottom of the screen to open the FaceTime tools palette. At the top of the palette, you see the same controls as on the FaceTime screen along with some additional options.

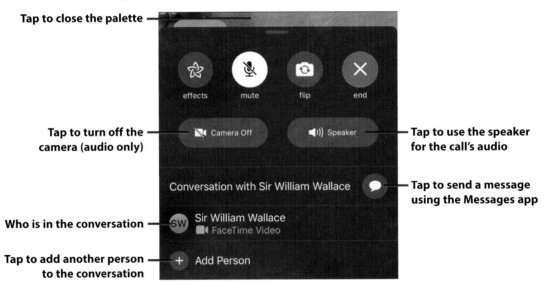

Tap to close the palette

effects   mute   flip   end

Tap to turn off the camera (audio only) — Camera Off

Speaker — Tap to use the speaker for the call's audio

Conversation with Sir William Wallace — Tap to send a message using the Messages app

Who is in the conversation — SW Sir William Wallace / FaceTime Video

Tap to add another person to the conversation — + Add Person

- Tap Camera Off to turn off the camera. The audio portion of the session continues.

- Tap Speaker to use the speakerphone for the audio part of the session.

- Tap the conversation button to move into the Messages app so that you can send messages to the other person in the session.

- Tap Add Person to add one or more people into the conversation (see "Making Group FaceTime Calls" later in this chapter).

- Tap above the palette or swipe down from the horizontal line at the top of the palette to close it and return to the FaceTime screen.

FaceTime works in landscape orientation, too

The preview shows the landscape orientation

- Rotate your iPhone to change the orientation to horizontal. This affects what the other person sees (as reflected in your preview), but you continue to see the other person in her iPhone's current orientation.

- Just like when you're in a voice call, you can move into and use other apps (if your provider's technology supports this functionality). A smaller FaceTime window floats on top of the other screens so you can continue the FaceTime conversation while doing other things. You can drag this screen around to change its location. Swipe it off the side of the screen to collapse it down to an arrow; tap this to open the FaceTime small window again. Tap the small window to move back into the Full Screen mode.

Making Group FaceTime Calls

You can add up to 32 people to a single FaceTime session (though you can expect the quality of the video and audio signal for that many people to degrade a bit unless everyone has very fast Internet connections). Working with group FaceTime calls is similar to having a FaceTime session with two people.

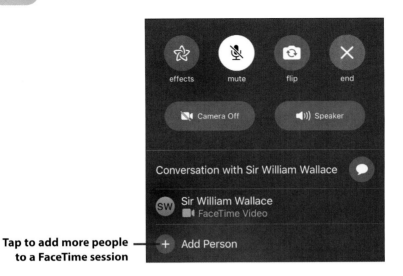

Tap to add more people to a FaceTime session

- Start a FaceTime session as described in the prior section. Open the FaceTime tools palette. Tap Add Person (+).

Enter the phone number or email address for the person you want to add

Tap the person you want to add

Tap to add the person to a FaceTime session

- Enter the contact information for the person you want to add to the conversation or tap her name on the list of matching contacts. Then tap Add Person to FaceTime; she receives an invitation on her device. If she chooses to join, she's added to the conversation.

Tap to focus

Each person has their own window

Your image

- You see each member of the conversation in a separate box. Your preview window becomes smaller and is locked into its current location. You can hear and see each person as you can in a two-way conversation. To focus on one person, tap her box.

- As more people are added to the conversation, each person's box becomes smaller. After three or four are added, the boxes start aligning themselves in different ways to make the most of the screen's space. You can work with each box in the same way no matter its size.

Person in focus ———   **Jill Godfrey**   ——— Tap to enlarge

effects   mute   flip   end

- When you tap in a box, you see the person's name and the expand icon. Tap that icon to enlarge the person's window. When you have only two other people in the call, this isn't so useful, but as you add more people, each person's box becomes smaller, so it's helpful to be able to enlarge a specific box to better see its image.

- If you have the Speaking switch set to on (green) when someone speaks, that person's box enlarges automatically.

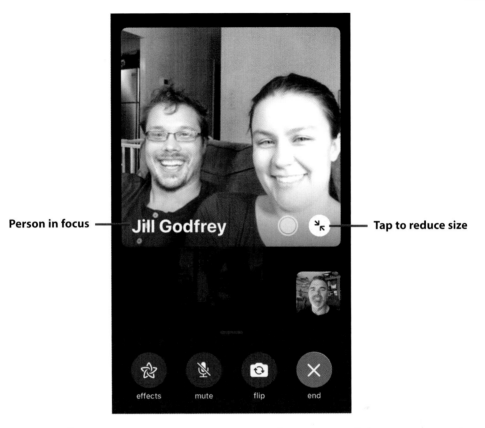

Person in focus ——— Jill Godfrey   ——— Tap to reduce size

effects   mute   flip   end

- When you focus on an image, it moves to the center of the screen, and you see the reduce icon; tap this to return to the previous configuration.

Tap to show controls

Tap to focus

- The FaceTime tools, such as mute or flip, work in the same ways as in a two-person conversation.

Tap to join a FaceTime group session

- When you are invited to a group conversation, you see the names of those involved at the top of the screen. Tap join to enter the group FaceTime conversation.

>>>Go Further

FACETIME FUN WITH ANIMOJIS AND EFFECTS

You can add a variety of effects to FaceTime conversations, including Animojis (which is where your image is replaced by an animated "cartoon" that matches your facial movements). To see what's available, tap effects. Above the tools, you see a palette of options. Tap this palette, and the tools enlarge and have captions added to them, which helps you know what they are. Swipe to the left and right on this palette to browse all of the options.

What you apply appears "on top" of your video image, either altering it, replacing it, or just floating on it. There are many possibilities; following are descriptions of a few to get you started (not all iPhone models support all of these features):

- **Animoji**—When you tap this, you see a selection of Animojis that you can apply. Tap one and place your face within the yellow markers on the screen. When you do so, your face is replaced by the Animoji. It mimics your facial movements; for example, when you smile, the animoji smiles too.

- **Filters**—When you tap filters, you see a number of different video effects that you can apply to your image. Swipe to the left or right to browse the filters available to you. As you move through the filters, the one currently applied is indicated by the yellow box, and you see your image as adjusted by the filter. Other people in the call also see the filtered image. When you find the filter you want to use, tap close (x) to return to the conversation. To remove the filter, open the Filters tool and select the Original filter.

- **Text**—Tap the Text tool and then tap Aa. Swipe up and down to see various styles you can apply; tap a style to apply it. A text placeholder appears on the screen. Type the text you want to show and tap in the video image. The text appears to everyone in the conversation. You can drag it around on the screen to change its location. To remove text, tap it and then tap delete (x).

Tap to configure email settings

Use email to share content, such as photos

Number of new emails

Tap to use email

In this chapter, you explore all the email functionality that your iPhone has to offer. Topics include the following:

→ Getting started
→ Working with email
→ Managing email
→ Finding email
→ Managing junk email

Sending, Receiving, and Managing Email

For most of us, email is an important way we communicate with others, both in our public and personal lives. Fortunately, your iPhone has great email tools, so you can work with email no matter where you are.

Getting Started

To work with email on your iPhone, you use the Mail app to access your email accounts—or accounts, if you have more than one.

Before you can start using the Mail app, you have to configure the email accounts you want to access with it. The iPhone supports many kinds of email accounts, including iCloud, Gmail, and many others. Setting up the most common types of email accounts is covered in Chapter 4, "Setting Up and Using an Apple ID, iCloud, and Other Online Accounts," so if you haven't done that already, go back to that chapter and get your accounts set up. Then come back here to start using those accounts for email.

If you have a lot of email activity, you'll want to configure the notifications associated with email. For a detailed explanation of configuring notifications, refer to Chapter 6, "Making Your iPhone Work for You."

Email is sent via the Internet, so you need to be connected to the Internet through a Wi-Fi or cellular data connection to send or receive email. You can compose new messages and read and reply to downloaded messages (though messages you write won't be sent until you have an Internet connection). (In Chapter 3, "Using Your iPhone's Core Features," you find out how to connect your iPhone to the Internet.) As you use email, it's helpful to understand that email isn't sent directly between devices—for example, from an email application on a computer to Mail on an iPhone. Rather, all email flows through email servers.

When you send an email, it moves from your iPhone to an email server. That server then routes it to various other servers until it reaches the one associated with the recipient's email account (for example, a Gmail account). The recipient(s) receive the email on their devices from that server.

Likewise, when someone sends email to you, it ends up on the server associated with your email account (for example, an iCloud account). The devices, such as an iPhone and a computer, that you use to work with your email account then receive the message from the server.

This means you can have the same email messages on all your devices at the same time. For example, you see the same messages on your iPhone and an iPad that's also configured to work with email associated with the same account.

There are three ways emails move from the server onto your iPhone: Manually, Fetch, and Push. You can select which method your iPhone uses for each email account; the details are provided in Chapter 4.

Working with Email

The Mail app is ideally suited for working with email on your iPhone. This app offers a consolidated Inbox, so you can view email from all your accounts at the same time.

You've got email
(well, actually I do)

When you move to a Home screen, you see the number of new email messages you have (if you have any) in the badge on the Mail app's icon; tap the icon to move to the app. Even if you don't have any new email, the Mail icon still leads you to the Mail app. Mail also notifies you of new messages by displaying visual notifications and the new mail sound if these features are enabled.

On Assumptions

The steps and figures in this section assume you have more than one email account configured and are actively receiving email to those accounts on your iPhone. If you have only one email account active, your Mailboxes screen contains that account's folders instead of mailboxes from multiple accounts and the Accounts sections that appear in these figures and steps. Similarly, if you enable the Organize by Thread setting, you see groups of messages based on the subject of the message. The examples in this chapter show Threading turned off, which means you work with messages individually. To learn about the settings, refer to the Go Further sidebar, "Mail Settings," at the end of this chapter.

The Mail app enables you to receive and read email for all the email accounts configured on your iPhone. The Mailboxes screen is the top-level screen in the app and is organized into two sections.

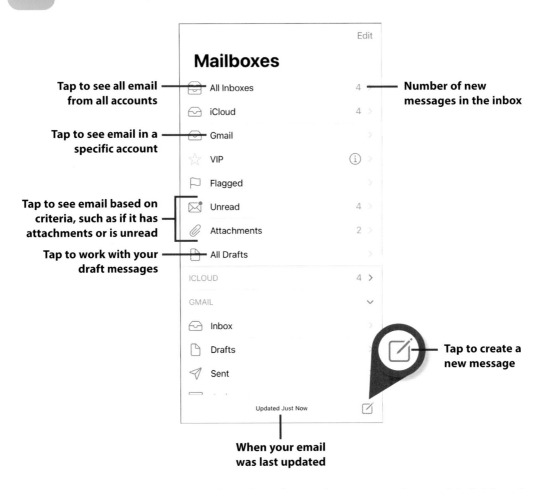

Tap to see all email from all accounts

Number of new messages in the inbox

Tap to see email in a specific account

Tap to see email based on criteria, such as if it has attachments or is unread

Tap to work with your draft messages

Tap to create a new message

When your email was last updated

The Inboxes section shows the inbox for each account along with folders for email based on various criteria, such as having attachments or being unread, along with a folder for your draft messages (those you've started but haven't sent yet). Next to each inbox or folder is the number of new emails in that inbox or folder. (A new message is simply one you haven't viewed yet.) At the top of the section is All Inboxes, which shows the total number of new messages to all accounts; when you tap this, the integrated inbox containing email from all your accounts is displayed.

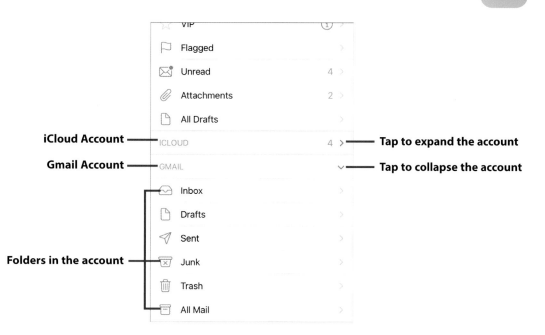

The Accounts section, which is underneath the Inboxes section on the screen, shows the set of inboxes and folders associated with each email account. The difference between these sections is that the Inbox options take you to just the inbox for one or all of your accounts or specific folders (such as the Attachments folder), whereas the Account options take you to all the folders under each account.

Tap the right-facing arrow for an account to expand it so you can see the folders it contains. You can tap any folder or inbox under an account to view the emails stored in that folder or inbox. Tap the downward-facing arrow to collapse an account so that you see only its name.

Receiving and Reading Email

To read email, perform the following steps:

(1) On the Home screen, tap Mail. When you open Mail (assuming you didn't use the App Switcher to quit the app when you left it), you move back to the screen you were last on; for example, if you were reading an email you return to it. If the Mailboxes screen isn't showing, tap Back (<) in the upper-left corner of the screen until you reach the Mailboxes screen.

(2) Tap the inbox that contains messages you want to read, or tap All Inboxes to see the messages from all your email accounts. Various icons indicate the status of each message, if it has attachments, if it is from someone you have designated as a VIP (explained in the Go Further sidebar, "VIPs," later in this chapter), or if it is part of a thread. A message is part of a thread when it has double right-facing arrows along the right side of the screen—individual messages have only one arrow.

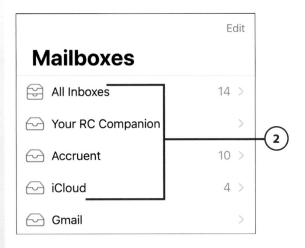

Edit

Mailboxes

All Inboxes	14	>
Your RC Companion		>
Accruent	10	>
iCloud	4	>
Gmail		>

(**3**) Swipe up or down the screen to browse the messages. You can read the preview of each message to get an idea of its contents.

(**4**) Tap the message you want to read. As soon as you open a message, it's marked as read and the new mail counter reduces by one. You see the message screen with the address information at the top, including whom the message is from and to whom it was sent. You also see the time and date it was sent. Under that section is the subject of the message in large, bold type; if the message is a reply to other messages, the subject has "Re" added to it. Below that is the body of the message. If the message has an attachment or is a reply to another message, the attachment or quoted text appears toward the bottom of the screen.

New message (**4**)

OX

Search 🎤

★ **The Bruce**   9:22 AM >
● Re: Flying Today?
Here is that brake document I was talking about. Regards, Robert

(**3**)

★ **The Bruce**   9:21 AM >
● Re: Flying Today?
I'm here to help. Maybe this photo will —— **Preview**
make you feel better. Regards, Robert

★ **Sir William Wallace**   9:19 AM >
↩ Re: Flying Today?
Sounds like a good time! I'm in. I'll plan to
be there about 10 AM. I'll bring my F-86....   **Message with attachment**

● **Micro-Mark**   9:19 AM >
Proxxon Precision Mini Power Tools
Please add Micro-Mark@e.micromark.com
to your email address book. If you are hav...

★ **The Bruce**   9:14 AM —— **When you received the message**
● Flying Today?
Fellas, It looks perfect outside for some RC
flying action. What say you? I'm planning...

Subject of the message   **Who sent the message**

5 Swipe up and down the screen to read the entire message.

See More

When all the information, such as for a quoted message, isn't being displayed, tap the related See More command. The section expands, and you can see the additional text.

Standard Motions Apply

You can use the standard gestures on email messages, such as unpinching or tapping to zoom, swiping to scroll, and so on. You also can rotate the phone to change the orientation of messages from vertical to horizontal.

6 If the message contains an attachment, swipe up the screen to get to the end of the message. Some types of attachments, most notably photos, appear directly in the message; you don't have to download them to the device. If an attachment hasn't been downloaded yet, it starts to download automatically (unless it's a large file). If the attachment hasn't been downloaded automatically, which is indicated by Tap to Download in the attachment icon, tap it to download it into the message. When an attachment finishes downloading, its icon changes to represent the type of file it is. If the icon remains generic, it might be of a type the iPhone can't display, and you need to open it on a computer or other device.

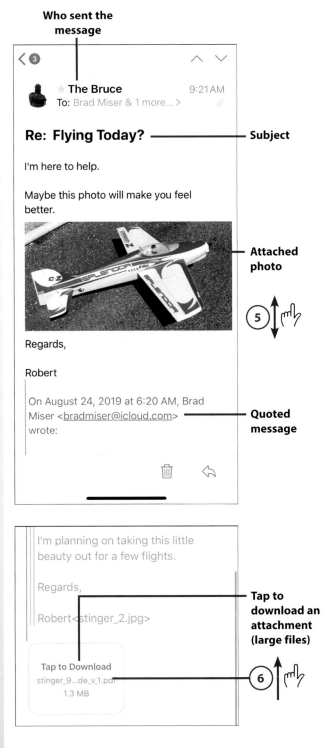

Who sent the message

Subject

Attached photo

Quoted message

Tap to download an attachment (large files)

(7) Tap the attachment icon to view it.

(8) Scroll the document by swiping up, down, left, or right on the screen.

(9) Unpinch to zoom in.

(10) Pinch to zoom out.

(11) Tap Done (depending on the type of attachment you were viewing, you might tap Back [<] instead).

(12) To view detailed information about the sender and recipients of the message, tap the top part of the message. The information expands and addresses turn blue indicating you can take action on them.

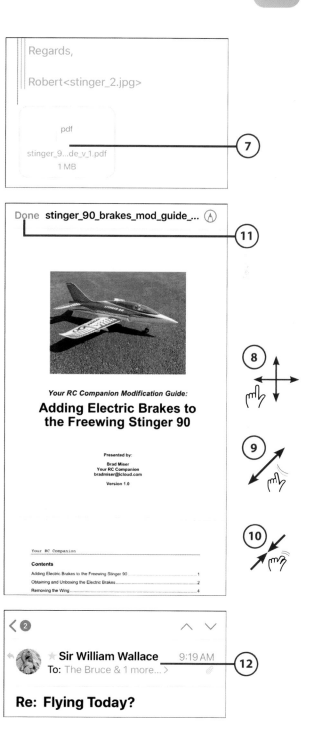

13 To get information about a specific person, tap his name or email address. The Info screen appears. On this screen, you see as much information for the person as is available. If it is someone in the Contacts app, you see all of the information stored there, and you can place a call, send a message, and perform other actions associated with her contact information. If it is not someone in the Contacts app, you see the person's email address along with actions you might want to perform, such as creating a contact for him or adding new information to an existing contact. (See Chapter 8, "Managing Contacts," for information about working with contacts.)

14 Tap Done to return to the message.

15 To read the next message in the current inbox or thread, tap the down arrow. (If the arrow is disabled, you are viewing the most recent email in the inbox or thread.)

16 To move to a previous message in the current inbox or thread, tap the up arrow. (If the arrow is disabled, you are viewing the oldest message in the inbox or thread.)

17 To move back to the inbox from where you came, tap back (<), which shows the number of unread messages in that inbox (all your inboxes if you were viewing them all).

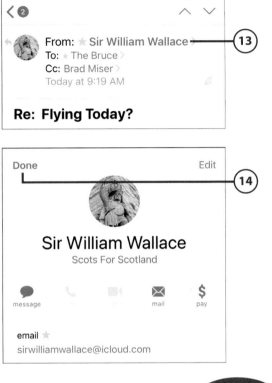

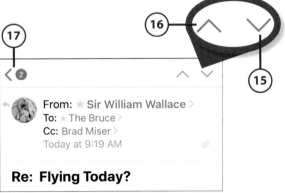

>>>Go Further

MORE ON RECEIVING AND READING EMAIL

Check out these additional pointers for working with email you receive:

- You can touch and hold on the Mail icon to see its Quick Action menu. Choose a command to use it. For example, tap All Inboxes to jump to the Inboxes screen or tap Search to find email messages.

- If a message includes a photo, Mail displays the photo in the body of the email message if it can (if the image is large, you might have to tap to download it to see it). You can zoom in or out and scroll to view it just as you can for photos in other apps.

- If more messages are available than are downloaded, tap the Load More Messages link. The additional messages download to the inbox you're viewing.

- If you choose to have messages grouped by threads, meaning Threading has been enabled for Mail (using the Settings app), emails are collected based on the title. As replies are sent to the original message, they're collected in a thread. You can tap the double arrows (>>) for a thread to open it so you can see its individual messages. Tap the arrows (<<) to collapse the thread. You can work with the messages inside a thread the same way as messages that aren't part of a thread.

Sending Email

You can send email from any of your accounts. Follow these steps for a walk-through of composing and sending a new email message:

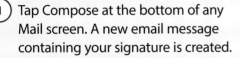

(1) Tap Compose at the bottom of any Mail screen. A new email message containing your signature is created.

Signing Off

The signature is a standard block of text automatically added at the bottom of every new message you create. It's used to end the message. A signature usually has your name and valediction; however, you can use any text you want. See the Go Further sidebar, "Mail Settings," to learn how to configure a signature for your email accounts.

(2) Tap the To field and type the first recipient's email address. As you type, Mail attempts to find matching addresses in your Contacts list, or in emails you've sent or received, and displays the matches it finds. These can include individuals or groups with which you've emailed. To select one of those addresses, tap it. Mail enters the rest of the address for you. Or just keep entering information until the address is complete.

(3) Address the email using your Contacts app by tapping Add (+).

(4) Use the Contacts app to find and select the contact to whom you want to address the message. When you tap a contact who has only one email address, that address is pasted into the To field and you return to the New Message window. When you tap a contact with more than one email address, you move to that contact's screen, which shows all available addresses; tap the address to which you want to send the message.

(5) Repeat steps 2–4 to add other recipients to the message.

(6) Tap the Cc/Bcc, From line. The Cc and Bcc lines expand. If you don't want to add these types of recipients, skip to step 9.

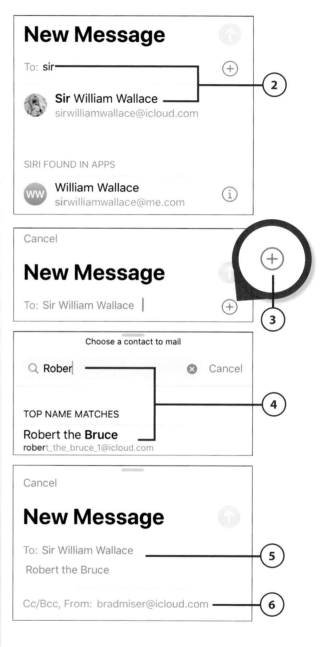

Removing Addresses

To remove an address, tap it so it's highlighted in a darker shade of blue; then tap Delete (x) on the iPhone's keyboard.

7 Follow the same procedures from steps 2–4 to add recipients to the Cc field. Use this field to include people who might benefit from reading the email, but don't have any responsibility for it (information only).

8 Follow the same procedures from steps 2–4 to add recipients to the Bcc field. Use the Bcc field for those people whom you want to receive the message but that you want to hide from others on the distribution list (hidden recipients).

9 If the account you want to send the email from is shown in the From section or if you have only one email account, skip to step 11; to change the account from which the email is sent, tap the From field. You see a list of all the accounts available for sending email. The current address is indicated by the check mark.

10 If necessary, swipe up or down the list until you see the address you want to use.

11 Tap the address you want to use. The account list closes.

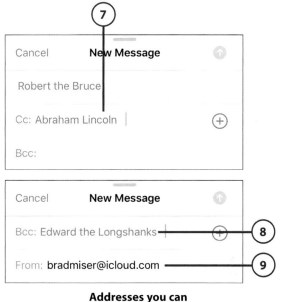

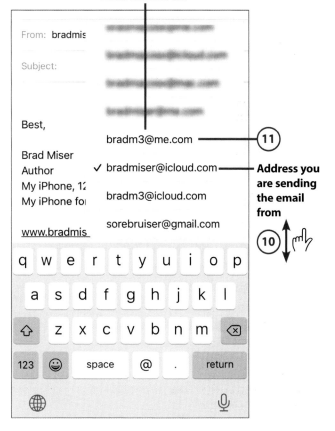

Addresses you can send email from

(12) Type the subject of the message.

(13) If you want to be notified when someone replies to the message you are creating, tap the bell; if not, skip to step 15.

(14) Tap Notify Me. When anyone replies to the message, you are notified.

(15) If you don't see the body of the message, swipe up the screen, and it appears.

(16) Tap in the body of the message, and type the message above your signature. Mail enables you to use all of the iOS's great text tools, such as Predictive text. Of course, you can tap the microphone to dictate your email, too. (Refer to Chapter 3 for the details of working with text.)

(17) To make the keyboard larger, rotate the iPhone so that it's horizontal.

(18) When you finish the message, tap Send. The progress of the send process is shown at the bottom of the screen; when the message has been sent, you hear the sent mail sound, which confirms that the message has been sent. If you enabled the reply notification for the message, you are notified when anyone replies to it.

>>>*Go Further*
MORE ON SENDING EMAIL

Following are several tips to help you send email:

- **Multiple Email Accounts**—If you have more than one email account, it's important to know from which account you're sending a new message. If you tap the Compose icon while you're on the Mailboxes screen or the Inboxes screen, the From address is the one for the account set as your default; otherwise, the From address is the email account associated with the inbox you're in. Of course, you can always use steps 9 and 10 to change the account from which you are sending a message regardless of where you start the message.

- **Draft Messages**—If you want to save a message you're creating without sending it, tap Cancel. A prompt appears; select Save Draft to save the message; or tap Delete Draft if you don't want to keep it. When you want to work on a draft message again, touch and hold down Compose. The Drafts screen appears. Here, you see your most recent draft messages; tap the draft message you want to work on. You can make changes to the message and then send it or save it as a draft again. You also can access your draft message in the Drafts folder under the account from which you are sending the message; moving among folders is covered later in this chapter.

- **Suggested Recipients**—As you create messages over time, Siri suggests recipients based on the new message's current recipients. For example, if you regularly send emails to the same group of people, when you add two or more people from that group, Siri suggests others (individually and as a group) that you might want to include just below the address bar you're working in (such as the To bar). You can tap these suggestions to quickly add recipients to a new message.

- **Including Photos in Email**—To add a photo or video to a message you create, tap twice in the body. Swipe to the left on the resulting toolbar (if you don't see it immediately) until you see the Insert Photo or Video command, and then tap it. Use the Photos tool (see Chapter 16, "Viewing and Editing Photos and Video with the Photos App," for information about this app) to move to and select the photos you want to send; the selected photos are marked with a blue circle. Tap Close (x). The photos you selected are added to the email.

- **Sending Email from All the Right Places**—You can send email from a number of apps on your iPhone. For example, you can share a photo with someone by viewing the photo in the Photos app, tapping Share, and then tapping Mail. Or you can tap a contact's email address to send an email from your contacts list. In all cases, the iPhone uses the Mail app to create a new message that includes the appropriate content, such as a photo or link; you use Mail's tools to complete and send the email as described in this chapter.

Replying to Email

Email is all about communication, and Mail makes it simple to reply to messages.

(1) Open the message you want to reply to.

(2) Tap Share (the arrow).

(3) Tap Reply to reply to only the sender or, if there was more than one recipient, tap Reply All to reply to everyone who received the original message.

(4) If the message to which you're replying has attachments, tap Include to include those in your reply or Don't Include to leave them out. If the message doesn't have attachments, you don't have this option.

The Re: screen appears showing a new message. Mail pastes the contents of the original message at the bottom of the body of the new message below your signature. The original content is in blue and is marked with a vertical line along the left side of the screen.

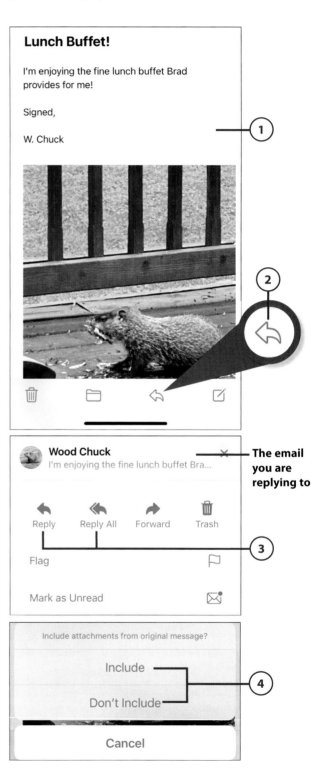

5. Add or change the To, Cc, or Bcc recipients and select the account from which you want to send the reply. These tasks work just like when you create a new email message.

6. Write or dictate your response.

7. Tap Send . Mail sends your reply.

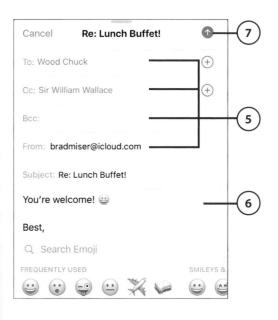

Forwarding Emails

When you receive an email you think others should see, you can forward it to them.

1. Read the message you want to forward.

2. If you want to include only part of the current content in the message you forward, tap where you want the forwarded content to start. This is useful (and considerate!) when only a part of the message applies to the people to whom you are forwarding it. If you want to forward the entire message, skip to step 4.

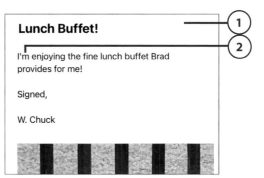

(**3**) Use the text selection tools to select the content you want to include in your forwarded message.

(**4**) Tap Share.

(**5**) Tap Forward.

(**6**) If the message includes attachments, tap Include at the prompt if you also want to forward the attachments, or tap Don't Include if you don't want them included. If it doesn't have attachments, you don't see this option.

The Forward screen appears. Mail pastes the contents of the message that you selected, or the entire content if you didn't select anything, at the bottom of the message below your signature. If you included attachments, they're added to the new message as well.

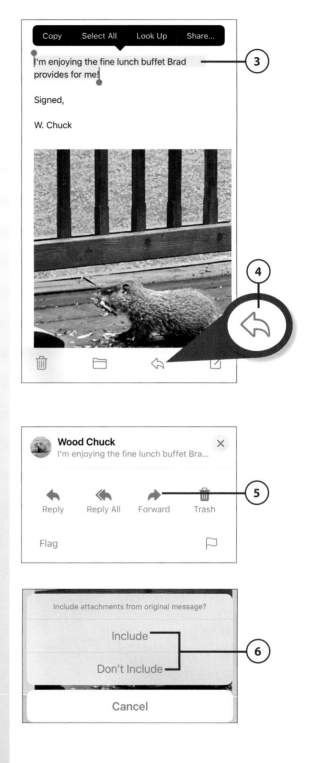

⑦ Address the forwarded message using the same tools you use when you create a new message.

⑧ Type your commentary about the message above your signature.

⑨ Tap Send. Mail forwards the message.

Large Messages

Some emails, especially HTML messages, are so large that they don't immediately download in their entirety. When you forward a message whose content or attachments haven't fully downloaded, Mail prompts you to download the "missing" content before forwarding. If you choose not to download the content or attachments, Mail forwards only the downloaded part of the message.

Cancel

Fwd: Lunch Buffet! ⬆

To: Crabby Hayes

Cc/Bcc, From: bradmiser@icloud.com

Subject: Fwd: Lunch Buffet!

At least someone is grateful for food!

Best,

Brad Miser
Author

Managing Email

Following are some ways you can manage your email. For example, you can check for new messages, see the status of messages, delete messages, and organize messages using the folders associated with your email accounts.

Checking for New Email

Edit

Mailboxes

Swipe down to update your email

📭 All Inboxes   12 >

📭 Your RC Companion   >

📭 Accruent   10 >

To manually retrieve messages, swipe down from the top of any Inbox or the Mailboxes screen. The screen "stretches" down and when you lift your finger, the Mail app checks for and downloads new messages.

Mail also retrieves messages whenever you move into the app or into any inbox or all your inboxes. Of course, it also retrieves messages according to the selected Fetch New Data option. It downloads new messages immediately when they arrive in your account if Push is enabled or automatically at defined intervals if you've set Fetch to get new email periodically. (Refer to Chapter 4 for an explanation of these options and how to set them.)

When your email was last updated

How many unread messages you have in the current inbox

The bottom of the Mailboxes or an Inbox screen always shows when email was most recently downloaded to your iPhone; on the bottom of inbox screens, you also see the number of new email messages (if there are any unread messages).

Understanding the Status of Email

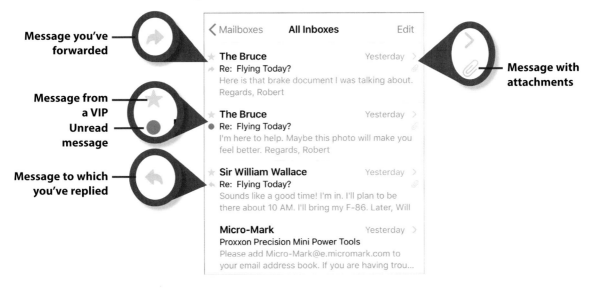

Message you've forwarded

Message from a VIP

Unread message

Message to which you've replied

Message with attachments

When you view an inbox or a message thread, you see icons next to each message that indicate its status. When a message doesn't have any icons, that means it doesn't have attachments, you've read it but haven't done anything else with it, or it isn't from a VIP. (VIPs are explained in the Go Further sidebar, "VIPs," later in this chapter.)

Managing Email from the Message Screen

Tap to delete a message

To delete a message while reading it, tap Trash. If the warning preference is enabled (see the Go Further sidebar, "Mail Settings," at the end of this chapter to learn how to enable this warning), confirm the deletion and the message is deleted. If the confirmation prompt is disabled, the message is deleted immediately.

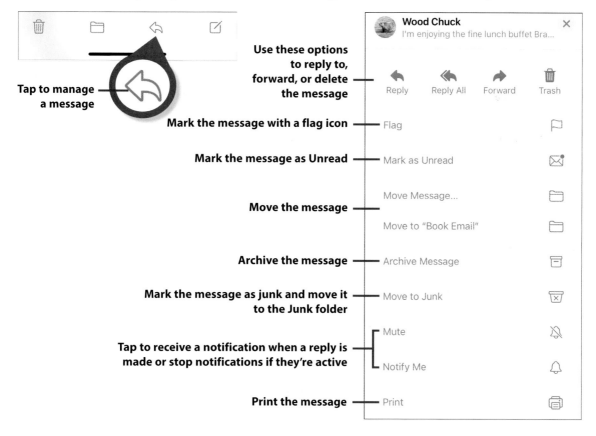

Tap to manage a message

Use these options to reply to, forward, or delete the message

Mark the message with a flag icon

Mark the message as Unread

Move the message

Archive the message

Mark the message as junk and move it to the Junk folder

Tap to receive a notification when a reply is made or stop notifications if they're active

Print the message

To take other action on a message you are reading, tap Share. On the Share sheet that opens, you can choose a number of commands. The action you select is performed on the message you're viewing.

>>>Go Further
SWIPE TO MANAGE

You can perform some of the actions described here on a message by swiping to the left or right on it when viewing its preview in an inbox. The direction you swipe determines which actions are available (you can configure these as described in the Go Further sidebar, "Mail Settings," at the end of this chapter). For example, when you swipe to the left on a message, you might see Trash, Flag, and More. Tap Trash to delete it, Flag to flag it, or More to see options like those shown in the previous figure. Swiping to the right on a message enables you to mark it as Read or Unread.

You can quickly delete a message from an inbox by quickly swiping all the way to the left on it.

Managing Multiple Emails at the Same Time

You can manage email by selecting multiple messages on an Inbox screen, which is more efficient because you can take action on multiple messages at the same time.

(1) Move to an Inbox screen showing messages you want to manage.

(2) Tap Edit. A selection circle appears next to each message.

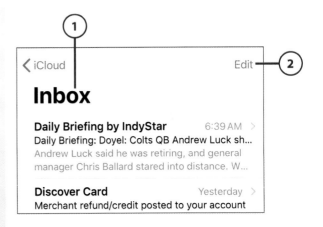

3 Select the message(s) you want to manage by tapping their selection circles. As you select each message, its selection circle turns into a white check mark in a blue circle. Selected messages are also highlighted in gray bars. At the top of the screen, you see how many messages you currently have selected. The actions you can perform become active at the bottom of the screen.

4 To delete the selected messages, tap Trash. Mail deletes the selected messages and exits Edit mode.

5 To change the status of the selected messages, tap Mark.

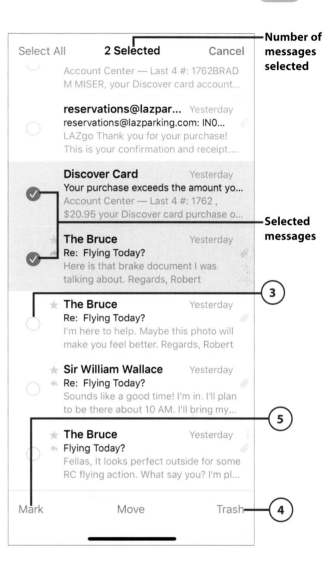

Number of messages selected

Selected messages

6 Tap the action you want to take on the selected messages. The action you select is performed and you return to the Inbox screen and exit Edit mode.

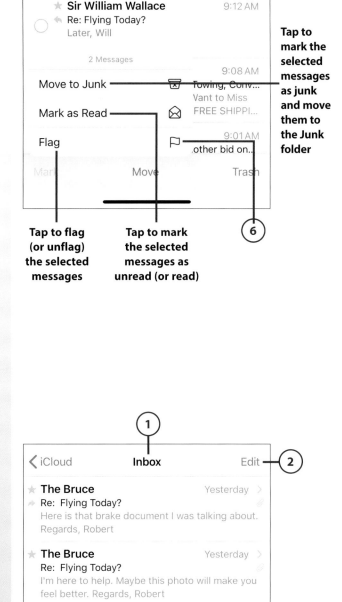

Tap to mark the selected messages as junk and move them to the Junk folder

Tap to flag (or unflag) the selected messages

Tap to mark the selected messages as unread (or read)

Storing Email in Specific Locations

Over time, you might end up with a lot of messages in your inboxes. You can keep the email you want to retain organized by storing it in folders, which work just like document folders on a desktop computer and are analogous to a file folder used to keep papers organized. To keep your inboxes neat and tidy, follow these steps:

1 Move to an Inbox screen showing messages you want to move to a folder.

What's in a Name?

Mail uses the term mailbox to refer to a location where messages are stored. A place to store files or documents is often called a folder (such as on a computer). For the purposes of this chapter, mailbox and folder are used interchangeably and refer to the same type of object.

2 Tap Edit. A selection circle appears next to each message.

3 Select the messages you want to move by tapping their selection circles. As you select each message, its selection circle turns into a white check mark on a blue circle. The actions available for what you select become active at the bottom of the screen.

4 Tap Move.

Straight to It

In some cases (such as when you haven't moved messages) after you tap Move, you skip steps 5 and 6 and move directly to the Mailboxes screen as shown for step 7.

5 If the location to which you want to move the messages appears with a Move command, such as Move to *Folder Name* (where *Folder Name* is the name of the folder), tap that location and the selected messages move there. Skip the rest of these steps.

6 If you want to choose a different location, tap Move *X* Messages, where *X* is the number of messages you have selected. The Mailboxes screen appears. At the top of this screen you see the number of messages you are moving. Under that are the mailboxes available under the current account.

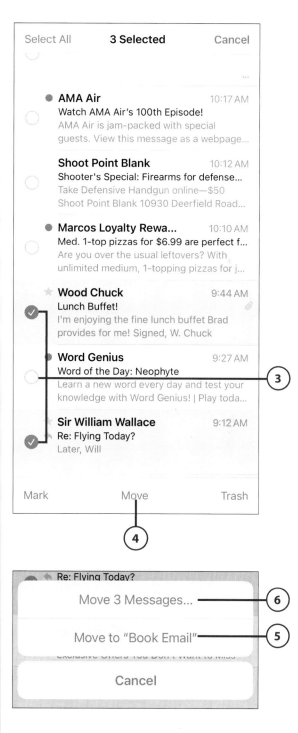

⑦ Swipe up and down the screen to browse the mailboxes available in the current account.

⑧ Tap the mailbox to which you want to move the selected messages. They are moved into that folder, and you return to the previous screen, which is no longer in Edit mode.

< Back	iCloud	Cancel
☰	Wood Chuck & 2 more... 3 messages	
◁	Sent	
☒	Junk	3
🗑	Trash	
🗄	Archive	
🗀	Book Email	
🗀	com.apple.Mail.ToDo	
🗀	Deleted Items	
🗀	Junk (bradmacosx)	
🗀	Junk E-mail	
🗀	Mail to File	

⑦

⑧

>>>Go Further
KEEPING MAILBOXES TIDY

If you use email frequently, you'll want to practice good email hygiene by keeping your email organized. Here are a few pointers:

- **Move to Other Accounts**—If you want to move selected messages to a folder under a different account, tap Accounts in the upper-left corner of the Move screen. Tap the account to which you want to move the messages (not all accounts will be available; if an account is grayed out, you can't move messages to it). Then tap the mailbox into which you want to move the messages. The messages move from the current account to the folder under the account you selected.

- **Makin' Mailboxes**—You can create a new mailbox to organize your email. Move to the Mailboxes screen and tap Edit. Then, tap New Mailbox located at the bottom of the screen. Type the name of the new mailbox. Tap MAILBOX LOCATION and then choose where you want the new mailbox located (for example, you can place the new mailbox inside an existing one). Tap Save. You can then store messages in the new mailbox.

- **Changin' Mailboxes**—You can change the mailboxes that appear on the Mailboxes screen. For example, you can display the Attachments mailbox to make messages with attachments easier to get to. Move to the Mailboxes screen and tap Edit. To cause a mailbox to be visible, tap it so that it has a check mark in its circle. To hide a mailbox, tap its check mark so that it shows an empty circle. Drag the Order icon for mailboxes up or down the screen to change the order in which mailboxes appear. Swipe up to the bottom of the screen and tap Add Mailbox to add a mailbox not shown on the list. Tap Done to save your changes.

Working with Messages in a Specific Mailbox

You can open a mailbox within an account to work with the messages it contains. For example, you might want to open the Trash mailbox to recover a deleted message.

1. Move to the Mailboxes screen.

2. If necessary, swipe up the screen to see the email accounts you're using. Each account has its own section showing the mailboxes stored on that account.

3. If you don't see an account's mailboxes, expand the account by tapping its right-facing arrow.

4. Swipe up and down to browse the mailboxes in the account.

5. Tap the folder or mailbox containing the messages you want to view. You see the messages it contains. In some cases, this can take a few moments for the messages to be downloaded if that folder or mailbox hasn't been accessed recently.

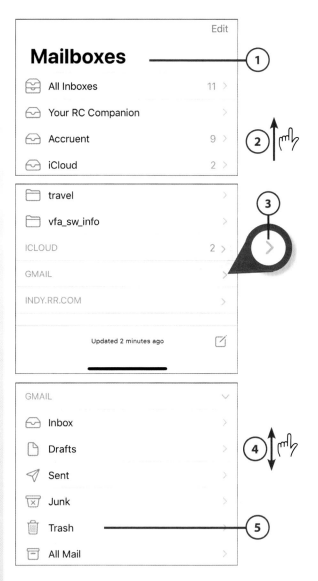

(6) Tap a message or thread to view it. Or tap Edit and select multiple messages to move them to a different location. For example, if you accidentally deleted messages, open the Trash mailbox for the account to which those messages were sent and move them back to the inbox for that account using the steps in "Storing Email in Specific Locations."

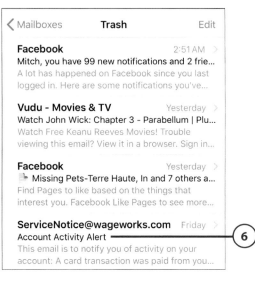

Saving Images Attached to Email

Email is a great way to share photos. When you receive a message that includes photos, you can save them on your iPhone.

(1) Move to the message screen of an email that contains one or more photos or images.

(2) Swipe up the message to see all the images it contains.

(3) Touch and hold on an image. The Share tools appear.

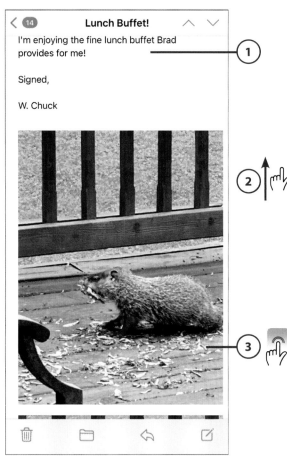

4. Tap Save Image to save just the image you touched or tap Save *X* Images, where *X* is the number of images attached to the message, to save all the photos. (If there is only one image, the command is just Save Image.) The images are saved in the Photos app on your iPhone. (See Chapter 16 for help working with the Photos app.)

Finding Email

As you accumulate email, you might want to find or focus on specific messages. You can do this by filtering an inbox or by searching for specific messages.

Filtering Email

You can quickly filter the email messages in an inbox as follows:

(1) Open the inbox you want to filter. Choose All Inboxes to filter all your email at once.

(2) Tap Filter. The contents of the inbox are filtered by the current criteria, which is indicated by the term under "Filtered by." The Filter icon is highlighted in blue to show the inbox is filtered.

(3) Tap the current filter criteria.

4 Set the criteria by which you want to filter the messages in the inbox; the current criteria are indicated by check marks or green switches. For example, tap To: Me to only show messages on which you are included in the To block.

5 Tap Done. You return to the inbox and only messages that meet your filter criteria are shown.

6 Tap Filter to display all the messages again.

Searching Email

You can find specific email messages by searching for them. For example, suppose you want to retrieve an email message that was related to a specific topic, but you can't remember where you stored it. Mail's Search tool can help you find messages like this quite easily.

1 Move to the screen you want to search, such as an account's inbox or a folder's screen. If you want to search all your mailboxes, this step is optional as you can select all your mailboxes later in the process.

2 If necessary, swipe down to move to the top of the screen to display the Search tool.

3 Tap in the Search tool.

Suggested Searches

When the Search screen first appears, you see the Suggested Searches list. If a search you want to perform is on this list, tap it to run the search.

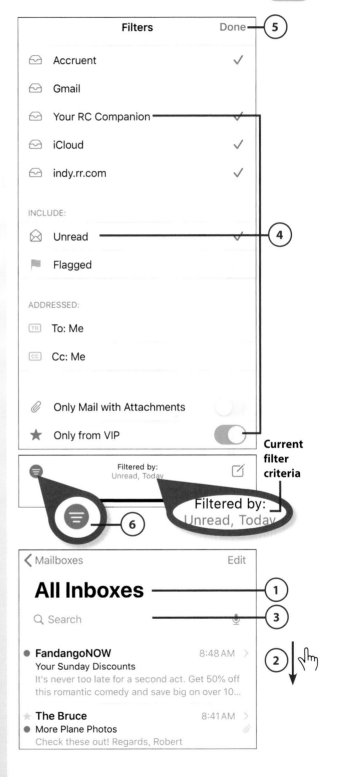

4 To search in all your mailboxes, tap All Mailboxes, or to search in only the current mailbox, tap Current Mailbox.

5 Enter the text for which you want to search. As you type, Siri makes suggestions about what you might be searching for. These appear in different sections based on the type of search Siri thinks you are doing, such as People, Subjects, and more. The Top Hits section shows messages that seem to be the best matches; tap one of these to open that message.

6 To use one of Mail's suggestions to search, such as a person, tap their name; or continue typing your search term and when you are done, tap search. Mail searches for messages based on your search criterion and you see the results.

7 Work with the messages you found, such as tapping a message to read it. When you're done with the message, tap back (<) in the upper-left corner of the screen to return to the search results.

8 To clear a search and exit Search mode, tap Cancel.

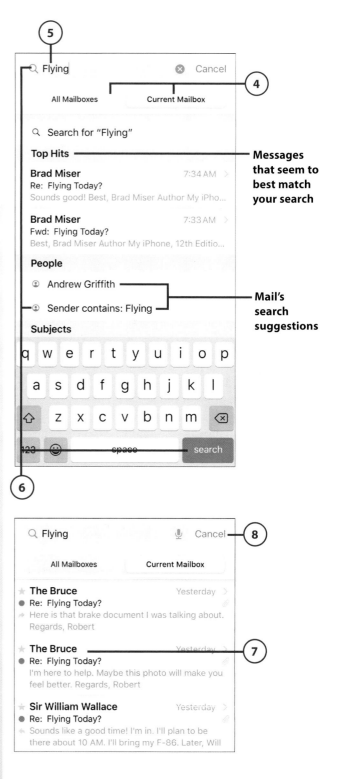

Messages that seem to best match your search

Mail's search suggestions

>>>Go Further

VIPS

The VIP feature enables you to designate specific people as your VIPs. When a VIP sends you an email, it's marked with a star icon and goes into the special VIP mailbox so you can access these important messages easily. You can also create specific notifications for your VIPs, such as a unique sound when you receive email from a VIP. (See Chapter 8 to learn how to mark a contact as a VIP and Chapter 6 to configure special notifications for VIPs.)

Here are a few tidbits on VIPs:

- Messages from VIPs are marked with the star icon no matter in which mailbox you see the messages.

- To see the list of your current VIPs, move to the Mailboxes screen and tap Info (i) for the VIP mailbox. You see everyone currently designated as a VIP. Tap Add VIP to add more people to the list. Tap VIP Alerts to create special notifications for VIPs.

- To return a VIP to normal status, view his contact information and tap Remove from VIP.

Managing Junk Email

Junk email, also known as *spam*, is an unfortunate reality. No matter what precautions you take, you're going to receive some spam emails. Of course, it's good practice to be careful about where you provide your email address to limit the amount of spam you receive.

Consider using a "sacrificial" email account when you shop, post messages, or provide your address in the other situations that might lead to spam. If you do get spammed, you can stop using the sacrificial account and create another one to take its place. Or you can delete the sacrificial account from your iPhone and continue to use it on your computer where you likely have better spam tools in place.

The Mail app on the iPhone includes a basic junk email tool. However, if you use an account or an email application on a computer that features a junk mail/ spam tool, it acts on mail sent to your iPhone, too. For example, if you configure spam tools for a Gmail account, those tools act on email before it reaches your iPhone. Similarly, if you use the Mail app on a Mac, its rules and junk filter work on email as you receive it; the results of this are also reflected on your iPhone. To change how you deal with junk email on your iPhone, change the junk email settings for your account online (such as for Gmail). Or change how an email app on a computer (for example, Mail on a Mac) deals with junk mail. The results of these changes are reflected in the Mail app on your iPhone.

Many email accounts, including iCloud and Gmail, have Junk folders; these folders are available in the Mail app on your iPhone. You can open the Junk folder under an account to see the messages that are placed there.

Marking Junk Email

You can perform basic junk email management on your iPhone by doing the following:

1. When you view a message that is junk, tap Share.

2. If needed, swipe up to see the Move to Junk command.

3. Tap Move to Junk. The message is moved from the inbox to the Junk folder for the account to which it was sent. Future messages from the same sender go into the Junk folder automatically.

Junk Them

You can move multiple messages to the Junk folder by tapping Edit on an inbox screen, selecting the messages you want to junk, tapping Mark, and then tapping Move to Junk.

Mark as Unread

Move Message

Archive Message

Move to Junk

Mute

Notify Me

Print

Junk It or Trash It?

The primary difference between moving a message to the Junk folder or deleting it is that when you mark a message as junk, future messages from the same sender are moved to the Junk folder automatically. When you delete a message, it doesn't change how future messages from the same sender are handled.

>>>Go Further
MAIL SETTINGS

There are a number of settings that affect how the Mail app works. The good news is that you can probably use the Mail app with its default settings just fine. You can change how the Mail app works by opening the Settings app and tapping Mail. The following table explains some of the more useful options that you might want to change.

Mail and Related Settings

Settings Area	Location	Setting	Description
Mail	ALLOW MAIL TO ACCESS	Notifications	Takes you to notifications settings for Mail; see Chapter 6 for details.
Mail	ALLOW MAIL TO ACCESS	Cellular Data	Setting this switch to on (green) enables Mail to send and receive email when your phone is connected to a cellular network (it's not using Wi-Fi).
Mail	N/A	Accounts	Enables you to configure online accounts; see Chapter 4 for the details.
Mail	MESSAGE LIST	Preview	Determines the number of lines you want to display for each email when you view the Inbox and in other locations, such as alerts. This preview enables you to get the gist of an email without opening it. More lines give you more of the message but take up more space on the screen.

Settings Area	Location	Setting	Description
Mail	MESSAGE LIST	Show To/Cc Labels	When enabled, Mail indicates if you are a To recipient or a Cc recipient with a small icon next to your name on email messages.
Mail	MESSAGE LIST	Swipe Options	Changes what happens when you swipe to the left or right on email when you are viewing an Inbox. You can set the Swipe Left motion to be None, Mark as Read, Flag, or Move Message. You can set the Swipe Right motion to be None, Mark as Read, Flag, Move Message, or Archive. Note that you can't have the same option configured for both directions.
Mail	MESSAGES	Ask Before Deleting	When this switch is on (green), you're prompted to confirm when you delete or archive messages. When this switch is off (white), deleting or archiving messages happens without the confirmation prompt.
Mail	THREADING	Organize By Thread	When this switch is on (green), messages in a conversation are grouped together as a "thread" on one screen. This makes it easier to read all the messages related to the same topic, which is called a thread. When this switch is off (white), messages are listed individually.
Mail	THREADING	Collapse Read Messages	When enabled and you read messages in a thread, the thread collapses so you see the thread rather than the individual messages in the thread.
Mail	THREADING	Most Recent Message on Top	With this switch set to on (green), the most recent message in a thread appears at the top of the thread and the messages move backward in time as you move down the list of messages.

Settings Area	Location	Setting	Description
Mail	THREADING	Complete Threads	With this switch enabled (green), all the messages in a thread are displayed when you view the thread, even if you've moved messages to a different folder (other than the inbox).
Mail	COMPOSING	Signature	Signatures are text that is automatically added to the bottom of new email messages that you create. For example, you might want your name and email address added to every email you create. You can configure the same signature for all your email accounts or have a different signature for each account.
Mail	COMPOSING	Default Account	Determines which email account is the default one used when you send an email (this setting isn't shown if you have only one email account).
Display & Brightness	N/A	Text Size	Changes the size of text in all apps that support Dynamic Type (Mail does). Drag the slider to the right to make text larger or to the left to make it smaller.
Display & Brightness	N/A	Bold Text	Changes text to be bold when the Bold Text switch is set to on (green).

Send messages from other apps too, such as to share photos

Tap to send and receive text messages, photos, video, and more

Tap to configure Messages

In this chapter, you explore the texting and messaging functionality your iPhone has to offer. Topics include the following:

→ Getting started
→ Preparing the Messages app for messaging
→ Sending messages
→ Receiving, reading, and replying to messages
→ Working with messages

Sending, Receiving, and Managing Texts and iMessages

You can use the iPhone's Messages app to send, receive, and converse; you also can send and receive images, videos, and links. You can maintain any number of conversations with other people at the same time, and your iPhone lets you know whenever you receive a new message via audible and visible notifications that you configure. In addition to conversations with other people, many organizations use text messaging to send updates, such as airlines and doctor's offices. You might find messaging to be one of the most used functions of your iPhone.

Getting Started

Texting, also called messaging, is an especially great way to communicate with others when you have something quick you want to say, such as an update on your arrival time. It's much easier to send a quick text, "I'll be there in 10 minutes," than it is to make a phone call or send an email. Texting/is designed for relatively short messages. It's also a great way to share photos and videos quickly and easily. And if you communicate with younger people, you might find they tend to respond more readily since texting is a primary form of communication for them.

There are two types of messages that you can send with and receive on your iPhone using the Messages app.

The Messages app can send and receive text messages via your cell network based on telephone numbers. Using this option, you can send text messages to and receive messages from anyone who has a cell phone capable of text messaging.

If you're communicating with people who have Apple devices (such iPhones or iPads), the second option is to use Apple's iMessage service. You can send and receive iMessage messages within the Messages app similar to text messages. Messages sent using iMessage travel over the Internet and are associated with an email account. Messages you send using iMessage don't have some of the limitations that text messages have; for example, they can be fairly long and can include photos, videos, and effects. (The recipients of your iMessage messages must have iMessages set up on their Apple devices, but as you will soon see, this isn't hard.)

You don't need to be overly concerned about which type is which because the Messages app makes it clear which type a message is by color and text. The app automatically uses iMessage when available; when iMessage can't be used, the Messages app automatically uses cellular texting instead.

You can configure your iMessage account on multiple devices, such as an iPhone and an iPad. This means you have the same text conversations on each device. So, you can start a conversation on your iPhone, and then continue it on an iPad at a later time.

Preparing the Messages App for Messaging

To use the Messages app, you need to configure the account you want to use for iMessage. There are a few other settings you should also configure so your Messages experience is as good as possible.

No Apple ID?

You need to have an Apple ID to use iMessage, although you can still use your iPhone for cellular text messages. Because iMessage offers more features than cellular text messages, I recommend you use iMessage whenever you can. You can obtain an Apple ID for free so you can give it a try with no risk. See Chapter 4, "Setting Up and Using an Apple ID, iCloud, and Other Online Accounts," for the steps to obtain and configure an Apple ID.

Perform the following steps to set up Messages on your iPhone:

(1) Move into the Settings app and tap Messages.

(2) If the iMessage switch is on (green), go right to step 3; if the iMessage switch is off (white), tap it to enable the iMessage feature; the switch becomes green to show it is active. When you turn iMessage on, your account is activated, and the Send & Receive option appears.

(3) Tap Send & Receive.

Apple ID Assumed

These steps assume you have already signed into an Apple ID on your iPhone (see Chapter 4). If you haven't, you're prompted to sign into your Apple ID when you perform step 2. Also, in some cases, you might see Use your Apple ID for iMessage at the top of the iMessage shown for step 4. If you do see this, tap it to activate Messages with your Apple ID. After you sign into the Apple ID, it can take a few moments for your iCloud addresses to appear on the iMessage screen.

(4) To prevent an email address from being available to others to send you messages, in the YOU CAN RECEIVE IMESSAGES TO AND REPLY FROM section, tap it so it doesn't have a check mark. Note that you can't disable your phone number since that is used for cellular messages.

(5) To enable an address so it can be used for messages, tap it so it does have a check mark.

(6) Tap the phone number or email address you want to use by default when you start a new text conversation in the START NEW CONVERSATIONS FROM section. Because there are no data or media limitations with iMessages, you usually want to choose your email address as the default way to start a conversation.

(7) Tap Messages (<).

(8) Tap Share Name and Photo.

Memoji Stickers

The first time you use the Share Name and Photo command, you're prompted to create a Memoji sticker to use with messages. If you want to do this, tap Get Started and follow the onscreen instructions to create it. Or tap Set Up Later in Messages, which is the option used for these steps (not shown in a figure).

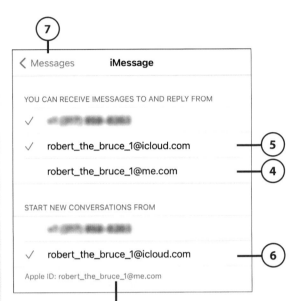

The Apple ID you are using for Messages

⑨ Tap Choose Name and Photo.

⑩ Tap Options (…).

⑪ Tap the option you want to use as the photo shared with others that you message. You can tap the Camera icon to take a photo, tap the Photos icon to use an existing photo, tap the smiley face to use an emoji, tap the pencil to use text, tap a Memoji to use it, or tap add (+) to create a new Memoji. If you swipe up the screen, you see even more options with icons.

These steps show using a recently used photo. The steps to use one of the other options are somewhat different.

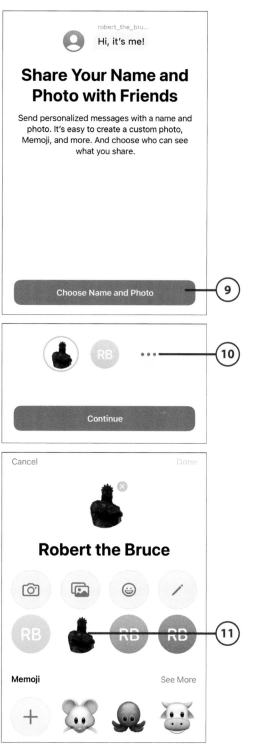

(12) Tap Edit.

(13) Pinch, unpinch, and drag the image until the part you want to display is shown in the circle.

(14) Tap Choose.

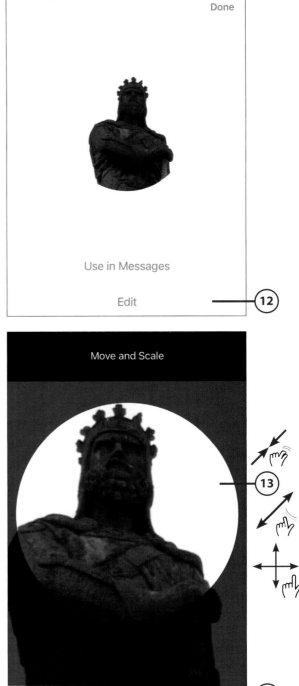

(15) Tap the filter you want to apply to the image.

(16) Tap Done.

(17) Tap Done.

(18) Tap Continue.

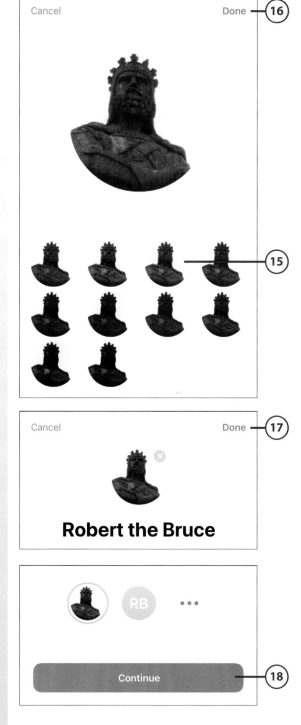

(19) If you want to use the selected image as your image for Apple ID and for your contact card, tap Use; if you want to use it only for Messages, tap Not Now.

(20) Enter the first and last name you want others to see when you message them.

(21) Tap Contacts Only if you want your name and image to be shared with people in your contacts list or tap Always Ask if you want to always be prompted to share your name and image when you message.

(22) Tap Back (<).

(23) Set the Show Contact Photos switch to on (green) if you want to see other people's images when you message.

(24) Slide the Send Read Receipts switch to on (green) to notify others when you read their messages. Be aware that receipts apply only to iMessages (not texts sent over a cellular network).

(25) Slide the Send as SMS switch to on (green) to send texts via your cellular network when iMessage is unavailable. If your cellular account has a limit on the number of texts you can send, you might want to leave this set to off (white) so you use only iMessage when you're texting. If your account has unlimited texting, you should set this to on (green).

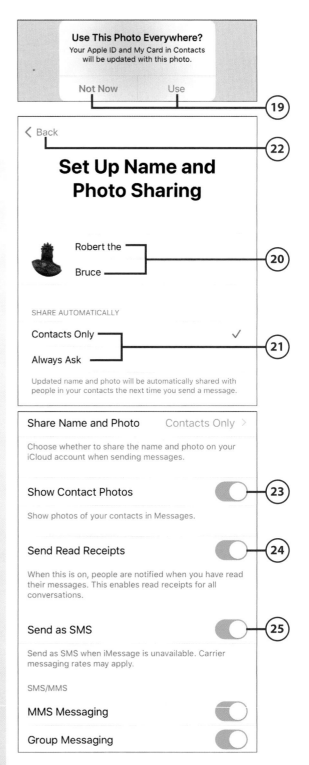

(26) Set the MMS Messaging switch to off (white) if you don't want to allow photos and videos to be included in messages sent via your phone's cellular network. You might want to disable this option if your provider charges more for these types of messages—or if you simply don't want to deal with anything but text in your messages.

(27) Set the Group Messaging switch to on (green) to keep messages you send to a group of people organized by the group. When enabled, replies you receive to messages you send to groups (meaning more than one person on a single message) are shown on a group message screen. Then, each reply from anyone in the group is included on the same screen. If this is off (white), when someone replies to a message sent to a group, the message is separated out as if the original message was just to that person. (The steps in this chapter assume Group Messaging is on.)

(28) Swipe up the screen.

(29) Tap Keep Messages.

(30) Tap the length of time for which you want to keep messages.

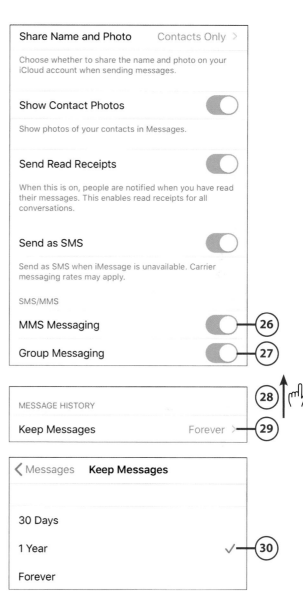

(31) If you tapped an option that made the length of time you're keeping messages shorter than it previously was, tap Delete (if it's longer, you skip this step). The messages on your iPhone older than the length of time you selected in step 30 are deleted.

(32) Tap Messages (<).

(33) Set the Notify Me switch to on (green) if you want to receive a notification when your name is mentioned in a message even when conversations are muted. This can help you know when you have something to respond to without being bothered with chatter in a group text conversation.

(34) Set the Filter Unknown Senders switch to on (green). When this switch is on, messages from unknown senders are filtered onto a dedicated screen so they don't clog up your active messages; notifications for such messages are disabled. (Using this feature is explained in the section "Working with Messages from People You Might Not Know" later in this chapter.)

Delete Older Messages?

This will permanently delete all text messages and message attachments from your device that are older than 1 year.

Delete ———(31)

Cancel

(32)

< Messages   **Keep Messages**

MENTIONS

Notify Me ———(33)

When this is on, you will be notified when your name is mentioned even if conversations are muted.

MESSAGE FILTERING

Filter Unknown Senders ———(34)

Sort messages from people who are not in your contacts into a separate list.

>>>Go Further

MORE ON MESSAGES CONFIGURATION

Following are a few Messages setting tidbits for your consideration:

- **Text Message Forwarding**—If you see this option on the Messages Settings screen, you can forward text messages to another device that is associated with your iCloud account, such as a Mac. This ensures messages sent to your phone as text messages will also be sent to those devices. Tap Text Message Forwarding and set the switches to on (green) for the devices that you want text messages to be sent to.

- **Show Subject Field**—This divides text messages into two sections; the upper section is for a subject, and you type your message in the lower section. This is not commonly used in text messages, and the steps in this chapter assume this setting is off.

- **Character Count**—This setting displays the number of characters you've written compared to the number that are allowed for a cellular text message (such as 59/160). These limits don't really apply these days so you can just leave this off.

- **Blocked Contacts**—This enables you to configure blocked contacts for Messages and other apps. Blocking contacts is explained in Chapter 18, "Maintaining and Protecting Your iPhone and Solving Problems."

- **Audio Messages**—The Messages app can record and send audio messages. The Expire setting determines how long audio messages that you receive are stored on your phone. The Raise to Listen setting determines whether those messages play automatically when you raise your phone.

- **Low Quality Image Mode**—To have the images in your messages sent at a lower quality level, set the Low Quality Image Mode switch to on (green). This can be a useful setting if you or the other recipients of your messages have limited data plans and aren't using a Wi-Fi connection, because lower quality images require less data to transmit and receive.

- **Messages in iCloud**—If you allow your messages to be stored in your iCloud space, you can access your messages from all the devices configured with your iCloud account. See Chapter 4 for details.

- **Notifications**—If you use Messages frequently, you should configure its notifications so that you're aware of activity without being distracted by excess notifications. Configuring notifications is explained in detail in Chapter 6, "Making Your iPhone Work for You."

Sending Messages

You can use the Messages app to send messages to people using a cell phone number (as long as the device receiving it can receive text messages) or an email address that has been registered for iMessage. If the recipient has both a cell number and iMessage-enabled email address, the Messages app assumes you want to use iMessage for the message.

When you send a message to more than one person and at least one of those people can use only the cellular network, all the messages are sent via the cellular network and not as an iMessage.

When you send messages to or receive messages from a person or a group of people, you see those messages in a conversation. Every message sent among the same people is added to that conversation (if the Group Messaging option is enabled). When you send a message to a person or group you haven't messaged before, a new conversation is created. If you send a message to a person or group you have messaged before, the message is added to the existing conversation.

Navigating in Messages

The Messages app has three major screens and it is useful to know how to get around them.

Tap to open Messages

Tap Messages on a Home screen to open the Messages app.

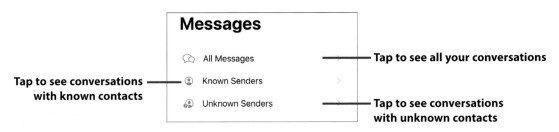

Tap to see conversations with known contacts

Tap to see all your conversations

Tap to see conversations with unknown contacts

Filtering Unknowns

This section assumes that the Filter Unknown Senders switch is on as described in the previous section. If it isn't, you only see the Messages screen that shows all conversations or individual conversations screens.

At the top level of the app is the Messages screen. Here, you see the conversations organized into three groups. All Messages contains all conversations. Known Senders leads you to conversations with people you "know," meaning those who are in your Contacts app or with whom you've previously communicated. Unknown Senders contains conversations with senders who aren't in your contacts app or with whom you haven't previously communicated.

Tap one of these to move to a conversation screen. For example, to see conversations with known people, tap Known Senders.

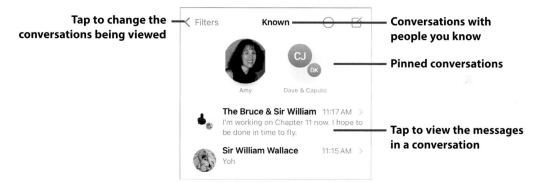

Tap to change the / **conversations being viewed** — (Filters)

Conversations with / **people you know** — (Known)

Pinned conversations — (CJ / DK)

Tap to view the messages / **in a conversation** — (The Bruce & Sir William)

When you move onto a conversations screen, such as Known conversations, you see all the conversations of that type. As you learn later in this chapter, you can browse conversations or tap a conversation to read its messages, among other things. To change the kind of conversations you are viewing, tap Filters. To move into a conversation, tap it.

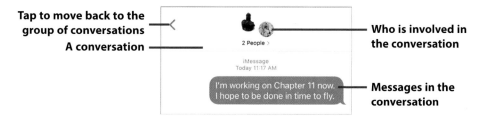

Tap to move back to the / **group of conversations**

A conversation

Who is involved in / **the conversation**

Messages in the / **conversation**

When you move into a conversation, you see all the messages it contains along with information about who is in the conversation. You can read the messages, add new messages, get information, or tap back (<) to move back to the conversations screen, such as the Known screen.

Creating a New Message and Conversation

You can send messages by entering a phone number or email address manually or by selecting a contact from your contacts list.

(1) On the Home screen, tap Messages.

(2) Move to a conversations screen, such as Known, and tap Compose.

(3) Type the recipient's name, email address, or phone number. As you type, the app attempts to match what you type with a saved contact or to someone you have previously messaged and shows you a list of suggested recipients. You see phone numbers or email addresses for each recipient on the list. Phone numbers or addresses in blue indicate the recipient is registered for iMessage and your message is sent via that means. Messages to phone numbers in green are sent as text messages over the cellular network. If a number or email address is gray, you haven't sent any messages to it yet; you can tap it to attempt to send a message. You also see groups you have previously messaged.

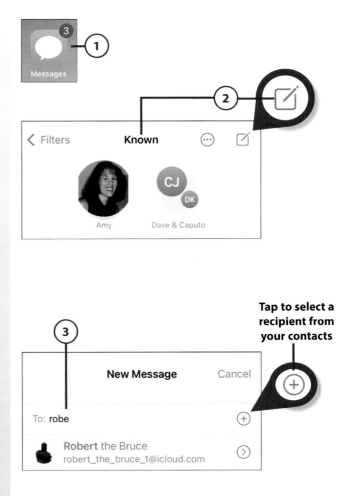

Tap to select a recipient from your contacts

Straight to the Source

You can tap Add (+) in the To field to use the Contacts app to select a contact to whom you want to address the message (see Chapter 8, "Managing Contacts," for the details of using the Contacts app).

4 If the contact has more than one way to message him, tap expand (>); if you don't see this, the contact has only one option, so tap the contact instead.

5 Tap the phone number, email address, or group to which you want to send the message. The recipients' names are inserted into the To field. Or if the information you want to use doesn't appear, just type the complete phone number (as you would dial it to make a call to that number) or email address.

6 If you want to send the message to more than one recipient, tap in the space between the current recipient and Add (+) and use steps 3 through 5 to enter the other recipients' information, either by selecting contacts using Add (+) or by entering phone numbers or email addresses. As you add recipients, they appear in the To field.

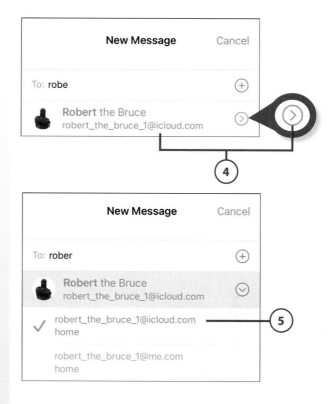

Change Your Mind?

To remove a contact or phone number from the To box, tap it once so it becomes highlighted in blue, and then tap delete (x) on the keyboard.

7 When you've added all the recipients, tap in the Message bar, which is labeled iMessage if you entered iMessage addresses or Text Message if you entered a phone number. The cursor moves into the Message bar, and you're ready to type your message.

8 Type the message you want to send in the Message bar.

9 Tap Send, which is blue if you're sending the message via iMessage or green if you're sending it via the cellular network. The Send status bar appears as the message is sent; when the process is complete, you hear the message sent sound and the status bar disappears.

To: Robert the Bruce Sir William Wallace (+)

iMessage — **7**

I'm finishing up Ch 11 tomorrow. After this, I should have time to fly. — **8**

9

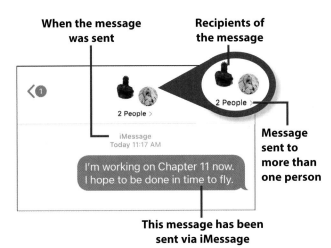

When the message was sent

Recipients of the message

2 People >

iMessage
Today 11:17 AM

I'm working on Chapter 11 now. I hope to be done in time to fly.

Message sent to more than one person

This message has been sent via iMessage

If the message is addressed to iMessage recipients, your message appears in a blue bubble in a section labeled iMessage. If the person to whom you sent the message has enabled his read receipt setting, you see when he reads your message.

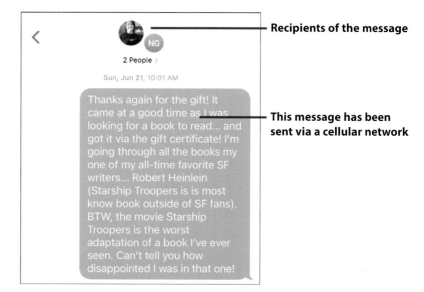

Recipients of the message

This message has been sent via a cellular network

If you sent the message via the cellular network instead of iMessage, you see your message in a green bubble in a section labeled Text Message.

When you send a message to people you haven't texted before, you see a new conversation screen. If you've texted them before, you move back to the existing conversation screen and your new message is added to that conversation instead.

In a group conservation, you see icons and the names of each person who received the message; if there are more than two or three people, the icons change size and move around to display as many as possible. (If you sent a message via a cellular text, you only see the person's icon with initials or number.)

>>>Go Further

TEXT ON

Following are some additional points to help you take your texting to the next level (where is the next level, anyway?):

- **Group messaging**—If you've enabled Group Messaging in the Messages setting, when you include more than one recipient, any messages sent in reply are grouped in one conversation. If this setting isn't enabled, each reply to your message appears in a separate conversation.

- **Larger keyboard**—Like other areas where you type, you can rotate the iPhone to be horizontal so the keyboard is larger as is each key. This can make texting easier, faster, and more accurate.

- **Recents**—When you enter To information for a new message, included on the list of potential recipients are people Siri suggests to you. When a suggested recipient has the Info (i) option, tap Info, tap Ignore Contact, and then tap Ignore at the prompt to prevent that person from being suggested in the future.

Sending Messages in an Existing Conversation

As you learned earlier, when you send a message to or receive a message from one or more people, a conversation is created. You can add new messages to an existing conversation as follows:

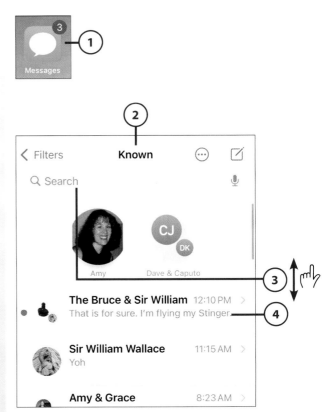

(**1**) On the Home screen, tap Messages.

(**2**) Move to a conversations screen, such as the Known screen. You see a list of conversations in the group you selected. The conversation containing the most recent message you've sent or received is at the top; conversations get "older" as you move down the screen.

(**3**) Swipe up or down the screen or tap in the Search bar and type names, numbers, or email addresses to find the conversation to which you want to add a message.

(**4**) Tap the conversation to which you want to add a message. At the top of the screen, you see the people involved in the conversation. Under that, you see the current messages in the conversation.

(5) Tap in the Message bar.

(6) Type your message.

(7) Tap Send. Your new message is added to the conversation and sent to everyone participating in the conversation.

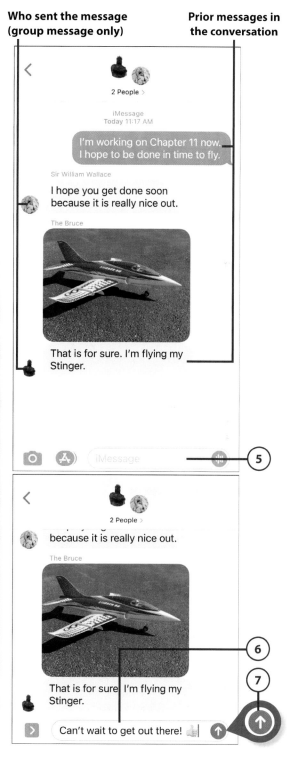

Who sent the message (group message only)

Prior messages in the conversation

Including Photos or Video You've Recently Taken in Messages You Send

It's easy to use Messages to quickly send photos or video you've taken recently as you see in the following steps:

1. Move into the conversation with the person or people to whom you want to send a photo, or start a new conversation.

2. Tap Photos. In the Photos pane, you see the photos and video you've taken recently.

3. Swipe to the left or right on the photo panel to browse the recent photos and videos.

4. Tap the first photo or video you want to send. It's marked with a check mark and is added to the message you're sending.

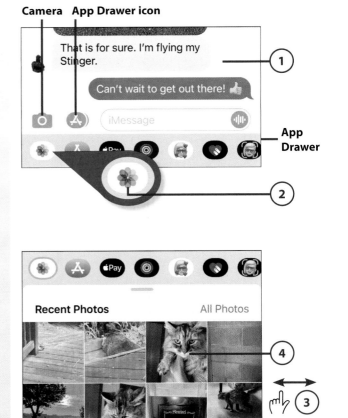

Camera   App Drawer icon

That is for sure. I'm flying my Stinger.

Can't wait to get out there! 👍

iMessage

App Drawer

Recent Photos   All Photos

5 Swipe left or right to review more recent photos and videos.

6 Tap the next photo or video you want to send. It's also marked with a check mark and added to the message.

7 Repeat steps 5 and 6 until you've added all the photos and videos you want to send.

8 Tap in the Message bar.

9 Type the comments you want to send with the photos or videos.

10 Tap Send. The photos, videos, and comments are added to the conversation.

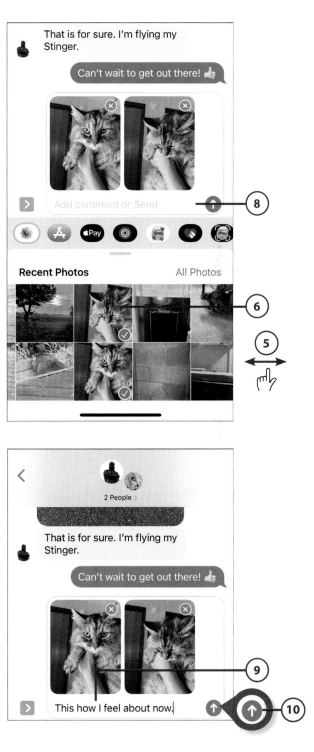

Including Photos or Video Stored in the Photos App in Messages You Send

You can add any image, photo, or video stored in Photos to a conversation by performing the following steps:

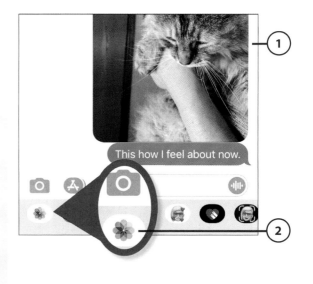

(1) Move into the conversation with the person or people to whom you want to send a photo or video, or start a new conversation.

(2) Tap the Photos icon.

(3) Tap All Photos.

4 Select the source containing the photo or video you want to send.

5 Swipe up or down the screen to find the photos or videos you want to send. (For more information about using the Photos app to find and select photos, see Chapter 16, "Viewing and Editing Photos and Video with the Photos App.")

6 Tap the photo or video you want to send.

7 Tap Add.

8 Tap in the Message bar and type the message you want to send with the photo or video.

9 Tap Send.

Commentary Not Required
You can send photos and other content without commentary. Just add the content and tap Send.

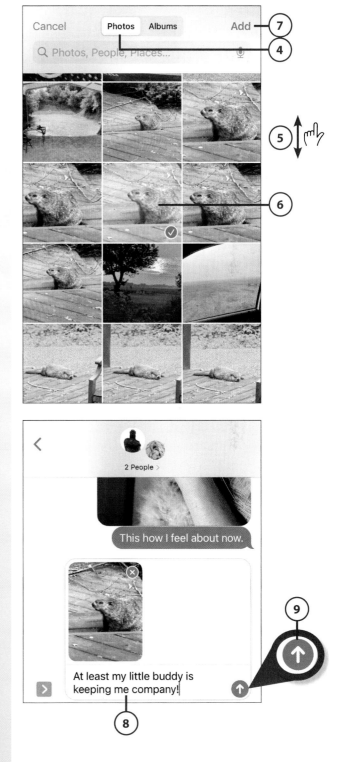

Taking Photos or Video and Sending Them in Messages

You can capture new photos and video using the iPhone's cameras and immediately send them to others via messages as follows:

1. Move into the conversation with the person or people to whom you want to send a photo or video, or start a new conversation.

2. Tap Camera.

3. Use the Camera app to take the photo or video you want to send. (For more information about taking photos or videos, see Chapter 15, "Taking Photos and Video with Your iPhone.")

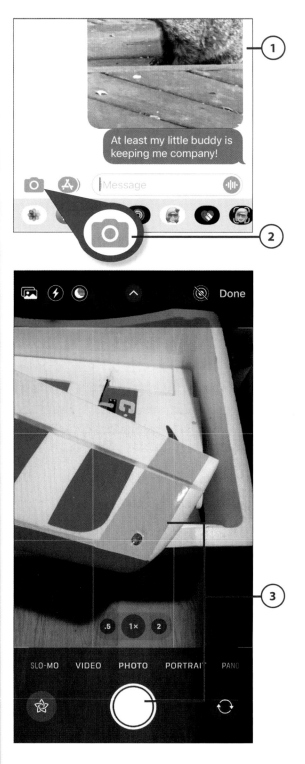

(4) To send the photo or video without comment, tap Done and skip the rest of these steps.

(5) To include a comment with the photo or video, tap Send.

(6) Tap in the Message bar and type the message, if any, you want to send with the photo.

(7) Tap Send. The photo (or video) and message are added to the conversation.

>>>*Go Further*

MORE TEXTING TIDBITS

Here are some other items that will help you with Messages:

- **Don't See the Photos or Camera Icons?**—If you don't see the Photos or Camera icons, tap the right-facing arrow that appears next to the Message bar. The App Drawer opens and you see the icons.

- **Improving Photos**—When you take a photo within Messages, you can use the Effects, Edit, and Markup tools to make changes to the photo before you send it. See Chapter 16 for information about these tools.

- **Problems Sending**—If a message you try to send is undeliverable or has some other problem, it's marked with an error icon (an exclamation point inside a red circle). The most common cause of this issue is a poor Internet connection. Or you might be trying to send something that is too large, such as a video file. Tap the error icon and then tap Try Again to attempt to resend the message. If it doesn't work again, you might need to wait until you have a better connection or shorten the video before you send it.

- **Sharing with Messages**—You can share all sorts of information via Messages from many apps, such as Photos, Safari, Contacts, Maps, and so on. From the app containing the information you want to share, tap Share. Then tap Messages; the information with which you are working is automatically added to a new message. Or tap a Siri suggestion from a recent conversation, and the content is added to that conversation. Use the Messages app to complete and send the message.

Adding Other Kinds of Content to Your Messages

You've seen how easy it is to send text messages, photos, or videos to others. With Messages enabled, you can send many other types of content too; you can also also enhance your messages with effects.

When you send photos stored in the Photos app, you are actually accessing the Photos app's tools within Messages. You can access many other apps within Messages in similar ways.

Tap to open or close the App Drawer

Photos

App Drawer

Apps within Messages are available on the App Drawer. You can open and close this by tapping the stylized "A" icon just to the right of the Camera icon. (If you don't see this, tap the right-facing arrow just to the left of the Message bar.) On the App Drawer, you see the apps you can work with; you've already learned how to work with the Photos app.

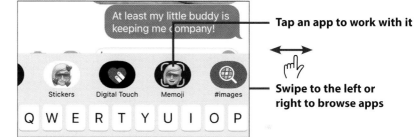

Tap an app to work with it

Swipe to the left or right to browse apps

When you touch the App Drawer, it enlarges to make the icons easier to see. Swipe left or right to browse the apps. When you see the app you want to use, tap it.

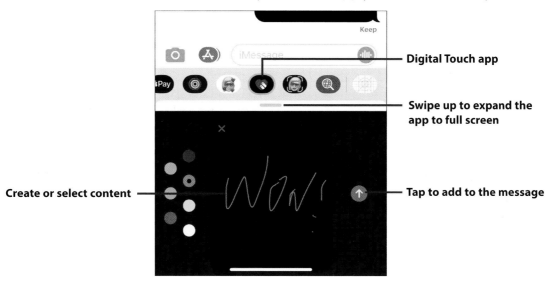

Digital Touch app

Swipe up to expand the app to full screen

Create or select content

Tap to add to the message

The selected app's tools open in the bottom of the window and you can use it to add content to your message. You can work with the app in the small window at the bottom of the screen or swipe up on the line just under the App Drawer to expand the app's window to full screen.

There are a number of apps provided by default. You can also download and add more apps for use within the Messages app. To get you started, following is a quick look at some of the default apps:

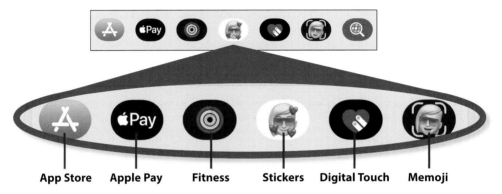

App Store   Apple Pay   Fitness   Stickers   Digital Touch   Memoji

- **App Store**—Use this app to add more apps to the App Drawer. This works very similarly to using the App Store app to download apps to your iPhone; see Chapter 5, "Customizing Your iPhone with Apps and Widgets," for the details. When you use this app, you are taken to apps designed to work within Messages. You can download and install them in the App Drawer with a few taps. You can use the apps you download just like the others already installed.

- **Apple Pay**—Use this to send money to others via the Messages app. Note that you can send money only when you're in a conversation with another individual; this doesn't work in a group conversation.

- **Fitness**—This app enables you to add various animated activity graphics to messages. For example, if you want to tell someone about a hike, you can add an animated hiker to the conversation.

- **Stickers**—This one provides static images, similar to emojis, that you can add to messages.

- **Digital Touch**—With this app, you can use your fingers to create content. You can use your fingers to "write" on the screen and you can choose from default touches, such as sending a fireball by pressing with one finger.

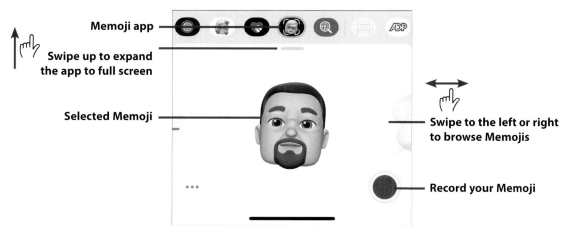

Memoji app

Swipe up to expand
the app to full screen

Selected Memoji

Swipe to the left or right
to browse Memojis

Record your Memoji

- **Memoji**—This one enables you to overlay images or recorded motion over your face. When you record them, they move with your facial movements; for example, if you smile, your Memoji smiles too. You can choose from default Memojis, and you can even create custom Memojis (tap Add [+] to get started). When you open the Memoji app, you choose the Memoji you want to use and then place your face within the yellow box. You can then record an animated message with sound that you can send via Messages.

No iOS 10 or later?

Recipients must be using Apple devices running iOS 10 or later, iPad OS, or Macs running macOS Sierra or later for much of the content you send to appear the same to the recipient as it does to you. If you send messages with effects and content to people not using a compatible device, the messages are delivered, but not the effects and additional content. Or the content might be static instead of having the motion you see.

In addition to using apps to send content, there are a couple of other options that might be of interest to you.

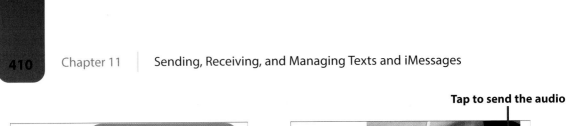

Tap to send the audio

Touch and hold to record
an audio message

Recording an audio message

You can record audio messages to send via Messages. Touch and hold on the Audio tool in the Message bar (this appears only when there is nothing in the Message bar). The audio recording palette opens and begins recording. Speak your message. When you're done, take your finger off the screen. You can play the message you want to send, and then send it.

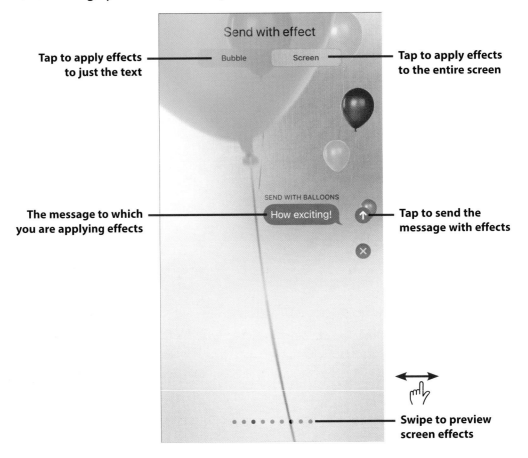

Tap to apply effects
to just the text

Tap to apply effects
to the entire screen

The message to which
you are applying effects

Tap to send the
message with effects

Swipe to preview
screen effects

You can add effects to your messages by touching and holding Send (which appears after you've added content in the Message bar). On the resulting screen, you can tap Bubble to apply effects to the bubble in which the message is sent. Or tap Screen to apply effects to the entire screen. With both options, you can preview the effects you're applying. When you're happy with the effect, tap Send to send the message. The recipients see the effects as soon as they view the message (assuming they are using a compatible device).

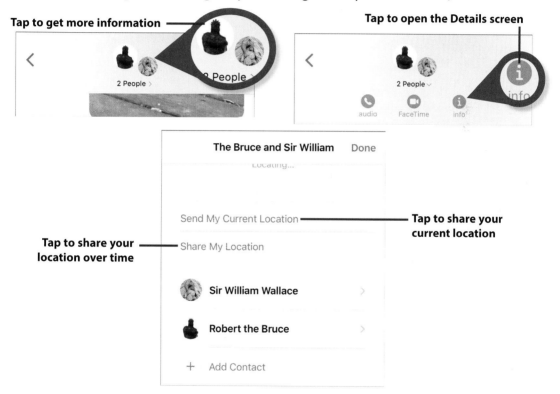

It can be useful to include location information in your messages. For example, if you're meeting someone at a location, you can quickly send your location via a message.

To share location information, tap the icons showing who is involved in the conversation at the top of the screen. Then tap Info (i). On the Details screen, tap Send My Current Location to share a one-time view of where you are. Or tap Share My Location and then choose a timeframe; recipients are able to track your location during the time you select.

When others send their locations to you, tap the message to open the location in the Maps app.

>>>*Go Further*

ORGANIZE THE APP DRAWER

You might want to organize the App Drawer to make using apps easier. Swipe all the way to the left on the App Drawer, and then tap More (…). You see the apps currently installed in two sections. FAVORITES shows the apps currently installed. MORE APPS shows apps that are available but not currently installed.

Tap Edit. Drag apps up or down to reorganize them on the FAVORITES list; apps higher on the list appear toward the left end of the App Drawer. Tap Unlock (–) and then tap Remove from Favorites to remove an app from the App Drawer; it moves to the MORE APPS list (you can add it again). Tap Add (+) to move an app from the MORE APPS list onto the FAVORITES list (and onto the App Drawer).

When you're done organizing the App Drawer, tap Done and then tap Done again.

Receiving, Reading, and Replying to Messages

Text messaging is about communication, so when you send messages you expect to receive responses. People can also send new messages to you. As you learned earlier, the Messages app keeps messages grouped as a conversation consisting of messages you send and replies you receive to the same person or group of people.

Receiving Messages

When you aren't currently using the Messages screen in the Messages app and receive a new message (as a new conversation or as a new message in an ongoing conversation), you see, hear, and feel the notifications you've configured for the Messages app. (Refer to Chapter 6 to configure your message notifications.)

If you're on the Messages screen in the Messages app when a new message comes in, you hear and feel the new message notification sound and/or vibration, but a notification doesn't appear. On the conversation list, any conversations containing a new message are marked with a blue circle showing the number of new messages in that conversation.

If a new message is from someone or a group with whom you've previously sent or received messages, the new message is appended to an ongoing conversation, and you see the prior messages in that conversation. That conversation then moves to the top of the list of conversations in the Messages app. If there isn't an existing message to or from the people involved in a new message, a new conversation is started and the message appears at the top of that list.

Prior Messages?

If you have manually deleted messages from a conversation or they were deleted automatically based on their age, you won't see them when you send a new message to someone or a group you've messaged with before.

Speaking of Texting

Using Siri to hear and speak text messages is extremely useful. Check out Chapter 13, "Working with Siri," for examples showing how you can take advantage of this great feature. One of the most useful Messages commands is to activate Siri and say, "Get new messages." Siri reads any new messages you have received.

Reading Messages

You can get to new messages you receive by doing any of the following:

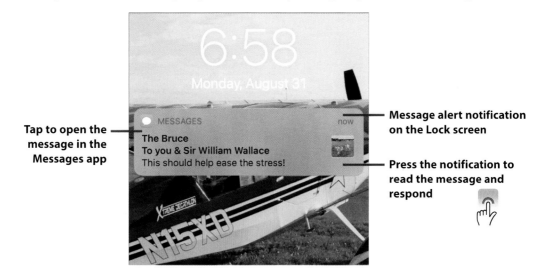

Tap to open the message in the Messages app

Message alert notification on the Lock screen

Press the notification to read the message and respond

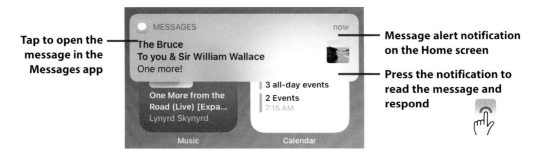

Tap to open the message in the Messages app

Message alert notification on the Home screen

Press the notification to read the message and respond

- Tap a banner alert notification from Messages to move into the message's conversation in the Messages app.

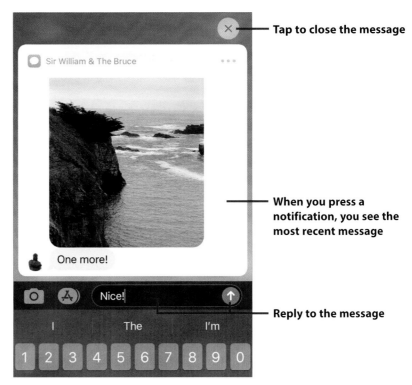

Tap to close the message

When you press a notification, you see the most recent message

Reply to the message

- Touch and hold on an alert to read and reply to the most recent message.

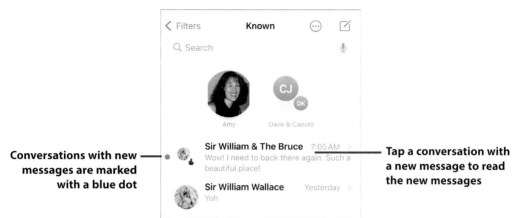

Conversations with new messages are marked with a blue dot

Tap a conversation with a new message to read the new messages

- Open the Messages app and tap a conversation containing new messages; these conversations appear at the top of the Messages list and are marked with a blue circle. The conversation opens and you see the new message.

- If you receive a new message in a conversation that you're currently viewing, you immediately see the new message.

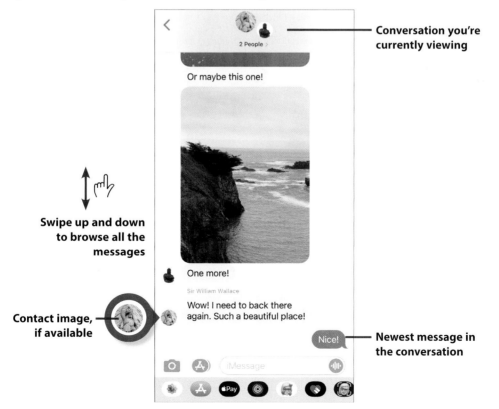

Conversation you're currently viewing

Swipe up and down to browse all the messages

Contact image, if available

Newest message in the conversation

However you get to a conversation, you see the new messages in either an existing conversation or a new conversation. The newest messages appear at the bottom of the screen. Swipe up and down the screen to see all the messages in the conversation. As you move up the screen, you move back in time.

Messages sent to you are on the left side of the screen and appear in gray bubbles. If the message is for a group, just above the bubble is the name of the person sending the message; if you have an image for the contact, that image appears next to his bubble.

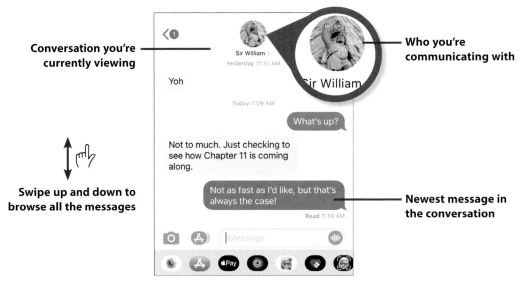

Conversation you're currently viewing

Who you're communicating with

Swipe up and down to browse all the messages

Newest message in the conversation

If the conversation is with just one other person, at the top of the screen you see her name and the image associated with her contact information or a circle with her initials if there isn't an image stored with her contact information. The color of the bubbles around your messages indicates how they were sent: blue indicates iMessages while green is used for cellular messages.

Viewing Photos or Video You Receive in Messages

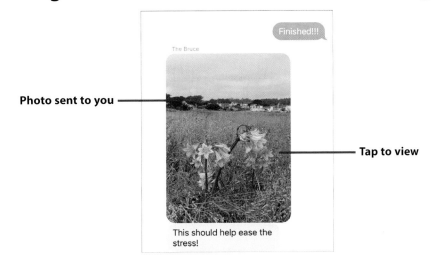

Photo sent to you ——

—— Tap to view

When you receive a photo or video in a message, it appears in a thumbnail along with the accompanying message.

Tap to return to the conversation ——

Number of photos in the conversation

Swipe to see other photos or videos in the conversation

Tap to bring up a list of recent photos in the conversation

Tap to share ——

To view a photo or video attachment, tap it. You see the photo or video at a larger size. You can rotate the phone, zoom, and swipe around the photo just like viewing photos in the Photos app (see Chapter 16 for details). You can watch a video in the same way, too.

Tap Share to share the photo with others via a message, email, tweet, Facebook, and so on.

If there is more than one photo or video in the conversation, you see the number of them at the top of the screen. Swipe to the left or right to move through the available photos.

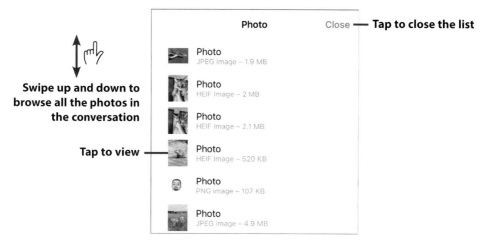

Swipe up and down to browse all the photos in the conversation

Tap to view

Tap to close the list

Tap List to see a list of the recent photos in the conversation (this only appears if there is more than one photo in the conversation).

Tap a photo on the list to view it. Tap Close to return to the photo you were viewing.

To move back to the conversation, tap Done.

The Bruce

This should help ease the stress!

Press to save a photo

Tap to save

Save

Reply

Copy

More...

To save a photo to your photo library in the Photos app, press on it. On the resulting menu, tap Save. The photo is saved to the Photos app and you can use it just like photos you take with the iPhone's cameras (see Chapter 16).

>>>Go Further

VIEWING OTHER TYPES OF CONTENT IN MESSAGES

Earlier, you learned about some of the content you can include in messages, such as Digital Touches, Memojis, and audio messages. Of course, other people who use an iOS or iPad OS device can send these types of content to you, too. Some of this content, for example Digital Touches and effects, plays automatically when you view the message with which it is sent. Other types of content have a Play icon you can tap to play that content. Some content (such as Digital Touches) plays only once but remains as a static image afterward, while other content (including effects) disappears after it plays.

Replying to Messages from the Messages App

To reply to a message, read the message and do the following:

① Read, watch, or listen to the most recent message.

② Use the photos, camera, or other tools to reply with content of that type as you learned about earlier in this chapter.

③ Tap in the Message bar if you want to reply with text.

④ Type your reply or use the Dictation feature to speak your reply.

⑤ Tap Send.

Replying to Messages from a Banner Alert

If you have banner alerts configured for your messages, you can reply directly from the alert from either the Home or Lock screens:

① Press on the notification. The conversation opens.

② Type your reply or use the other tools to add content to your response.

③ Tap Send. Your message is added to the conversation.

Having a Messages Conversation

Messaging is all about the back-and-forth communication with one or more people. You've already learned the skills you need, so put them all together. You can start a new conversation by sending a message to one or more people with whom you don't have an ongoing conversation; or you can add to a conversation already underway.

(1) Send a new message to a person or add a new message to an existing conversation. You see when your message has been delivered. If you sent the message to an individual person via iMessage and he has enabled his Read Receipt setting, you see when he has read your message, and you see a bubble as he is composing a response. (If you're conversing with more than one person, the person doesn't have her Read Receipt setting enabled, or the conversation is happening via the cellular network, you don't see either of these.)

As the recipient composes a response, you see a bubble on the screen where the new message will appear when you receive it (again, only if it's an iMessage with a single individual). Of course, you don't have to remain on the conversation's screen waiting for a response. You can move to a different conversation or a different app. When the response comes in, you're notified per your notification settings.

A new message, ready to send

Your message has been delivered

Your message has been read

The recipient is composing a response

2. Read the response.

3. Write and send your next message.

4. Repeat these steps as long as you want. Conversations remain in the Messages app until you remove them. Messages within conversations remain forever (unless you delete them, for one year, or for 30 days depending on your Keep Messages setting as shown earlier in this chapter).

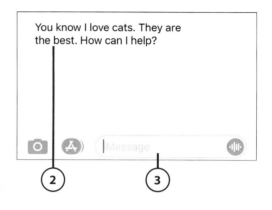

The iMessage Will Be With You...Everywhere

Messages that are sent with iMessage move with you from device to device, so they appear on every device configured to use your iMessage account. Because of this, you can start a conversation on your iPhone while you're on the move and pick it up on your iPad or Mac later.

Working with Messages

As you send and receive messages, the interaction you have with each person or group becomes a separate conversation. A conversation consists of all the messages that have gone back and forth. You manage your conversations from the Messages screen.

Multiple Conversations with the Same People

The Messages app manages conversations based on the phone number or email address associated with the messages in that conversation rather than the people (contacts) involved in the conversation. So, you might have multiple conversations with the same person if that person used a different means, such as a phone number and an email address, to send messages to you.

Managing Messages Conversations

Use the Messages screen to manage your messages.

(1) On the Home screen, tap Messages.

The Messages screen showing conversations you have going appears, or you move into the conversation you were most recently using (tap the left-facing arrow at the top of the screen to move back to the conversation list). Conversations containing new messages appear at the top of the list. The name of the conversation is the name of the person or people associated with it, or it might be labeled as Group if the app can't display the names. If a contact can't be associated with the person, you see the phone number or email address you are conversing with instead of a name.

(2) Swipe up and down the list to see all the conversations.

(3) Tap a conversation you want to read or reply to. The conversation screen appears. If it's a group conversation, you see the number of people involved and an icon for each at the top of the screen. If there is only one other person involved, you see his name and image (if you have an image for the contact, if you don't have an image, you see a bubble with his initials) at the top of the screen instead. Note that if you use one of the smaller models, such as the iPhone SE, the images won't appear as described and shown; you'll just see people's names or conversation labels.

If you have the badge enabled, you see the number of new messages on the Messages icon

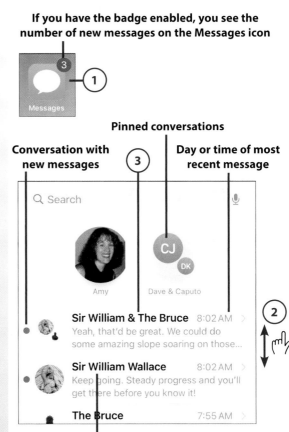

Conversation with new messages

Pinned conversations

Day or time of most recent message

Who is involved in the conversation

4 Read the new messages or view other new content in the conversation. Your messages are on the right side of the screen in green (cell network) or blue (iMessage), whereas the other people's messages are on the left in gray. Messages are organized so the newest message is at the bottom of the screen.

5 Swipe up and down the conversation screen to see all the messages it contains.

6 To add a new message to the conversation, tap in the Message bar, type your message or use an app to add content, and tap Send.

7 Swipe down to scroll up the screen and move back in time in the conversation.

8 To see details about the conversation, tap X People, where X is the number of people in the conversation or the person's name if there is only one other person involved.

9 Tap audio to place a FaceTime audio call to the person or people involved in the conversation (refer to Chapter 9, "Communicating with the Phone and FaceTime Apps," for details).

10 Tap FaceTime to place a FaceTime call to the person or people involved in the conversation (refer to Chapter 9).

11 Tap Info (i). The Details screen appears.

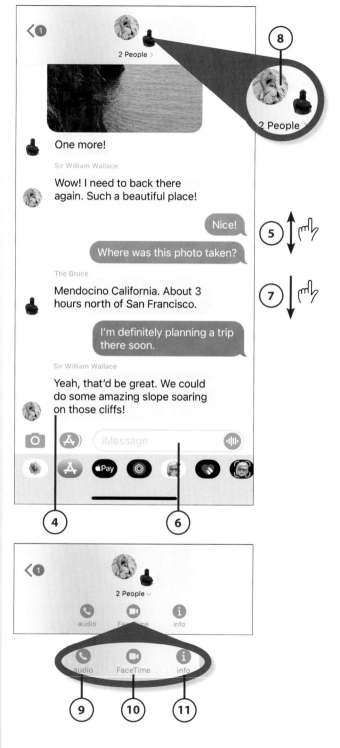

(12) Swipe up and down the Details
screen. On this screen, you can see
a variety of information about the
conversation, can access some of its
content, and perform a variety of
actions, which include the following:

- **Name a Group**—You can name
  a group conversation by tapping
  Change Name and Photo, which
  is located at the top of the Details
  screen just under the icons for
  the participants. On the resulting
  sheet, you can name the conversation, such as giving a group of
  people with whom you communicate a recognizable name, or
  change a person's image.

- **Send Location Information**—
  Tap the location options to send
  your current location or to share
  your location for a specific period
  of time.

- **Work with Participants**—You
  can tap people to see their contact information or tap one of the
  icons to communicate with them
  individually. Tap Add Contact to
  add others to the conversation.

- **Hide Alerts**—Set Hide Alerts
  to on (green) to prevent
  notifications associated with the
  conversation.

- **Exit the Conversation**—Tap
  Leave this Conversation to with-
  draw from a Messages conversation (you can't do this if you started the conversation). You won't
  receive any future messages.

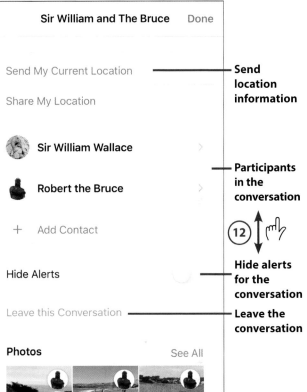

Sir William and The Bruce   Done

Send My Current Location — **Send location information**

Share My Location

Sir William Wallace >

Robert the Bruce > — **Participants in the conversation**

+ Add Contact — **(12)**

Hide Alerts — **Hide alerts for the conversation**

Leave this Conversation — **Leave the conversation**

Photos   See All

- **View Photos**—In the Photos section, you see the photos recently included in the conversation. Tap See All to see all of the photos in the conversation.

- **View Locations**—In the LOCATIONS section at the bottom of the screen, you see location information that has been included in the conversation. Tap a location to see it in the Maps app.

(13) Tap Done to return to the conversation.

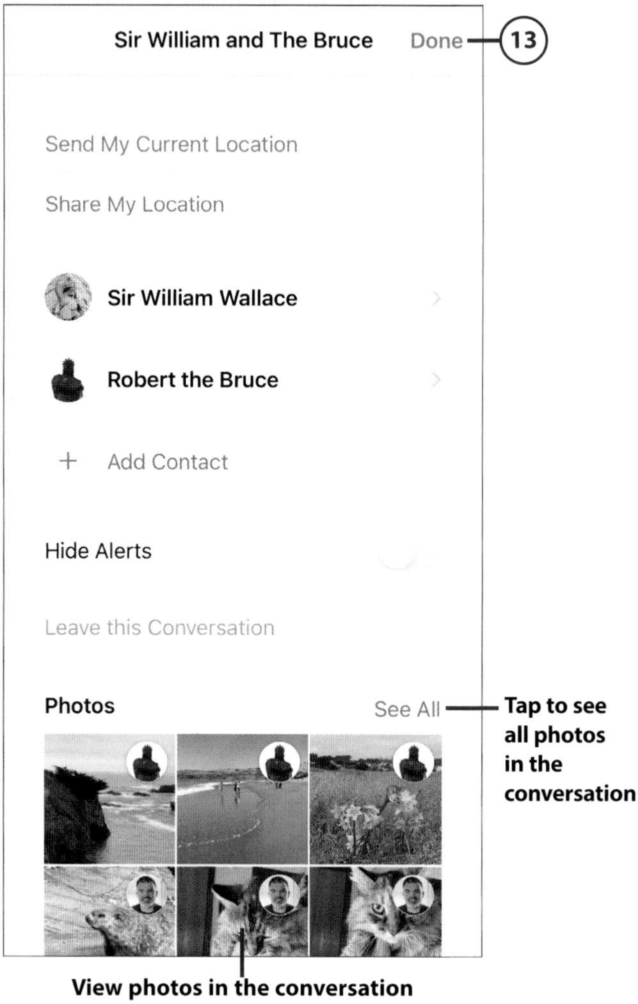

| | Sir William and The Bruce | Done ——(13) |

Send My Current Location

Share My Location

Sir William Wallace   >

Robert the Bruce   >

+   Add Contact

Hide Alerts

Leave this Conversation

Photos   See All ——— **Tap to see all photos in the conversation**

View photos in the conversation

Pinning Conversations

Over time, you'll likely have people or groups of people with whom you communicate regularly. You can pin these conversations to the top of the screen so that they are always available for you.

(1) Move to a list of conversations, such as the Known conversation list.

(2) Tap Options (…).

(3) Tap Edit Pins.

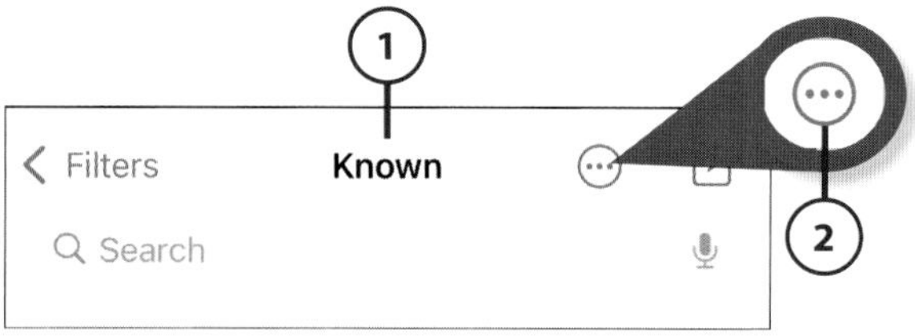

(1)
< Filters   Known   …   (2)
Q Search   🎤

< Filters   Known   …   ✏️
Q Se   Select Messages   ⊘   🎤
Edit Pins   📌 ——(3)
Edit Name and Photo   ⊙

4 Tap a pin to pin the associated conversation. That conversation moves to the top of the screen and becomes pinned there.

5 To unpin a conversation, tap remove (–). The conversation moves back to the list where it is still available for you.

6 Tap Done when you are done pinning or unpinning conversations. You see the pinned conversations at the top of the screen. As new messages come in, the conversations are marked with the blue circle, and you see a preview of the most recent message.

7 Tap a pinned conversation to work with it.

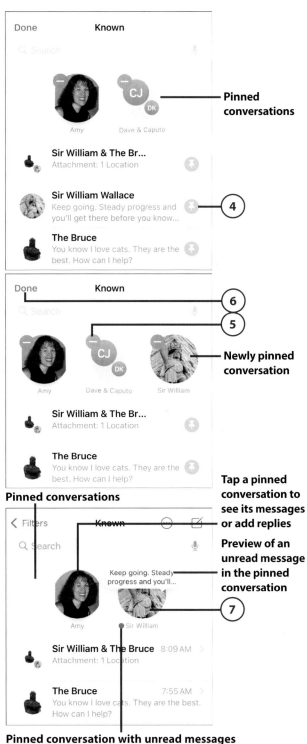

Pinned conversations

Newly pinned conversation

Pinned conversations

Tap a pinned conversation to see its messages or add replies

Preview of an unread message in the pinned conversation

Pinned conversation with unread messages

>>>Go Further

TELL ME MORE

Messages has lots of great tools you can use to communicate with others and be informed about important events. Here are a few more pointers:

- **Whose Got the Time?**—To see the time or date associated with every message in the conversation being displayed, swipe to the left and hold your finger down on the screen. The messages shift to the left and the time of each message appears along the right side of the screen. The date associated with each message appears right before the first message on that date.

- **Information Messages**—Many organizations use messages to keep you informed. Examples include codes you need to authenticate your credentials when you sign into an account, airlines that send flight status information, retailers that use messages to keep you informed about shipping, and and health care offices that confirm appointments. These messages are identified by a set of numbers that don't look like a phone number. You can't send a response to most of these messages; they're one-way only. In some cases, you can issue commands related to the texts from that organization, such as "Stop" to stop further texts from being sent.

- **Disconnected?**—My Acquisitions Editor Extraordinaire pointed out that people can't receive messages when their phones aren't connected to the Internet (via Wi-Fi or cell). You don't see a warning in this case; you can only tell the message wasn't delivered because the Delivered status doesn't appear under the message. The message is delivered as soon as the other person's phone is connected to the Internet again, and its status is updated accordingly on your phone. Also, when an iMessage can't be delivered, you can tap and hold on it; then tap Send as Text Message. The app tries to send the message via the cellular network instead of iMessage.

Viewing Photos in Conversations

As photos are added to a conversation, they're collected so you can browse and view them at any time:

1. Move to the conversation in which you want to browse attachments.

2. Tap the participants at the top of the screen.

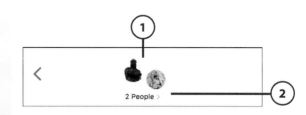

3. Tap info (i).

4. Swipe up until you see Photos.

5. Tap one of the recent photos to view it and skip to step 9.

6. Tap See All. You see all the photos included in the conversation.

7. Swipe up and down to browse the photos.

8. Tap a photo you want to view.

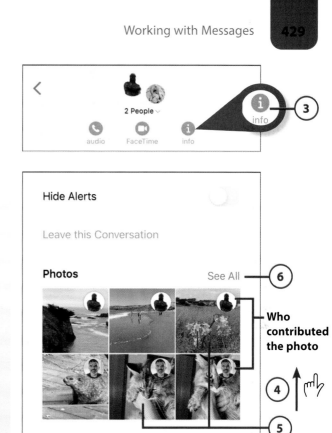

Who contributed the photo

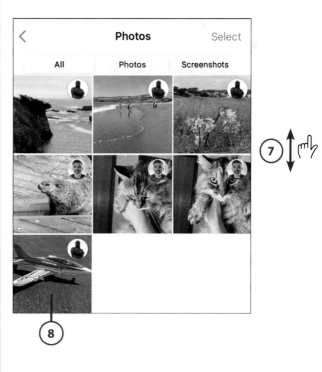

(9) View the photo.

(10) Swipe to the left or right to view other photos in the conversation.

(11) To save the photo in the Photos app, tap Share; to just view the photos, skip to step 13.

(12) Tap Save Image and skip the rest of these steps. The image is saved to your Photos app, and you move back to the conversation.

(13) Tap Photos (<) when you're done viewing the photos.

(14) Tap back (<) to return to the Details screen.

(15) Tap Done to return to the conversation.

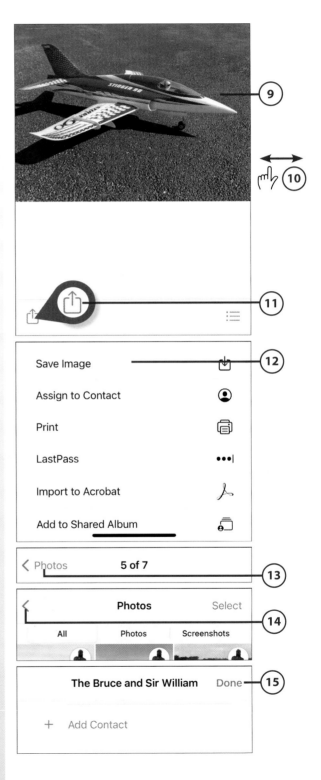

Working with Messages from People You Might Not Know

As you use Messages, it's likely you'll receive messages from people who aren't in your contacts or whom you haven't sent messages to before. Some of these will be legitimate contacts with whom you haven't messaged previously, whereas other messages will come from a person who made a mistake, such as not typing the phone number correctly; others might be sent for nefarious purposes. The Messages app can filter messages from unknown people for you and put them on the Unknown Senders list so you can easily identify messages that are from people whom you might not know. You won't receive any notifications for messages that end up on the Unknown Senders list.

To use this functionality, the Filter Unknown Senders switch on the Messages Settings screen must be turned on (green). (The details of configuring Messages settings are provided in "Preparing the Messages App for Messaging" earlier in this chapter.) When this switch is on, the Unknown Senders list contains conversations from people who aren't contacts or with whom you haven't communicated previously.

Unknown But Useful

Most of the messages you receive from organizations for things such as shipping updates or confirmation codes end up on the Unknown Senders list. You should review this list regularly to make sure you aren't missing important information.

Too Much Bother?

If dealing with the Unknown Senders list is more trouble than it is worth, you can view the All Messages list, which shows you all messages regardless of whether the sender is known or unknown. You still won't receive notifications for messages from unknown senders, but at least you can see all the messages on one screen.

You can work with "suspicious" messages as follows:

(1)   Open the Messages app and move to the Messages screen (tap back [<] from a conversation list to move here).

(2)   Tap Unknown Senders. You move to the Unknown contacts conversation list. You can work with conversations on this list just as you can with the Known Senders list.

(3)   Swipe up and down to review the conversations on the list.

(4)   Tap a message to read it so you can determine if it's a legitimate message for you.

(5)   Review the identification of the sender as best Messages can determine it, such as an email address.

(6)   Read the message.

(7)   If the message is from someone you don't recognize and don't want to receive future messages from, tap Report Junk. The sender's message is deleted from the Messages app and the associated email address or phone number is blocked so you won't receive any more messages from the sender. Skip the rest of these steps.

(8)   If you want to take other action on the sender, such as to create a contact so he will become a known sender or block future messages, tap the label under the image at the top of the screen (which is a silhouette in most cases).

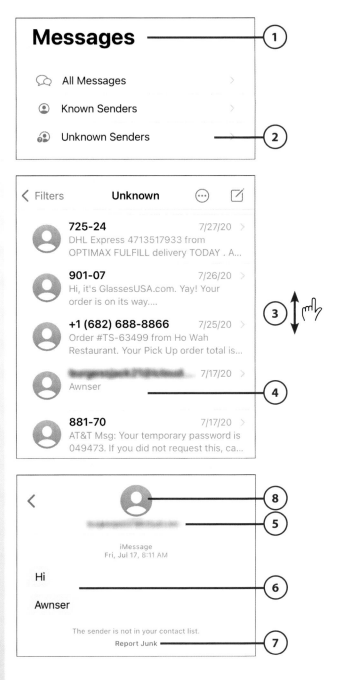

9 Tap info (i).

10 Tap info.

11 If you recognize the person, want to communicate with her, but don't have a contact configured for her, tap Create New Contact to create a new contact for the person. (For information about creating or updating contacts, see Chapter 8.) The conversation moves to the Known Senders list and you receive notifications for future messages.

12 If you recognize the person, want to communicate with her, and have a contact configured for her, tap Add to Existing Contact to update a current contact with new information. The conversation moves to the Known Senders list and you receive notifications for future messages.

13 If you don't want to receive future messages from the sender, tap Block this Caller.

14 Tap Block Contact. The contact is blocked and the conversation disappears.

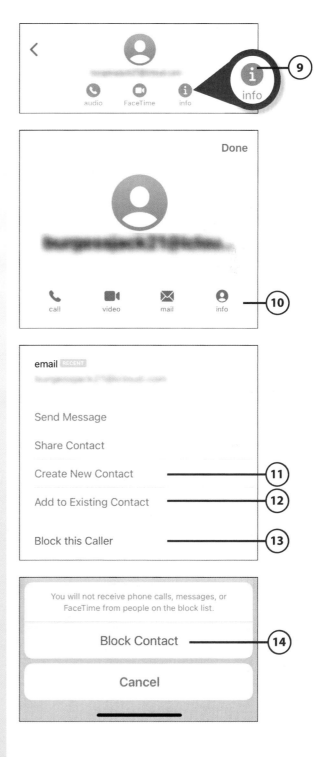

Junk or Block?

When you report a message as junk, the contact is blocked from communicating with you, but their information is not stored on your iPhone. When you block the contact instead, messages don't reach you, but the contact information is stored on the list of people you have blocked. You can unblock people from the Blocked Contacts list in the Messages setting screen if you want to resume communicating with them. You can't resume communicating with someone whose messages you have marked as junk.

No Notifications

Remember that you won't receive notifications for messages from unknown senders. So, it's a good idea to review the Unknown Senders list periodically to make sure you aren't missing messages that you don't want to miss. Messages from anyone who isn't a contact or with whom you haven't communicated go to this list automatically. It can be easy to miss these messages because you don't receive notifications when they come in.

Responding to a Message with Icons

You can quickly respond to messages with an icon as follows:

(1) View a conversation.

(2) Press and hold on a message you want to respond to. You can respond to any message, even if it's one you sent.

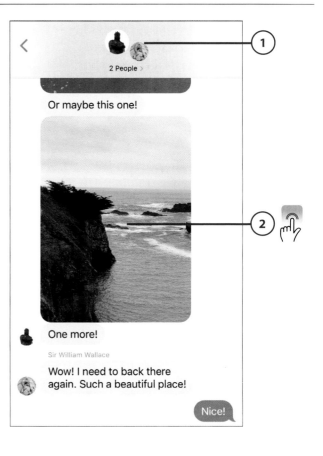

3) Tap the icon you want to add to the message; for example, tap the thumbs-up icon to indicate you like a message. The icon you select is added to the message. Others in the conversation receive notifications about what happened. For example, if you added a thumbs-up to a message, others receive a notification indicating that you liked it.

Your input on the message

Or maybe this one!

Deleting Messages and Conversations

Old text conversations never die, nor do they fade away (if you have your Keep Messages setting set to Forever, that is). All the messages that you receive from a person or that involve the same group of people stay in the conversation. Over time, you can build up a lot of messages in one conversation, and you can end up with lots of conversations. (If you set the Keep Messages setting to be 30 Days or 1 Year, messages older than the time you set are deleted automatically.)

Long Conversation?

When a conversation gets very long, the Messages app might not display all its messages. It keeps the more current messages visible on the conversation screen. To see earlier messages, swipe down on the screen to move to the top and earlier messages appear.

When a conversation gets too long, if you just want to remove specific messages from a conversation, or, if you want to get rid of messages to free up storage space, take these steps:

1. Move to a conversation containing an abundance of messages.

2. Press and hold on a message you want to delete.

3. Tap More. The message on which you tapped is marked with a check mark to show it's selected.

④ Tap other messages you want to delete. They're marked with a check mark to show you have selected them.

⑤ Tap Trash.

⑥ Tap Delete *X* Messages, where *X* is the number of messages you have selected. The messages are deleted and you return to the conversation.

Tap to delete all the messages in a conversation

Delete All   2 People ›   Cancel

Nice!

Where was this photo taken?

The Bruce

Mendocino California. About 3 hours north of San Francisco.

I'm definitely planning a trip there soon. ④

Sir William Wallace

Yeah, that'd be great. We could do some amazing slope soaring on those cliffs!

The Bruce

Location from 8/31/20   ›

⑤

⑥

Delete 2 Messages

Cancel

>>>Go Further

MORE ON CONVERSATIONS

There are a few more things you should know about conversations:

- **Delete Them All**—To delete all the messages in the conversation, instead of performing step 4, tap Delete All, which appears in the upper-left corner of the screen. Tap Delete Conversation in the confirmation box. The conversation and all its messages are deleted.

- **Copy Messages**—You can copy a message by pressing on it. On the resulting menu, tap Copy. You can paste the copied message in other messages or in other apps.

- **Reply to a Specific Message**—You can reply to a specific message in a conversation by pressing on it and tapping Reply on the resulting menu. After you send your reply, it is added immediately under the message you replied to, and your reply is indicated by the Reply link. Tap the Reply link to read the replies. The original message and its replies fill the screen. You can work with these just like the messages in a conversation. Tap the reply screen to return to the conversation.

- **Pass It On**—If you want to send one or more messages to someone else, perform steps 1–4 to select the messages you want to forward to someone else. Tap Share (curved arrow) that appears in the lower-right corner of the screen. A new message is created, and the messages you selected are pasted into it. Select or enter the recipients to whom you want to send the messages, and tap Send.

Deleting Conversations

If a conversation's time has come, you can delete it.

(1) Move to the Messages screen.

(2) Swipe to the left on the conversation you want to delete.

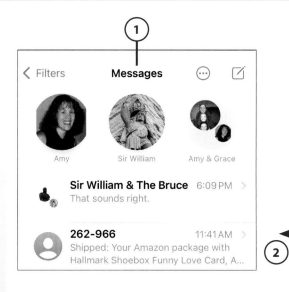

③ Tap Trash.

④ Tap Delete at the prompt
(not shown on the figure).
The conversation and all the
messages it contains are deleted.

	Sir William & The Bruce 6:09 PM >
	That sounds right.

	11:41 AM >		
	Your Amazon package with	🔕	🗑 |—③
	Shoebox Funny Love Card, A...		

| | The Bruce | 7:55 AM > |

Gone, but Not Forgotten?

When you delete an iMessage conversation, it's removed from your iPhone and other devices.
However, if you enable Messages for your iCloud account, the conversation remains in the cloud.
If you send a message to the same person or people who were on the conversation you deleted,
it's restored to your iPhone. (If your messages are set to be kept for 30 days or 1 year, they are
automatically deleted after that amount of time passes and so won't reappear in conversations.)

>>>Go Further

TEXTING LINGO

People frequently use shorthand when they text. Here is some of the more common shorthand
you might see. This list is extremely short, but there are many websites dedicated to providing
this type of information if you're interested. One that boasts of being the largest list of text
message acronyms is www.netlingo.com/acronyms.php.

- FWIW—For What It's Worth
- LOL—Laughing Out Loud
- ROTFL—Rolling On the Floor Laughing
- CU—See You (later)
- PO—Peace Out
- IMHO—In My Humble Opinion
- TY—Thank You
- RU—Are You
- BRB—Be Right Back
- CM—Call Me
- DND—Do Not Disturb
- EOM—End of Message

- FSR—For Some Reason
- G2G—Got to Go
- ICYMI—In Case You Missed It
- IDK—I Don't Know
- IKR—I Know, Right?
- ILU—I Love You
- NM or NVM—Never Mind
- OMG—Oh My God
- OTP—On the Phone
- P911—Parent Alert
- PLZ—Please

Go here to figure out where you're supposed to be and when to be there

In this chapter, you explore calendar functionality your iPhone has to offer. Topics include the following:

→ Getting started
→ Viewing calendars and events
→ Navigating calendars
→ Adding events to calendars
→ Searching calendars
→ Sharing calendars
→ Managing calendars and events

Managing Calendars

When it comes to time management, your iPhone is definitely your friend. Using the iPhone's Calendar app, you can view calendars that are available on all your devices. You can also make changes to your calendars on your iPhone, and they appear on the calendars on your other devices so you have consistent information no matter which device you happen to be using at any time.

Getting Started

The Calendar app lets you record meetings, doctor appointments, dinner reservations, and similar events—and reminds you when those times and dates are approaching. This app has lots of features designed to help you work with multiple calendars and accounts, manage events that other people are invited to, and more. You don't have to use all these features, and you might just want to use the Calendar app's core functionality, such as to record doctor appointments, dinner reservations, and similar events for which it is important to know the time and date (and be reminded when those times and dates are approaching).

To make your calendars accessible on multiple devices, store your calendars on the Internet cloud, using iCloud, Google, or similar service. If you don't have an online account, go to Chapter 4, "Setting Up and Using an Apple ID, iCloud, and Other Online Accounts," and set up at least one account to work with calendar information. When that's done, you're ready to continue in this chapter.

When you set up an online account for calendars, you can use multiple calendars at the same time. For example, you can use one calendar for personal events and another for club activities. You can work with all your events at the same time while displaying events in different colors to distinguish the calendar they are on.

You also can show or hide different calendars. You might want to see only your personal calendar, so you can hide other calendars to focus on it.

Speaking of online accounts, you can enable calendars on multiple accounts at the same time. For example, you can have a set of calendars managed in iCloud and another set stored in Google. It's easy to add events to your calendar, and you can use the app to search for specific events so you can find them quickly.

It can also be useful to share calendars. For example, you can create a calendar to share with a group of friends so that everyone can see that calendar on their devices.

Viewing Calendars and Events

When you use the Calendar app to view and work with your calendars, you can choose how you view them, such as by month, week, or day. You can also view a list of your calendar events.

To get into your calendars, move to the Home screen and tap the Calendar app (which shows the current day and date in its icon). The most recent screen you were viewing appears.

What's the Meaning of This?

The Calendar badge (the red circle with a number in it) appears when you've been invited to at least one event; after you respond, the counter in the badge decreases by 1 and disappears entirely if you don't have any new invitations. You learn about invitations in the Go Further sidebar, "Invitations," later in this chapter.

Most of the time, you'll be looking at your calendars in various views, such as showing a month, week, day, or event. You can easily get more information about your events, set alarms, and update event details.

Configuring Calendars

You can configure the calendars you see within the Calendar app. For example, you might want to see only your personal calendar or you might want to see all of your calendars at the same time. You also can change certain aspects of a calendar, such as the color with which its events are highlighted.

To configure the calendar information you see in the app, perform the following steps:

1 On a Home screen, tap the Calendar app or widget.

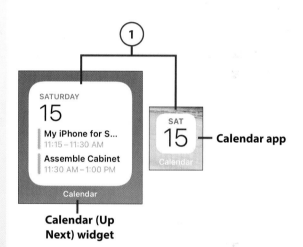

Calendar app

Calendar (Up Next) widget

Widget?

You can add widgets directly to the Home screens. I find the Calendar widget particularly useful. There are several versions of this widget. The smallest shows only your upcoming events. Another shows those events plus a month-at-a-glance calendar. The third shows a larger list of events on your calendars. Tapping one of these widgets takes you into the Calendar app just like tapping the app's icon does. To learn how to configure widgets on your Home screens, see "Customizing How Your iPhone Works with Widgets" in Chapter 5, "Customizing Your iPhone with Apps and Widgets."

(2) If you have only one calendar, skip to step 3. Otherwise, tap Calendars to see all your calendars. If you don't see this at the bottom of the screen, you're already on the Calendars screen (look for "Calendars" at the top of the screen) or you are on the Inbox; in that case, tap Done and then tap Calendars.

The Calendars screen displays the calendars available, organized by the account from which they come, such as ICLOUD or GMAIL (the names you see are the descriptions of the accounts you created on your iPhone). By default, all your calendars are displayed, which is indicated by the circles with check marks next to the calendars' names. Any calendars that have an empty circle next to their names are not displayed when you view your calendars.

(3) Tap a calendar with a check mark to hide it. The check mark disappears and the calendar is hidden. (The calendar is still available in the app, you just don't see it when you're viewing your calendars.)

(4) To show a calendar again, tap its name. It's marked with a check mark and appears when you're viewing calendars.

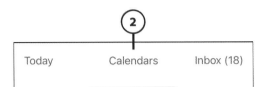

Calendars without check marks are hidden

Calendars with check marks are displayed

5 Tap Info (i) to see or change a calendar's settings. Not all types of calendars support this function, and those that do can offer different settings. The following steps show an iCloud calendar; if you're working with a calendar of a different type, such as a Google calendar, you might not have all of the same options as those shown here. In any case, the steps to make changes are similar across all available types of calendars.

6 Change the name of the calendar by tapping it and then making changes on the keyboard; when you're done making changes, swipe down the screen or tap Done to close the keyboard.

7 To share the calendar with someone, tap Add Person, enter the email address of the person with whom you're sharing it, and tap Add. (Sharing calendars is explained in more detail later in this chapter.)

8 If the calendar is shared and you don't want to be notified when shared events are changed, added, or deleted, set the Show Changes switch to off (white). When this switch is enabled (green) and someone makes a change to a shared calendar, you receive notifications about the changes that were made. If the calendar is not shared, you don't see this switch.

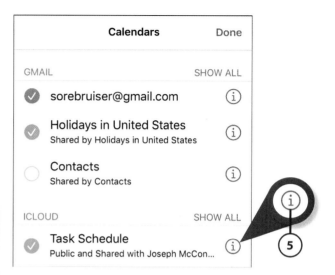

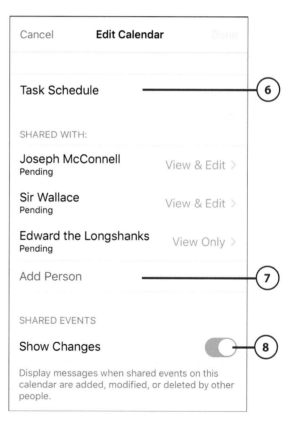

9 Swipe up the screen.

10 Tap the color you want events on the calendar to appear in.

11 If you want alerts to be enabled (active) for the calendar, set the Event Alerts switch to on (green). This is helpful because you're notified about events as they get close.

12 To make the calendar public so that others can subscribe to a read-only version of it, set the Public Calendar switch to on (green), tap Share Link, and then use the resulting Share tools to invite others to subscribe to the calendar (more on this later in this chapter).

13 To remove the calendar entirely (instead of hiding it from view), tap Delete Calendar and then tap Delete Calendar at the prompt. The calendar and all its events are deleted. (It's usually better just to hide a calendar as described in step 3 so you don't lose its information.)

14 Assuming that you didn't delete the calendar, tap Done to save the changes you made.

15 Edit other calendars as needed.

16 Tap Done. The app moves into viewing mode, and the calendars you enabled are displayed.

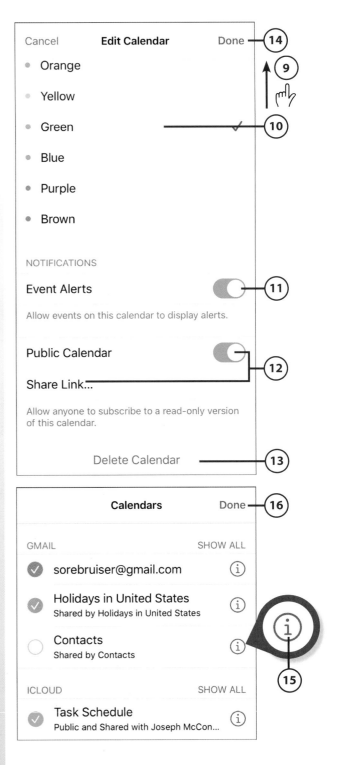

>>>*Go Further*

TAKING CHARGE OF YOUR CALENDAR

Here are few more calendar managing tidbits:

- **All or Nothing**—You can make all your calendars visible by tapping Show All at the bottom of the screen; tap Hide All to do the opposite. After all the calendars are shown or hidden, you can tap individual calendars to show or hide them. You can show or hide all the calendars from the same account by tapping the SHOW/HIDE ALL command at the top of each account's calendar list.

- **Creating Calendars**—To create a new calendar, move to the bottom of the Calendars screen and tap Add Calendar. Name the calendar, choose the account it is associated with, and choose the color in which its events will be shown. When you're done, tap Done. You can work with the new calendar as described in this section.

- **Show Declined Events**—By default, when you decline an event invitation, it disappears from the Calendar app. If you want to see these events on your calendars, tap Show Declined Events on the Calendars screen so that it is marked with a check mark. Declined events appear on your calendar.

- **Pass the Buck**—The Delegate Calendars option on the Calendars screen enables you to give access to your calendars by others. For example, if you wanted someone else to be able to add, change, or delete events on your calendar, you can delegate that permission to them.

- **Changes Here Are Made There**—When you change your calendars in the Calendar app, those changes are reflected in your calendars on other devices; this is a big benefit of storing your calendars online. For example, when you change the color associated with a calendar on your iPhone, that change is reflected on your iPad and in the Calendar app on a Mac computer.

Navigating Calendars

The Calendar app uses a hierarchy of detail to display your calendars. The highest level is the year view that shows the months in each year. The next level is the month view, which shows the days of the month (days with events are marked with a dot). This is followed by the week/day view that shows the days of the week and summary information for the events on each day. The most detailed view is the event view that shows all the information for a single event.

Viewing Calendars

You can view your calendars from the year level all the way down to the week/day. It's easy to move among the levels to get to the time period you want to see. Here's how:

1. Starting at the year view, swipe up and down until you see the year in which you're interested. (If you aren't in the year view, keep tapping back [<] located in the upper-left corner of the screen until that icon disappears.)

2. Tap the month in which you're interested. The days in that month display, and days with events are marked with a dot.

3. Swipe up and down the screen to view different months in the year you selected.

4. To see the detail for a date, tap it. There are two ways to view the daily details: the Calendar view or the List view. Steps 5 through 8 show the Calendar view, whereas steps 9 through 11 show the List view. Each of these views has benefits, and, as you can see, it is easy to switch between them.

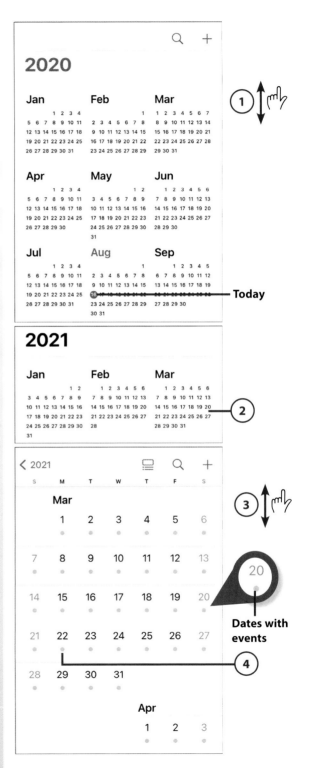

5 To see the Calendar view, ensure the List icon is not selected (isn't highlighted). At the top of the screen are the days of the week you're viewing. The date in focus is highlighted with a red circle when that day is today or a black circle for any other day. Below this area is the detail for the selected day showing the events on that day.

6 Swipe to the left or right on the dates to change the date for which detailed information is being shown in the lower part of the screen. As you change the date being displayed, the black (or red when the events are for the current day) circle indicates the date for which detail is shown.

7 Swipe up or down on the date detail to browse all its events.

8 Tap an event to view its detail and skip to step 12.

9 See the events in List view by tapping the List icon so it's highlighted.

10 Swipe up and down to see the events for each day.

11 Tap an event to view its detail.

The date being displayed below

Tap to jump to the current day

(12) Swipe up and down the screen to see all of the event's information.

(13) View information about the event, such as its location, repetition, and relationship to other events on the calendar.

(14) Tap the Calendar or Alert fields to change these settings or tap Invitees to see information about others invited to the event (more on this later in this chapter).

(15) To change the alerts for the event, tap Alert to change the first alert or Second Alert to change it (configuring alerts is covered in the next section).

(16) Tap any links or documents attached to the event to move to information related to the event.

(17) Read notes associated with the event.

(18) Tap a location to get directions to it.

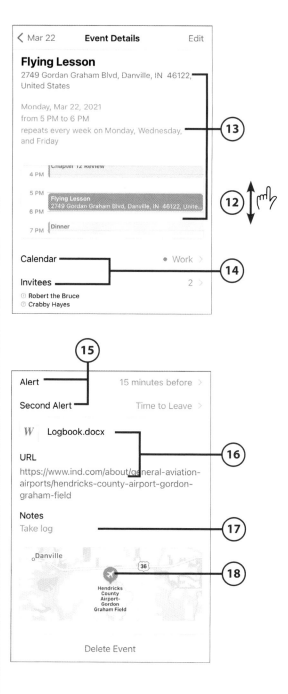

19 Use the Maps app to get directions.

20 Tap Calendar to return to the Calendar app.

21 Tap the date to move back to the week/day view.

22 Tap back (<, which is also labeled with the month and year you are viewing) to move back to the month view.

(23) To view your calendars in the multiday view, rotate your iPhone so it is horizontal. You can do this while in the week/day view or the month view.

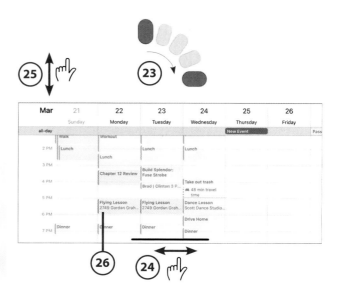

(24) Swipe left or right to change the dates being displayed.

(25) Swipe up or down to change the time of day being displayed.

(26) Tap an event to see its details.

Current Time or Event

When you're viewing today in the Calendar view, the red line stretching horizontally across the screen indicates the current time, which is shown at the left end of that line. When you are viewing today in the List view, the current event is indicated by the text "Now" in red along the right edge of the screen.

What's Next?

To quickly see the next event on your calendar, move to a Home screen or the Lock screen and swipe to the right to open the Widget Center. Find the UP NEXT widget, which shows you the next event on your calendar; you might want to move this to a Home screen so you can see this information even more easily. (For more on configuring your widgets, see Chapter 5.)

What's Next? But Even Faster

You can find out what's next even faster by activating Siri and saying, "What's my next event?" Siri shows you your next event and speaks the time and name of the event. Siri also can help with many other calendar-related tasks. (See Chapter 13, "Working with Siri," for the details.)

Adding Events to Calendars

There are a number of ways you can add events to your calendar. You can create an event in an app on a computer, website, or other device and sync that event to the iPhone through an online account. You also can manually create events in the Calendar app on the iPhone. Your events can include a lot of information, or they can be fairly basic. You can choose to create the basic information on your iPhone while you are on the move and complete it later from a computer or other device, or you can fill in all the details directly in the Calendar app. You can also add an event by accepting an invitation (covered later in this chapter).

(1) Tap Add (+), which appears in the upper-right corner of any of the views when your phone is vertical (except when you are viewing an event's details). The initial date information is taken from the date currently being displayed, so you can save a little time if you are on the date of the event before tapping Add.

⟨ August				☰	Q	+
S	M	T	W	T	F	S
16	17	18	19	20	21	22

Sunday August 16, 2020

(2) Tap in the Title field and type the title of the event. As you type, events that are similar to the one you are creating are suggested; tap this to create the new event based on the suggestion. Then you can change the details to be specific to the event you are creating. If you use a suggested event that includes a location, skip to step 6. If you don't want to use a suggested event, finish typing the event's name and continue to step 3.

Cancel	New Event	Add

Fly|

● **Fly at Plainfield**
501 W Main St, Plainfield, IN, 46168-1250 —— **Suggested event**
Today from 7 AM to 11 AM

Location

(3) Tap the Location bar and type the location of the event; if you allow the app to use Location Services, you're prompted to find and select a location; if not, just type the location and skip to step 6.

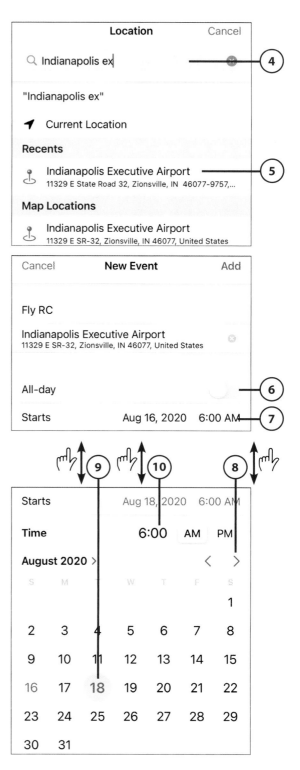

(4) Type the location in the Search bar. Sites that meet your search are shown below. The results screen has several sections including Recents, which shows locations you've used recently, and Map Locations, which are sites that the app finds that match your search criteria.

(5) Tap the location for the event; tap Current Location if the event takes place where you are. You return to the New Event screen and see the location you entered.

(6) To set the event to last all day, set the All-day switch to the on position (green); when you select the All-day option, you provide only the start and end dates (you don't enter times as described in the next several steps). To set a specific start and end time, leave this setting in the off position (white); you set both the dates and times as described in the following steps.

(7) To set a timeframe for the event, tap Starts. The date and time tool appears. By default, the date on which you create the event is selected.

(8) To change the month in which the event occurs, tap < to move to a previous month or tap > to move to a later month. If the event is in the current month, skip this step.

(9) Tap the day on which the event occurs. It is highlighted in red.

(10) Tap the start time. The time selection tool appears.

(11) Use the number keys to enter the start time of the event.

(12) Tap AM or PM.

(13) Swipe up the screen to see the Ends section.

Time Zones

The Calendar app assumes the event you are creating is in your current time zone. If you want to set the time in a different time zone, tap Time Zone and choose a city in the time zone that you want the starting time to be in. (It doesn't matter if you can't find the exact city for the event; you just need one in the desired time zone.) You can also set a different time zone for the end time.

(14) Tap Ends.

| Time | | 3·00 | AM | PM | (12) |

August 2020 >   < >

S	M	T	W	T	F	S
						1
2	3	4	5	6	7	8
9	10	11	12	13	14	15
16	17	18	19	20	21	22
23	24	25	26	27	28	29
30	31					

(13)

1	2 ABC	3 DEF	(11)
4 GHI	5 JKL	6 MNO	
7 PQRS	8 TUV	9 WXYZ	
	0	⊗	

30	31

Time Zone   America/Indiana/Indianapolis >

Ends   4:00 PM   (14)

15 Use steps 8 through 12 to set the end date and time for the event.

16 Swipe up the screen.

17 To make the event repeat, tap Repeat and move to Step 18; for a one-time event, keep the default, which is Never, and skip to step 19.

Your Results Might Vary

The fields and options available when you create a new event are based on the calendar with which the event is associated. For example, an iCloud calendar might have different options than a Google calendar does. If you aren't creating an event on your default calendar, it's a good idea to associate the event with a calendar before you fill in its details (to do that, perform step 25 before you do step 2).

No Changes?

When you change a selection on most of the screens you see, you automatically return to the previous screen; for example, when you select an alert time, you immediately return to the New Event screen. If you don't make a change, you can return to the previous screen by tapping New Event (<) in the upper-left corner of the screen.

18 Tap the frequency at which you want the event repeated, such as Every Day, Every Week, and so on; if you want to use a repeat cycle not shown, tap Custom and create the frequency with which you want the event to repeat.

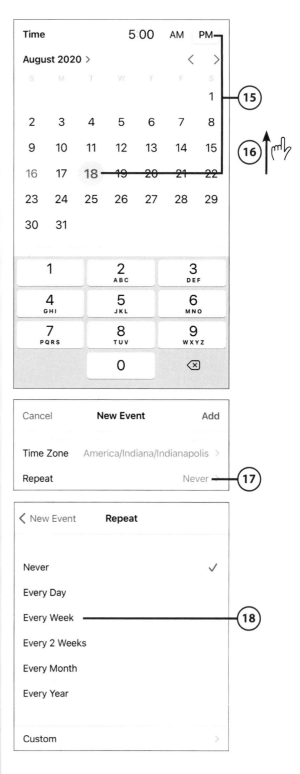

Custom Repeat

To configure a custom repeat cycle for an event in step 18, such as the first Monday of every month, tap Custom on the Repeat screen. Then use the Frequency (Monthly for the example) and Every (for example, On the first Monday) settings to configure the repeat cycle. Tap Repeat to return to the Repeat screen and then tap New Event to get back to the event you are creating.

Stop It!

If you configure an event to repeat and want it to stop repeating at some point in the future, tap End Repeat. Use the resulting screen to choose when you want the event to stop repeating. Beyond the end repeat date, the event is no longer on the calendar. If you don't choose an end repeat date, the event continues forever.

(19) To configure travel time for the event, tap Travel Time; if you don't want to configure this, skip to step 25.

(20) Set the Travel Time switch to on (green).

(21) Tap Starting Location to build a travel time based on a starting location and skip to step 23.

(22) To manually set a travel time for the event, tap it and skip to step 24.

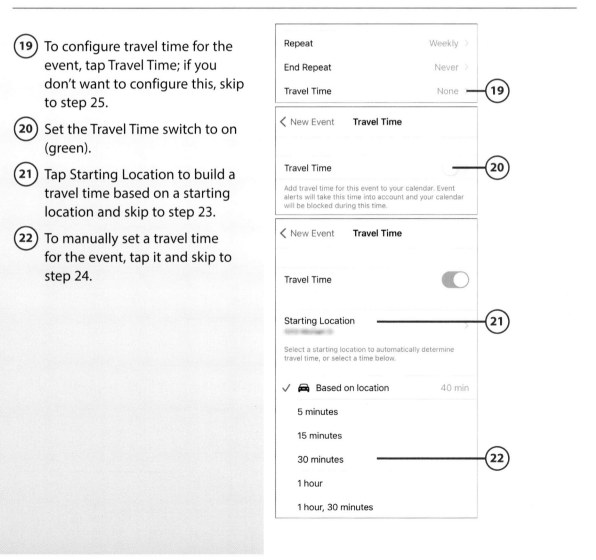

(23) To use your current location as the starting point, tap Current Location. Alternatively, use the Search tool to find a starting location, and then tap it to select that location. This works just like setting a location for the event. Or you can choose a recently used location. The travel time is calculated based on the starting and ending locations entered for the new event. Most of the time, the travel mode is assumed to be by car. Depending on where the start and end locations are, you might also see different options based on how you will travel. For example, if the two locations are relatively close, the travel time might be based on walking. Or, you might see public transportation instead. You can configure alerts based on this travel time as you see in later steps.

(24) Tap New Event.

(25) To change the calendar with which the event is associated, tap Calendar (to leave the current calendar selected, skip to step 27).

(26) Tap the calendar with which the event should be associated.

	Location	Cancel

🔍 ▓▓▓▓▓▓▓▓▓▓ ⊗

▓▓▓▓▓▓▓▓▓▓

➤ Current Location ——————— (23)

Recents

📍 Indianapolis Executive Airport ——
11329 E State Road 32, Zionsville, IN 46077-9757,...

(24)

‹ New Event **Travel Time**

Travel Time ⬤

Starting Location ›
▓▓▓▓▓▓▓

Select a starting location to automatically determine travel time, or select a time below.

✓ 🚗 Based on location 40 min

End Repeat Never ›

Travel Time 40 min ›

Calendar • Task Schedule ⟶ (25)

‹ New Event **Calendar**

~ United States Holidays

GMAIL

• sorebruiser@gmail.com

ICLOUD

• Robert's Calendar

• Calendar ——————— (26)

• Task Schedule ✓
Public and Shared with Edward the Longshanks...

27 To invite others to the event, tap Invitees; if you don't want to invite someone else, skip to step 31.

28 Enter the email addresses for the people you want to invite; as you type, the app tries to identify people who match what you are typing. You can tap a person to add him to the event or keep entering the email address until it is complete. You can also tap Add (+) to choose people in your Contacts app (see Chapter 8, "Managing Contacts," for help using that app). Repeat this step for each person you want to invite.

29 Tap Done. You move to the Invitees screen and see those whom you invited.

30 Tap New Event (<).

31 To set the amount of warning time at which the alert plays before the event that is different than the default, tap Alert; if you want to use the default alert, skip to step 33.

32 Tap when you want to see an alert for the event. You can choose a time relative to the time you need to start traveling or relative to the event's start time.

Calendar	● Calendar >
Invitees	None

— 27

Cancel	**Add Invitees**	Done

— 29

To: Wood Chuck   Sir William Wallace   ⊕

— 28

‹ New Event   **Invitees**

— 30

Add invitees   ›

NO RESPONSE

🌰 Wood Chuck   (?)

👤 Sir William Wallace   (?)

Calendar	● Calendar >
Invitees	2 >
Alert	15 minutes before

— 31

‹ New Event   **Alert**

None

At start of travel time

5 minutes before travel time

10 minutes before travel time

15 minutes before travel time   ✓

30 minutes before travel time

1 hour before travel time

— 32

2 hours before travel time

(33) To set the amount of warning time for the second alert that is different than the default, tap Second Alert; to use the default, skip to step 35.

(34) Tap when you want to see a second alert for the event. If you have included travel time in the event, the At or before start of travel time options are useful because they alert you relative to when your journey should begin.

Are You Available?

The Show As field indicates how other people who try to schedule events with you will see your calendar. In most cases, you want the default Busy status so people don't try to schedule your time when you're already busy.

Attachments

You can add documents or other files to an event by tapping Add attachment and selecting the file you want to add. Before you can do this, however, the files you want to add need to be stored on your iCloud Drive.

(35) Tap in the URL field to enter a URL associated with the event; if you don't need to add this, skip to step 38.

(36) Type the URL.

(37) Tap done.

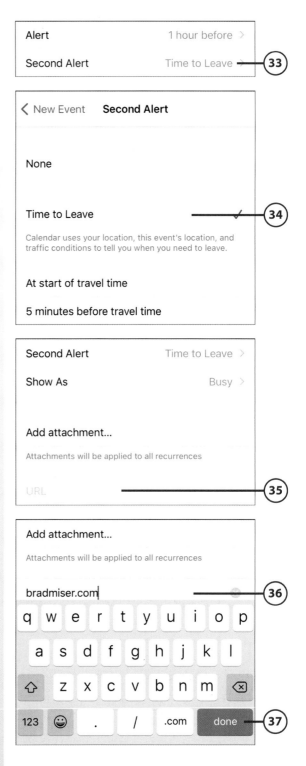

38 Tap in the Notes field.

39 Type the information you want to associate with the event.

40 Tap Add. The event is added to the calendar you selected, and invitations are sent to its invitees. Alerts trigger according to the event's settings.

bradmiser.com

Notes ——————————— **38**

Cancel   **New Event**   Add — **40**

Add attachment...

Attachments will be applied to all recurrences

bradmiser.com

Take:
-Batteries
-Transmitter ——————————— **39**
-Charger|

‹ August   ☰ 🔍 +

S   M   T   W   T   F   S
16   17   **18**   19   20   21   22
Tuesday August 18, 2020

Lunch

Noon |

Workout
1 PM |

2 PM |

🚗 40 min travel time
3 PM ⋮
Fly RC
Indianapolis Executive Airport
11329 E State Road 32, Zionsville, IN 46077-9757, United
4 PM States ——————— **New event on the calendar**

A Better Way to Create Events

Adding a lot of detail to an event in the Calendar app can be a bit tedious. One effective and easy way to create events is to start with Siri. You can activate Siri and say something like "Create meeting with William Wallace in my office on November 15 at 10:00 a.m." Siri creates the event with as much detail as you provided (and might prompt you to provide additional information, such as which email address to use to send invitations). When you get to a computer or iPad, edit the event to add more information, such as website links. When your calendar is updated on the iPhone, via syncing, the additional detail for the event appears in the Calendar app, too. (See Chapter 13 for detailed information about using Siri.)

Searching Calendars

You can search for events to locate specific ones quickly and easily. Here's how:

1 Tap the magnifying glass.

2 Type your search term. The events shown below the Search bar are those that contain your search term.

3 Swipe up or down the list to review the results.

4 Tap an event to see its detail.

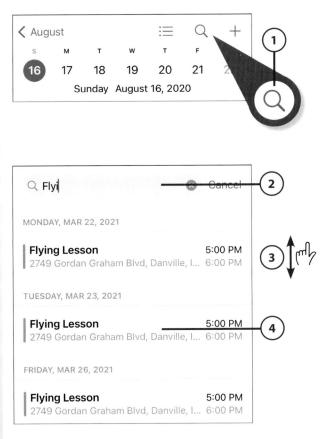

5 Swipe up or down the event's screen to review its information.

6 Tap Back (<) to return to the results.

7 Continue reviewing the results until you get the information for which you were searching.

8 Tap Cancel to exit Search mode.

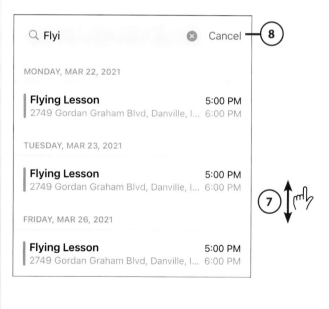

>>>Go Further
INVITATIONS

When someone invites you to an event, you receive an invitation notification on your iPhone, as an email, and in the Calendar app. From any of these locations, you can accept the invitation, at which point the event is added to your calendar. You can tentatively accept, in which case the event is added to your calendar with a tentative status, or you can decline the event if you don't want it added to your calendar.

After you've accepted or tentatively accepted an invitation, that event is added to your calendar. You can make some changes to it, such as alert times, but you can't change information set by the creator, such as the time and date.

You can work with your invitations by moving to the Calendars screen and tapping Inbox.

Sharing Calendars

Sharing your calendars with other people enables them to both see and change your calendar, according to the permissions you provide. If you set the View & Edit permission, the person is able to both see and change the calendar. If you set someone's permission to View Only, she can see, but not change, the calendar.

(1) Move to the Calendars screen. Calendars you're sharing are indicated by the Shared with *name* text just under the calendar name, where *name* is the name of the person with whom you are sharing the calendar.

(2) To see who is currently sharing a calendar, or to share a calendar, tap Info (i) for the calendar to be shared. If the calendar is currently being shared, the SHARED WITH section appears on the Edit Calendar screen. This section contains the names of and permissions granted to the people who are sharing the calendar. The status of each person's acceptance is shown under his name (Accepted, Pending, or Declined).

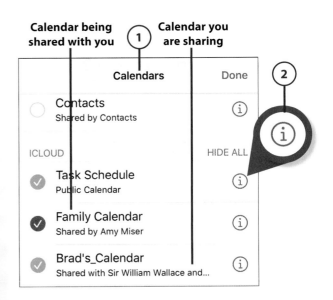

Calendar being shared with you — (1) — Calendar you are sharing

3 Tap Add Person to share a calendar (whether or not it's currently shared) with someone else.

4 Type a person's name or email address, tap a name or email address that the app suggests based on what you are typing, or use the Contacts app to choose the people with whom you want to share the calendar. (You can add multiple people at the same time.)

5 Tap Add. You return to the Edit Calendar screen, and see the person you added on the SHARED WITH list. The person you invited receives an invitation to join the calendar you're sharing. If he accepts, the shared calendar becomes available in the calendar app he uses. As people make decisions about the calendar you're sharing, the status below each invitee's name changes to reflect the current status. You also see notifications when an invitee responds if you allow the app to provide notifications to you.

6 If you don't want a person to be able to change the calendar, tap View & Edit.

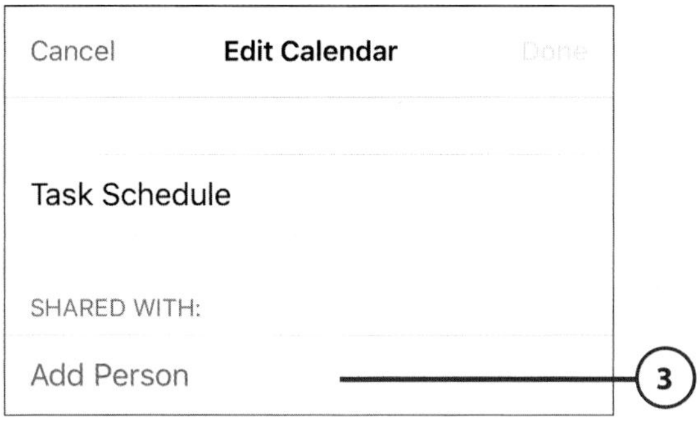

Cancel	**Edit Calendar**	Done

Task Schedule

SHARED WITH:

Add Person ————————— **3**

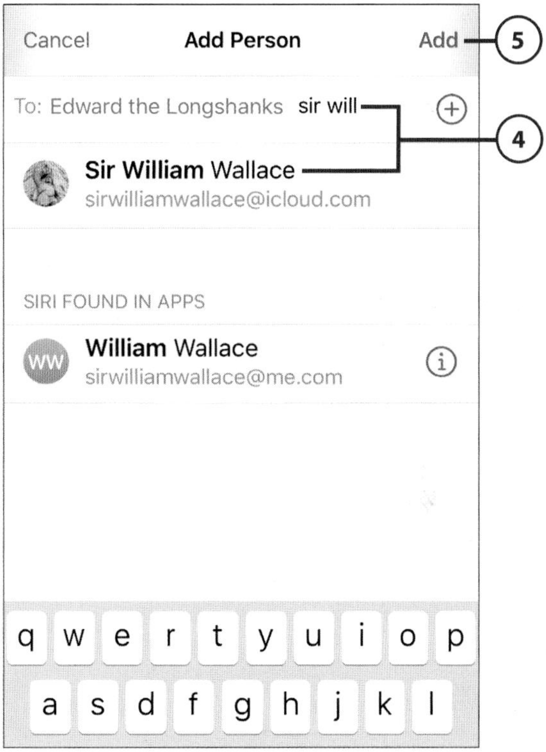

Cancel	**Add Person**	Add —— **5**

To: Edward the Longshanks   sir will ——
⊕ —— **4**

Sir William Wallace ——
sirwilliamwallace@icloud.com

SIRI FOUND IN APPS

William Wallace
sirwilliamwallace@me.com   ⓘ

q w e r t y u i o p

a s d f g h j k l

SHARED WITH:

Sir William Wallace
Pending   View & Edit → **6**

Edward the Longshanks
Pending   View & Edit ›

Add Person

(7) Slide the Allow Editing switch to off (white); the person will be able to view but not change the calendar.

(8) Tap Back (<).

(9) When you're done sharing the calendar, tap Done.

(10) When you're finished configuring calendar sharing, tap Done.

8

< Back   **Sir William Wallace**

Sir William Wallace
sirwilliamwallace@icloud.com   >

Allow Editing   — 7

Allow this person to make changes to
the calendar.

Resend Sharing Invitation

Cancel   **Edit Calendar**   Done — 9

Task Schedule

SHARED WITH:

Sir William Wallace   View Only >
Pending

Edward the Longshanks   View & Edit >
Pending

Calendars   Done — 10

○ Contacts
  Shared by Contacts   (i)

ICLOUD   HIDE ALL

✓ Task Schedule
  Public and Shared with Sir William Wall...   (i)

Managing Calendars and Events

Following are some more points about the Calendar app you might find helpful:

- You can change an existing event by viewing it and tapping Edit. Use the controls on the Edit Event screen to change the event; you have the same options as when you create an event.

- To remove an event from your calendar, view it, tap Edit, swipe up from the bottom of the screen, tap Delete Event, and then tap Delete Event at the confirmation. The event, and all the information it contains, is removed from your calendar.

- You can use the List view when you are viewing the calendar in Month view. Tap the List icon (just to the left of the magnifying glass). The list opens at the bottom of the screen and shows the events on the day currently selected (the date is in a black circle unless the day selected is today, in which case the circle is red).

- When an event's start time matches the alert time, an onscreen notification appears (according to the notification settings for the Calendar app) and the calendar event sound you've selected plays. To take action on a calendar's notification, you can press the notification to view its details; you can tap the details to move into the Calendar app or close the notification by tapping Close (x). You can swipe up from the bottom of a notification to manually close it without taking further action on it.

- When you share a calendar with someone, he receives a notification sent to the email address you used. If the recipient chooses to accept the invitation, he can view (and edit) the calendar in his calendar app. If not, he doesn't see the calendar. You receive notifications in both cases.

- When you move to a calendar's information screen, you see the status of people you've invited, such as Pending if they haven't made a decision or Accepted if they have accepted the invitation.

- When you receive an invitation, the event is tentatively added to your calendar until you make a decision about it. The status changes as you accept, tentatively accept (Maybe), or decline invitations.

- If an event gets canceled after you accept it or indicated maybe, you receive a notification in your Inbox that displays a strikethrough through the event's title. Tap Delete to remove the event from your calendar.

- Siri is useful for working with calendars, as discussed in the sidebar "A Better Way to Create Events" earlier in this chapter, especially for creating events. See Chapter 13 for more detailed information about using Siri.

- You can publish a calendar by making it public. When you do this, anyone who can access the shared calendar on the Web can view, but not change, the published calendar. This puts the calendar on the Web, where it can be viewed with a web browser. This is unlike a calendar you share with someone, which adds your calendar to her calendar app.

   To make a calendar public, move to its Edit Calendar screen and slide the Public Calendar switch to on (green). Tap Share Link. Then tap Mail to send the link via email, tap Message to send it via the Messages app, or tap Copy to copy the link so you can paste it elsewhere. Tap AirDrop, Twitter, or Facebook to share the link in those ways. A person can use the link to view your calendar or, if the person uses a compatible application, to subscribe to it so it appears in her calendar application.

>>>*Go Further*

OTHER USEFUL APPS FOR MANAGING YOUR TIME

Your iPhone includes some other apps that can help you manage your time. The Clock app enables you to easily see the current time in multiple locations, set alarms, set a consistent time for sleep every day, use a stopwatch, and count down time with a timer. The Reminders app enables you to create reminders for events, tasks you need to perform, or just about anything else. You can learn a bit more about these apps in Chapter 17, "Working with Other Useful iPhone Apps and Features."

>>>Go Further
CALENDAR AND DATE & TIME SETTINGS

If you open the Calendar option in the Settings app, you see a number of options that change how the Calendar app works. In most cases, you can use the Calendar app just fine without changing any of these. After you've used the Calendar app for a bit, you should check out these settings to see if there are any you want to change. For example, you might want to change the Default Alert Times to be more applicable to you than the defaults configured, well, by default.

To configure how your iPhone displays and manages the date and time, open the Settings app, tap General, tap Date & Time, and change the settings described in the following list:

- **24-Hour Time**—Turning this switch on (green) causes the iPhone to display 24-hour instead of 12-hour time.

- **Set Automatically**—When this switch is on (green), your iPhone sets the current time and date based on the cellular network it is using. When it is off (white), controls appear that you use to manually set the time zone, time, and date. You usually want this on so your iPhone always has the correct date, but if it isn't getting the date and time correctly for some reason, turn your phone off and on. If it's still incorrect, you can disable this and set the time zone, time, and date manually.

Go here to change
how Siri works

Speak to Siri to
send and hear
messages, create
and manage
events, make calls,
and much more

Working with Siri

Siri is Apple's name for the iPhone's voice recognition feature. This technology enables your iPhone to "listen" to words you speak so that you can issue commands just by saying them, such as "Send text message to Sam," and the iPhone accomplishes the tasks you speak.

Getting Started

Siri gives you the ability to talk to your iPhone to control it, get information, and to dictate text. Siri also works with lots of iPhone apps—this feature enables you to accomplish many tasks by speaking instead of using your fingers on the iPhone's screen. For example, you can hear, compose, send, and reply to text messages; reply to text messages and emails; make phone and FaceTime calls; create and manage calendar events and reminders; and well, the list is practically endless. And Siri can make suggestions as you perform tasks, such as searches, based on what you've done before.

In fact, Siri does so many things, it's impossible to list them all in a short chapter like this one; you should give Siri a try for the tasks you perform and to get the information you need, and, in many cases, Siri can handle what you want to do.

Think of Siri as your personal, digital assistant to help you do what you want to do more quickly and easily (especially when you are working in handsfree mode).

You don't have to train Siri very much to work with your voice, either; you can speak to it normally, and Siri does a great job of understanding what you say. Also, you don't have to use any specific kind of phrases when you have Siri do your bidding. Simply talk to Siri like you talk to people (well, you probably won't be ordering other people around like you do Siri, but you get the idea).

Your iPhone has to be connected to the Internet—using a Wi-Fi network or its cellular data connection—for Siri to work. That's because the words you speak are sent over the Internet, transcribed into text, and then sent back to your iPhone. If your iPhone isn't connected to the Internet, this can't happen, and if you try to use it, Siri reports that it can't complete its tasks.

Because your iPhone is likely to be connected to the Internet most of the time (via Wi-Fi or a cellular network when you have cellular data enabled), this really isn't much of a limitation—but it is one you need to be aware of.

Just start speaking to your iPhone and be prepared to be amazed by how well it listens! You'll find many examples in this chapter to get you going with specific tasks; from there, you can explore to learn what else Siri can do for you.

Configuring Siri

You probably configured Siri when you first turned your iPhone on. The iPhone's software guides you through the process of enabling Siri and training it to recognize your voice. It's a good idea to review Siri's current settings. You might want to make the settings on your phone match the ones I used when creating the figures in this chapter. When you're comfortable using Siri, you can make adjustments to suit your preferences.

To access Siri's settings, tap Settings on the Home screen and then tap Siri & Search. There are quite a few options on the Siri & Search screen, but for our purposes, the settings in the following table are the ones you should check.

Siri & Search Settings

Setting	Description
Listen for "Hey Siri"	When this switch is on (green), you can activate Siri by saying, "Hey Siri" in addition to using controls on the phone or earbuds. When this switch is off (white), you can only activate Siri with the other options.
Press Side Button for Siri (iPhones without a Home button)/Press Home for Siri (iPhones with a Home button)	When enabled, you can use the identified button to activate Siri. When disabled, you have to use one of the other methods, such as saying "Hey, Siri."
Allow Siri When Locked	When this switch is on (green), you can activate and use Siri without unlocking your iPhone. In some cases, your iPhone needs to be unlocked to complete the task you are doing, but in other cases, Siri can do it for you with the phone locked.
Language	Set the language you want Siri to use to speak to you.
Siri Voice	Choose the accent and gender of the voice that Siri uses to speak to you (the options you see depend on the language you have selected).
Siri Responses Siri Feedback	Use these settings to configure how Siri responds to you. To match the figures in this chapter, configure the following settings: • **SPOKEN RESPONSES**—This determines when Siri provides spoken responses to you. The options are Always, When Silent Mode is Off, or Only with "Hey Siri." This chapter assumes When Silent Mode is Off. When your phone is muted, Siri won't respond verbally. • **Always Show Siri Captions**—Enable this to have Siri show you what it says using a caption on the screen. This is useful because you can see what Siri is doing. • **Always Show Speech**—When enabled, you see captions on the screen showing what you have said. This is helpful because you see what Siri heard you say.
My Information	Choose your contact information in the Contacts app, which Siri uses to address you by name, take you to your home address, etc.
Siri & Dictation History	You can use this to delete the record of any Siri interactions or the dictation you have done.

The rest of the settings on the Siri & Search screen enable you to configure Siri Suggestions. You can determine when Siri presents suggestions, such as on the Lock or Home screens. You can also configure suggestions and shortcuts for individual apps.

Understanding Siri's Personality

Siri works in two basic modes: when you ask it to do something or when it makes suggestions to you.

Siri's Personality Shift with iOS 14

In prior versions of iOS, Siri had a dominant personality. When activated, Siri took over the screen and you couldn't see anything but Siri while it was working. With iOS 14, Siri isn't so pushy. Instead, Siri is visible at the bottom of the screen, and you continue to see the screen you were on when you activated Siri.

Telling Siri What to Do

When you ask Siri to do something, Siri follows a consistent pattern, and it always prompts you for input and direction when needed.

Activate Siri using one of the following methods:

- On the iPhones without a Home button, press and hold the Side button for a couple of seconds until Siri's Listening symbol appears at the bottom of the screen.

- On iPhones with a Home button, press and hold the Touch ID/Home button until Siri's Listening symbol appears at the bottom of the screen.

- Press and hold the center part of the buttons on the EarPods until Siri's Listening symbol appears at the bottom of the screen and you hear the Siri tone.

- Double-tap an AirPod until Siri's Listening symbol appears at the bottom of the screen and you hear the Siri tone.

- Say "Hey Siri;" you see Siri's Listening symbol at the bottom of the screen and hear Siri respond with something like "Uh huh." You also see the Hey Siri command on the screen.

"Hey Siri" was used to activate Siri

Siri is listening

Siri Feedback

Siri's feedback described here and shown in the figures throughout this chapter assume the Siri Responses or Siri Feedback settings are set as described in the Siri Responses section of the table at the beginning of the chapter. If the settings on your phone don't match those, you will have a different experience than described here.

Activating Siri puts it in "listening" mode. Siri displays the Listening symbol at the bottom of the screen and provides audible feedback if you used "Hey Siri" to indicate that it's ready for your commands.

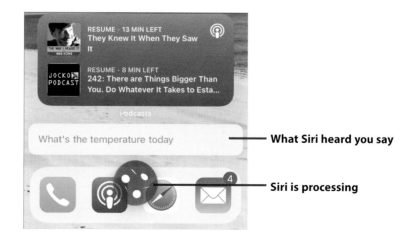

What Siri heard you say

Siri is processing

Speak your command or ask a question. As you speak, the symbol at the bottom of the screen spins to show you that Siri is hearing your input, and Siri displays its interpretation of what you've said. When you stop speaking, Siri goes into processing mode.

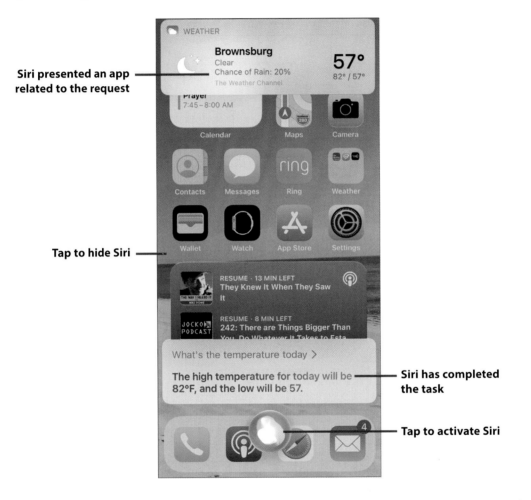

Siri presented an app related to the request

Tap to hide Siri

Siri has completed the task

Tap to activate Siri

Siri then tries to do what it thinks you've asked and speaks and shows you the outcome.

If it needs more input from you, Siri prompts you to provide it. Siri moves into "listening" mode automatically and prompts you to speak by playing a tone. If Siri asks you to confirm what it's doing or to make a selection, do so. Siri completes the action and displays what it has done; it also audibly confirms the result.

When Siri has completed the current task, you can

- **Tap a related app to work with it**—If Siri presents an app as part of the response, tap the app to use it.

- **Use Siri again**—To ask Siri to do something else, tap the Siri symbol at the bottom of the screen and speak your command.

- **Hide Siri**—When you're done with Siri, tap the screen anywhere outside of Siri windows. Siri moves into the background where it waits patiently for your next command.

Also, how Siri interacts with you can depend on how it was activated. For example, if you started the interaction using the verbal "Hey Siri" option, Siri assumes you want to interact verbally and might respond with other options than you would see or hear when you activate Siri manually. For example, when you ask Siri to show you your appointments for the day in this mode, you see the summary, and Siri reads the first few to you and then asks if you want to hear more. When you activate Siri by using the Side or Touch ID/Home button with the same request, Siri stops after showing you the summary.

When you're done with Siri, touch outside a Siri window. You move back to where you were.

Siri uses this pattern for all the tasks it does, but often Siri needs to get more information from you, such as when there are multiple contacts that match the command you've given. Siri prompts you for what it needs to complete the work. Generally, the more specific you make your initial command, the fewer steps you have to work through to complete it. For example, if you say "Meet Will at the park," Siri might require several prompts to get you to tell it who Will is and what time you want to meet him at the park. If you say, "Meet William Wallace at the park on 10/17 at 10:00 a.m.," Siri can likely complete the task in one step.

The best way to learn how and when Siri can help you is to try it—a lot. You find a number of examples in the rest of this chapter to get started.

Following are some other Siri tidbits:

• If Siri doesn't automatically quit "listening" mode after you've finished speaking, tap the Siri symbol. This stops "listening" mode and Siri starts processing your request. You need to do this more often when you're in a noisy environment because Siri might not be able to accurately discern the sound of you speaking versus the background noise.

• If Siri has trouble understanding your commands, speak a bit more slowly and clearly enunciate and end your words. If you tend to have a very short pause between words, Siri might run them all together, making them into something that doesn't make sense or that you didn't intend.

• However, you can't pause too long between words or sentences because Siri interprets pauses of a certain length to mean that you're done speaking, and it goes into processing mode. Practicing with Siri helps you develop a good balance between speed and clarity.

• If Siri doesn't understand what you want, or if you ask it a general question, it often performs a web search for you. You then see the results page for the search Siri performed, and you might have to manually open and read the results by tapping the listing you want to see. It opens in the Safari app. In some cases, Siri reads the results to you.

• When Siri presents information to you on the screen, you can often tap that information to move into the app with which it is associated. For example, when you tap an event that Siri has created, you move into the Calendar app, where you can add more detail using that app's tools, such as inviting people to an event, or changing the calendar it's associated with.

• When Siri needs direction from you, it presents your options on the screen, including Yes, Cancel, Confirm, or lists of names. You can speak these items or tap them to select them.

- Siri might ask you to help it pronounce some terms, such as names. When this happens, Siri asks you to teach it to pronounce the phrase. If you agree, Siri presents a list of possible pronunciations, which you can preview. Tap Select for the option you want Siri to use.

- To use Siri effectively, you should experiment with it by trying to say different commands or similar commands in different ways. For example, when sending email, you can include more information in your initial command to reduce the number of steps because Siri doesn't have to ask you for more information. Saying "Send an email to Wyatt Earp home about flying" requires fewer steps than saying "Send email" because you've given Siri most of the information it needs to complete the task, so it won't have to prompt you for who you want to send it to, which address you want to use, or what the subject of the email is.

- When Siri can't complete a task that it thinks it should be able to do, it usually responds with "I can't connect to the network right now," or "Sorry, I don't know what you mean." This indicates that your iPhone isn't connected to the Internet, the Siri server is not responding, or Siri just isn't able to complete the command for some other reason. If your iPhone is connected to the Internet, try the command again or try rephrasing the command.

- When Siri can't complete a task that it knows it can't do, it responds by telling you so. Occasionally, you can get Siri to complete the task by rephrasing it, but typically you have to use an app directly to get it done.

- Siri might not be able to complete some tasks because the phone is locked. If that happens, Siri prompts you to unlock your phone (which you can do by using Face ID, touching the Touch ID/Home button, or entering your passcode) and continue with what you were doing.

- Siri is really good at retrieving all sorts of information for you. This can include schedules, weather, directions, unit conversions, and so on. When you need something, try Siri first, as trying it is really the best way to learn how Siri can work for you.

- Siri sees all and knows all (well, not really, but it sometimes seems that way). If you want to be enlightened, try asking Siri questions, such as these:

  What is the best phone?

  Will you marry me?

  What is the meaning of life?

  Tell me a joke.

  Some of the answers are pretty funny, and you don't always get the same ones so Siri can keep amusing you. I've heard it even has responses if you curse at it, though I haven't tried that particular option.

Working with Siri Suggestions

Through Siri Suggestions, Siri becomes proactive and provides information or suggestions for you based on what you're doing and what you have done in the past. Over time, Siri "learns" more about you and tailors these suggestions to better match what you typically do. For example, when you create a text message and start to input a name, Siri can suggest potential recipients based on prior texts you've sent. Similarly, when you perform a search, Siri can tailor the search based on your history.

Because Siri works proactively in this mode, you don't do anything to cause Siri to take action. It works in the background for you and presents information or options at the appropriate times.

Learning How to Use Siri by Example

As mentioned earlier in this chapter, the best way to learn about Siri is to use it. Following are a number of tasks for which Siri is really helpful. Try these to get some experience with Siri and then explore on your own to make Siri work at its best for you.

Using Siri to Make Voice Calls

You can use Siri to make calls by speaking. This is especially useful when you're using your iPhone in handsfree mode.

(1) Activate Siri (such as by pressing and holding the Side button [iPhones without a Home button] or pressing and holding the Touch ID/Home button [iPhones with a Home button]).

Speeding Up Siri

You can combine these steps by saying "Hey Siri, call Sir William Wallace iPhone." This is an example where providing Siri with more information when you speak gets the task done more quickly.

(2) Say "Call *name*," where *name* is the person you want to call. Siri identifies the contact you named. If the contact has only one number, Siri places the call and you move into the Phone app. If the person has multiple numbers, Siri lists the numbers available and asks you which number to use.

(3) Speak the label for the number you want to call or tap it. Siri dials the number for you and you move to the Phone app as if you had dialed the number yourself.

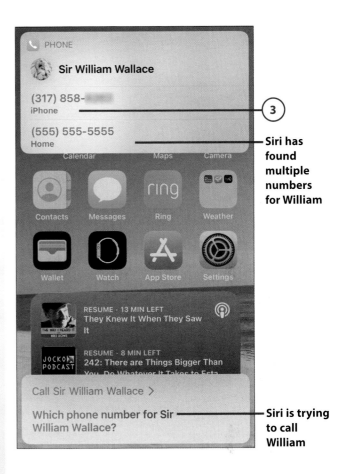

3

Siri has found multiple numbers for William

Siri is trying to call William

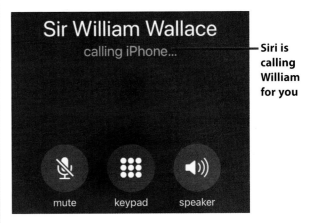

Siri is calling William for you

Siri Is Pretty Sharp

Siri can work with all kinds of variations on what you say. For example, if a contact has a nickname configured for him, you can use the nickname or the first name. If you want to call Gregory "Pappy" Boyington, you can say, "Call Pappy Boyington," or, "Call Gregory Boyington." If you say, "Call Pappy," and there is only one contact with that nickname, Siri calls that contact. If more than one contact has "Pappy" as part of their contact information, Siri presents a list of contacts and prompts you to select one.

Placing FaceTime Calls

You can also use Siri to make FaceTime calls by saying, "FaceTime *name.*"

Having Messages Read to You

The Messages app is among the best to use with Siri because you can speak the most common tasks you normally do with messages, including reading messages you receive. When you receive new text messages, do the following to have Siri read them to you:

(**1**) When you receive a text notification, activate Siri.

(**2**) Speak the command "Read text messages." (You can combine steps 1 and 2 by saying, "Hey Siri, read text messages.") Siri reads all the new text messages you've received, announcing the sender before reading each message. You have the option to reply (covered in the next task) or have Siri read the message again.

Siri reads each new message in turn until it has read all of them and then announces, "That's it," to let you know it has read all of them.

Siri only reads new text messages when you aren't on the Messages screen. If you've already read all your messages and you aren't in the Messages app, when you speak the command "Read text messages," Siri tells you that you have no new messages.

 If you're done with Siri, tap outside its windows or tap the Siri symbol to activate it again.

Siri is reading a text message

Reading Old Messages

To read an old message, move back to the conversation containing the message you want to hear. Activate Siri and say the command "Read text message." Siri reads the most recent text message to you.

Replying to Messages with Siri

You can also use Siri to speak replies to messages you've received. Here's how:

(1) Listen to a message.

(2) At the prompt asking if you want to reply, say, "Yes." Siri prepares a reply to the message.

3 Speak your reply. Siri displays your reply.

4 At the prompt, say "Send" to send the message, "Cancel" to delete it, "Review" to have your message read back to you, or "Change" to replace it. If you tell Siri that you want to send the message, Siri sends it and then confirms that it was sent.

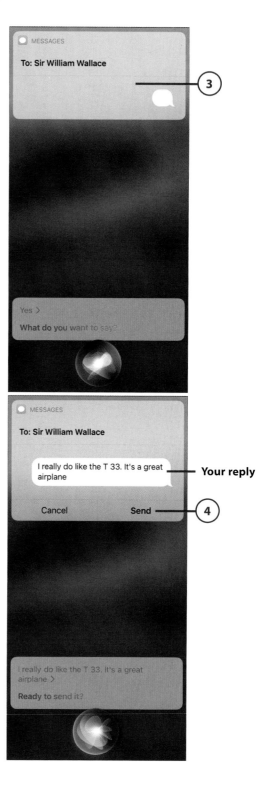

Your reply

Sending New Messages with Siri

To send a new text or iMessage message, do the following:

(1) Say "Hey Siri, send text message to *name*," where *name* is the person you want to text. Siri confirms your command and prepares to hear your text message.

(2) Speak your message. Siri listens and then prepares your message.

(3) If you want to send the message, say "Send." Siri sends the message.

(4) If you're done with Siri, tap outside its windows or tap the Siri symbol to activate it again.

>>>*Go Further*

DOING MORE MESSAGING WITH SIRI

Following are some other ways to use Siri with messaging:

- To send a text message to more than one recipient, say "and" between each name, as in, "Send text to William Wallace and Edward Longshanks." You can send a text message to as many recipients as you want.

- You can speak punctuation, such as "period" or "question mark" to add it to your message.

- Messages you receive or send via Siri appear in the Messages app just like messages you receive or send by tapping and typing.

Using Siri to Create Events

Siri is useful for capturing meetings and other events you want to add to your calendars. To create an event by speaking, use the following steps:

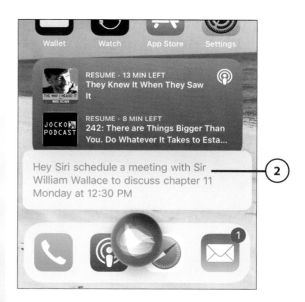

(1) Activate Siri.

(2) Speak the event you want to create. There are a number of variations in what you can say. Examples include, "Set up a meeting with William Wallace on Friday at 10:00 a.m." or "Doctor appointment on Thursday at 1:00 p.m." and so on. If you have any conflicts with the event you're setting up, Siri lets you know about them and asks you if you want to schedule the new event anyway.

3 Say "Confirm" if you don't have any conflicts or "Yes" if you do and you still want to have the appointment confirmed; you can also tap Confirm. Siri adds the event to your calendar. Say "Cancel" to cancel the event.

4 If you're done with Siri, tap outside its windows or tap the Siri symbol to activate it again.

Invitees

If you include the name of someone for whom you have an email address, Siri automatically sends invitations. If you include a name that matches more than one contact, Siri prompts you to choose the contact you want to invite. If the name doesn't match a contact, Siri enters the name but doesn't send an invitation.

Siri has created the event

Using Siri to Create Reminders

Creating reminders can be another useful thing you do with Siri, assuming you find reminders useful, of course. Here's how:

 1 Activate Siri.

2 Speak the reminder you want to create.

Here are some examples: "Remind me to buy the A-10 at Motion RC," "Remind me to finish Chapter 10 at 10:00 a.m. on Saturday," or "Remind me to buy milk when I leave work."

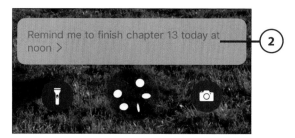

Siri provides a confirmation of what you asked. If you didn't mention a time or date when you want to be reminded, Siri prompts you to provide the details of when you want to be reminded.

3 Speak the date and time when you want to be reminded. If you included a date and time in your original reminder request, you skip this step. Unlike some of the other tasks, Siri creates the reminder without confirming it with you.

4 To add detail to the reminder, tap it. You move into the Reminders app and can add more information to the reminder, as you can when you create one manually.

Siri has created
the reminder

>>>Go Further
GOING FURTHER WITH SIRI TO MANAGE TIME

Following are some other ways to use Siri with the Calendar, Reminders, and Clock apps:

- You can change events with Siri, too. For example, if you have a meeting at 3 p.m., you can move it by saying something like, "Move my 3 p.m. meeting to Friday at 6 p.m."

- You can get information about your events with Siri by saying things such as:

  Show me today's appointments.

  Do I have meetings on November 3?

  What time is my first appointment tomorrow?

  What are my appointments tomorrow?

  Siri tells you about the events and shows you what they are on the screen. You can tap any event to view it in the Calendar app.

- You can speak to your iPhone to set alarms. Tell Siri what you want and when you want the alarm to be set. For example, you can say something like, "New alarm *alarmname* 6:00 a.m.," where *alarmname* is the label of the alarm. Siri sets an alarm to go off at that time and gives it the label you speak. It displays the alarm on the screen along with a status icon so you can turn it off if you change your mind. You don't have to label alarms, and you can just say something like, "Set alarm 6 a.m." However, a label can be useful to issue other commands. For example, if an alarm has a name, you can turn it off by saying, "Turn off *alarmname*." You can manage any alarms you create with Siri just like you would manage alarms you create directly in the Clock app. Note that alarms don't have dates associated with them so you can't set an alarm for a specific time that is more than 24 hours in the future; if you request one further out than that, Siri offers to create a reminder for you instead. You can create an alarm that activates at the same time every day by saying, "Hey Siri, create an alarm for 6 a.m. every day."

- To set a countdown timer, tell Siri to "Set timer for *x* minutes," where *x* is a number of minutes (you can do the same to set a timer for seconds or hours, too); to make it even easier, you can just say, "Hey Siri, *x* minutes." Siri starts a countdown for you and presents it on the screen. You can continue to use the iPhone however you want. When the timer ends, you see and hear an alert. You can also reset the time, pause it, and so on by speaking.

- You can get information about time by asking questions, such as "What time is it?" or "What is the date?" You can add location information to the time information, too, as in "What time is it in London, England?"

- Tapping any confirmation Siri displays takes you back into the related app. For example, if you tap a clock that results when you ask what time it is, you can tap that clock to move into the Clock app. If you ask about your schedule today, you can tap any of the events Siri presents to move into the Calendar app to work with them.

- When you use Siri to create events and reminders, they're created on your default calendar (events) or reminder list (reminders).

Using Siri to Get Information

Siri is a great way to get information about lots of different topics in many different areas. You can ask Siri for information about a subject, places in your area, unit conversion (such as inches to centimeters), and so on. Just try speaking what you want to learn to best get the information you need. Here's an example looking for Chinese restaurants in my area:

If you like Chinese food (or just about anything else), Siri can help you find it.

 (1) Activate Siri.

(2) Say something like, "Show me Chinese restaurants in my area." (Or a faster way is to combine steps 1 and 2 by saying, "Hey Siri, show me Chinese restaurants in my area.") Siri presents a list of results that match your query and even provides a summary of reviews at the top of the screen.

Siri prompts you to take action on what it found. For example, if you asked for restaurants, Siri tells you about the closest one and then asks if you want to try it. If you say yes, Siri offers to call it for you or gives you directions. You can tap other items on the list to get more information or directions.

If the result you find isn't the one you want, say no, and Siri presents the next option to you and asks if that's the one you want.

When you say yes, Siri offers to call or get directions.

(3) If you want to go to the location say, "Yes" when Siri asks if you want directions or say "Directions." Maps opens and generates directions.

(**4**) Tap GO to start navigating.

Siri is also useful for getting information about topics. Siri responds by conducting a web search and showing you the result. For example, suppose you want to learn about the F-15 fighter plane. Activate Siri and say, "Tell me about the F-15 Eagle." Siri responds with information about your topic. You can have Siri read the information by activating Siri and saying "Read." Siri reads the results (this doesn't always work; it works best when the results are presented via Wikipedia or something similar).

If you want to learn about something, just ask Siri

Using Siri to Play Music

You also can play music by telling Siri which music you want to hear.

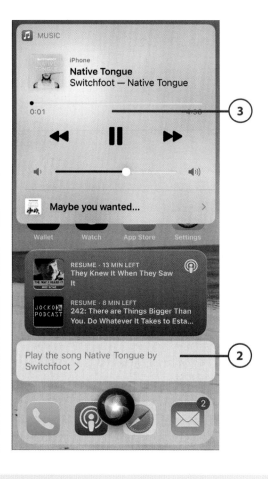

(1) Activate Siri.

(2) Tell Siri the music you want to hear. There are a number of variations in what you can say. Examples include:

Play album Time of My Life.

Play song "Gone" by Switchfoot.

Play playlist Jon McLaughlin.

Siri provides a confirmation of what you asked and begins playing the music.

(3) Use the Audio Player to control the music with your fingers.

>>>Go Further

MORE SPOKEN COMMANDS FOR MUSIC

There are a number of commands you can speak to find, play, and control music (and other audio). If you are currently playing music or other audio, you can say "Hey Siri" and the music pauses so you can speak your command. "Play *artist*" plays music by the artist you speak. "Play *album*" plays the album you name. In both cases, if the name includes the word "the," you need to include "the" when you speak the command. "Shuffle" plays a random song. "Play more like this" finds songs similar to the one playing and plays them. "Previous track" or "next track" does exactly what they sound like they do. To hear the name of the artist for the song currently playing, say "Who sings this song?" You can shuffle music in an album or playlist by saying "Shuffle playlist *playlistname*." You can stop the music, pause it, or play it by speaking those commands.

Using Siri to Get Directions

With Siri, it's easy to get directions—you don't even have to stop at a gas station to ask.

(1) Activate Siri.

(2) Say something like, "Give me directions to the airport." If you want to find a specific location, include the details, such as, "Hey Siri, give me directions to the Indianapolis International Airport." If you want directions starting from someplace other than your current location, include that in the request, such as, "Get directions from the Eagle Creek Airpark to the Indianapolis International Airport."

(3) If Siri asks if the found airport is the one you want, say "Yes." Siri uses the Maps app to generate directions.

(4) Tap Go to start turn-by-turn directions.

What's That Now?

Sometimes, Siri gets confused about what you said. In some of these cases, Siri presents a screen with suggestions about what you might have been trying to say. If one of those is what you intended to say, tap it. Siri proceeds based on what you tap. This can be faster than starting over.

Using Siri to Open Apps

As you accumulate apps on your iPhone, it can take several taps and swipes to get to a specific app, such as one that is stored in a folder that isn't on the page of the Home screen you're viewing. With Siri, you can open any app on your phone with a simple command.

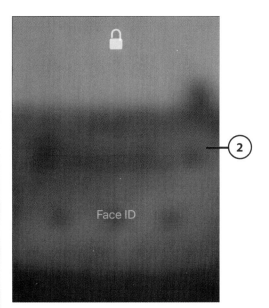

(1) Say "Hey Siri, open *appname*," where *appname* is the name of the app you want to open.

(2) If your phone needs to be unlocked to open the app, Siri prompts you to unlock it (such as by using Face ID or touching the Touch ID/Home button). Siri then opens the app for you, and you move to the last screen in the app you were using.

It's Not All Good

When Siri Misunderstands

Voice commands to Siri work very well, but they aren't perfect. Make sure you confirm your commands by listening to the feedback Siri provides when it repeats them or reviewing the feedback Siri provides on the screen. Sometimes, a spoken command can have unexpected results, which can include making a phone call to someone in the Contacts app whom you didn't intend to call. If you don't catch such a mistake before the call is started, you might be surprised to hear someone answering your call instead of hearing music you intended to play. You can put Siri in listening mode by tapping the Siri symbol, and then saying "no" or "stop" to stop Siri should a verbal command go awry.

Using Siri to Translate

Siri can translate words and phrases from the language you are using into other languages. For example, Siri supports the translation of English into French, German, Russian, Italian, Mandarin, and Spanish among others. Try this translating function to see if it supports the languages you need. Here's an example of translating an English phrase into Italian:

(1) Say "Hey Siri, translate How do I get to the airport into Italian." Siri does the translation for you and speaks the translated phrase.

(2) Tap Play to have Siri speak the translation again.

No Can Do
If Siri is unable to complete the translation you requested, it tells you that it can't speak that language yet. Apple adds language support to Siri over time so you can try again in the future to see if support for the language you need has been added.

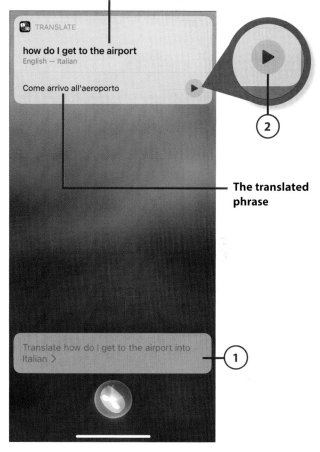

The phrase Siri is translating

The translated phrase

Translate with Authority
Doing one-off translations with Siri is useful, but if you need to do many translations, check out the Translate app. Select the language you want to translate from on the left side. Select the language you want to translate to on the right side. Tap the Microphone and speak the phrase you want to translate. The app performs the translation and speaks the translated phrase.

Tap to configure Safari

Tap to have the World Wide Web in the palm of your hand

In this chapter, you explore the amazing web browsing functionality your iPhone has to offer. Topics include the following:

→ Getting started
→ Visiting websites
→ Viewing websites
→ Working with multiple websites at the same time
→ Searching the Web
→ Working with bookmarks
→ Signing in to websites automatically

Surfing the Web

The Web has become an integral part for much of our lives. It is often the first step to search for information, make plans (such as travel arrangements), conduct financial transactions, shop, and so much more. Safari on the iPhone puts the entire Web in the palm of your hand.

Getting Started

The World Wide Web, more commonly called the Web, is a great resource for finding information, planning travel, keeping up with the news, and just about anything else you want to do.

You can use the iPhone's Safari web browser app to take advantage of all the Web has to offer. Safari is a full-featured web browser so that you don't have to make any compromises when using the Web on your iPhone. For example, you can use bookmarks to make returning to the same web pages easy, you can perform web searches, and do all the other things for which the Web has become an essential element of modern life.

Visiting Websites

If you've used a web browser on a computer before, using Safari on an iPhone is a familiar experience. If you've not used a web browser before, don't worry; using Safari on an iPhone is simple and intuitive.

Syncing Bookmarks

Using iCloud, you can synchronize your Internet Explorer favorites on a Windows PC or Safari bookmarks on a Mac to your iPhone so you have the same set of bookmarks available on your phone that you do on your computer and other devices, and vice versa (refer to Chapter 4, "Setting Up and Using an Apple ID, iCloud, and Other Online Accounts"). You can enable this functionality before you start browsing on your iPhone to avoid typing URLs or re-creating bookmarks. To do so, open the Settings app, tap your information at the top of the screen, tap iCloud, and ensure that the Safari switch is on (green).

Using Bookmarks to Move to Websites

Using bookmarks you've synced via iCloud or created on your iPhone (you learn how later in this chapter) makes it easy to get to websites.

(1) On the Home screen, tap Safari.

(2) Tap Bookmarks.

(3) Tap the Bookmarks tab if it isn't selected already. (If you don't see this tab, tap back [<], which is labeled with the name of the folder from which you moved to the current screen, in the upper-left corner of the screen until you see Bookmarks at the top of the screen.)

(4) Swipe up or down the list of bookmarks to browse the bookmarks and folders containing bookmarks available to you.

(5) To move to a bookmark, skip to step 10; to open a folder of bookmarks, tap it.

(6) Swipe up or down the folder's screen to browse the folder and bookmarks it contains.

(7) Tap a folder to see the bookmarks it contains.

Change Your Mind?

If you decide not to visit a bookmark, tap Done. You return to the website you were previously viewing.

(8) To return to a previous screen, tap back (<) in the upper-left corner of the screen, which is labeled with the name of the folder you previously visited (the parent folder); this disappears when you are at the top-level Bookmarks screen.

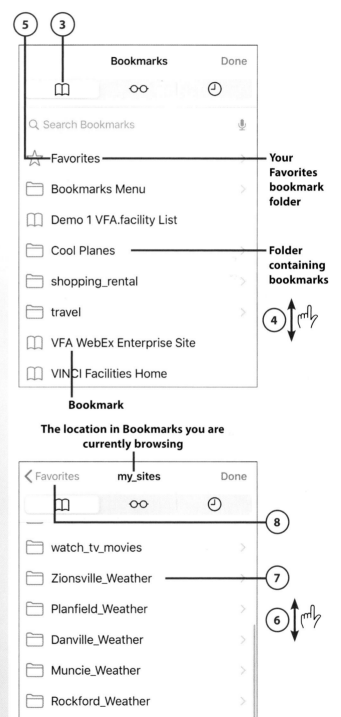

(5) (3)

Bookmarks   Done

Q Search Bookmarks

Favorites —————————— Your Favorites bookmark folder

Bookmarks Menu

Demo 1 VFA.facility List

Cool Planes ——————— Folder containing bookmarks

shopping_rental

travel (4)

VFA WebEx Enterprise Site

VINCI Facilities Home

Bookmark

The location in Bookmarks you are currently browsing

< Favorites   my_sites   Done

(8)

watch_tv_movies

Zionsville_Weather ——————— (7)

Planfield_Weather (6)

Danville_Weather

Muncie_Weather

Rockford_Weather

RC_sites

9 Repeat steps 5–8 until you see a bookmark you want to visit.

10 Tap the bookmark you want to visit. Safari moves to that website.

11 Use the information in the section "Viewing Websites" later in this chapter to get information on viewing the web page.

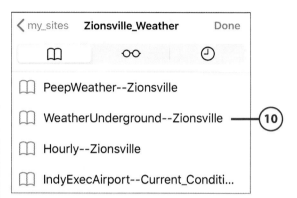

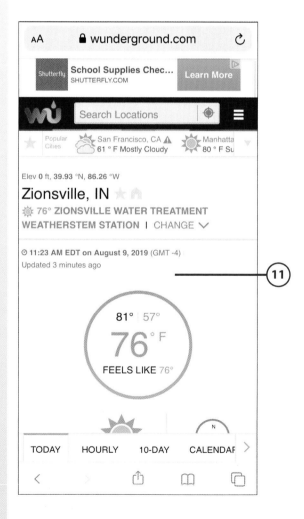

Using Your Favorites to Move to Websites

Using the Safari settings described in the Go Further sidebar "Safari Settings" at the end of this chapter, you can designate a folder of bookmarks as your Favorites. You can get to the folders and bookmarks in your Favorites folder more quickly and easily than navigating to it as described in the previous section. Here's how to use your Favorites:

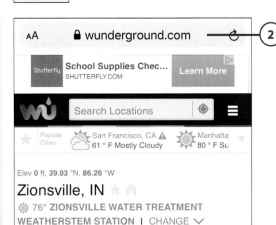

1. On the Home screen, tap Safari. (If you are in Safari and have the Bookmarks screen open, tap Done to close it.)

2. Tap in the Address/Search bar (if you don't see the Address/Search bar, tap at the top of the screen to show it). After you tap in the Address/Search bar, you see your Favorites (bookmarks and folders of bookmarks) immediately under the Address/Search bar. The keyboard opens at the bottom of the screen.

Show More/Show Less

You might see only a subset of your favorites. Tap Show More to see all your favorites or Show Less to display fewer of them.

Frequently Visited

Safari tracks sites you visit frequently and lists them in the Frequently Visited section at the bottom of the Favorites screen. Tap a site to return to it.

3 Swipe up and down on your Favorites. The keyboard closes to give you more room to browse.

4 To move to a bookmark, tap it and skip to step 8.

5 Tap a folder to move into it.

6 Continue browsing your Favorites until you find the bookmark you want to use. Like using the Bookmarks screen, you can tap a folder to move into it, tap a bookmark to move to its website, or tap back (<) to move to a previous screen.

7 Tap the bookmark for the site you want to visit.

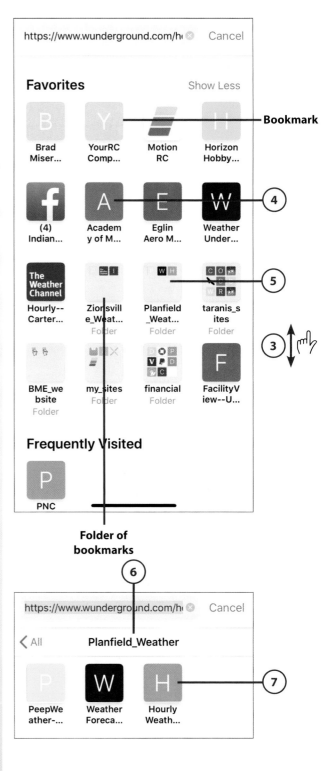

Bookmark

Folder of bookmarks

8 Use the information in the section "Viewing Websites" later in this chapter to view the web page.

forecast.weather.gov

Typing URLs to Move to Websites

A Uniform Resource Locator (URL) is the Internet address of a web page. URLs can be relatively simple, such as www.apple.com, or they can be quite long and convoluted. The good news is that by using bookmarks, you can save a URL in Safari so you can get back to it using its bookmark (as you learned in the previous two tasks) and thus avoid typing URLs more than once. To use a URL to move to a website, do the following:

1 On the Home screen, tap Safari. (If you are in Safari and have the Bookmarks screen open, tap Done to close it.)

2 Tap in the Address/Search bar (if you don't see the Address/Search bar, tap at the top of the screen). The URL of the current page becomes highlighted, or if you haven't visited a page, the Address/Search bar is empty. Just below the Address/Search bar, your Favorites are displayed. The keyboard appears at the bottom of the screen.

3 If an address appears in the Address/Search bar, tap Clear (x) to remove it.

4 Type the URL you want to visit. If it starts with www (which almost all URLs do), you don't have to type "www." As you type, Safari attempts to match what you are typing to a site you have visited previously and completes the URL for you if it can. Just below the Address/Search bar, Safari presents a list of sites that might be what you're looking for, organized into groups, such as Top Hits.

5 If one of the sites shown is the one you want to visit, tap it. You move to that web page; skip to step 8.

6 If Safari doesn't find a match, continue typing until you enter the entire URL (not shown).

7 Tap go. You move to the web page.

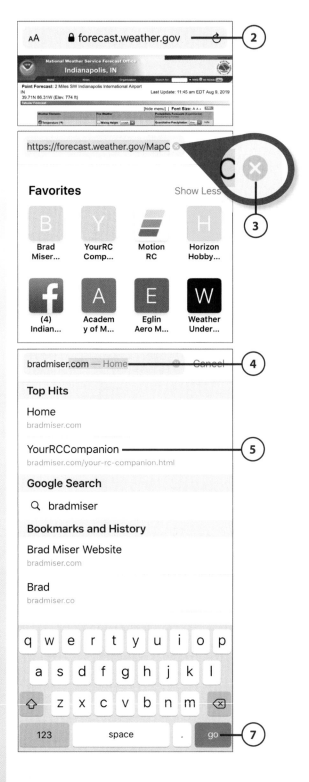

8 Use the information in the section "Viewing Websites" to view the web page.

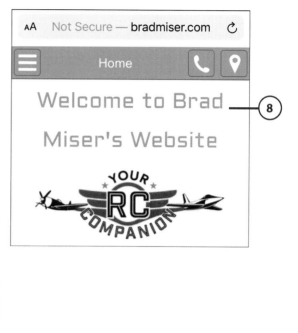

Shortcut for Typing URLs

URLs include a top-level domain code that represents the type of site (theoretically anyway) that URL leads to, such as .com (commercial sites), .edu (educational sites), and .gov (government). To quickly enter a URL's code, tap and hold the period key to see a menu from which you can select other options, such as .net or .edu. Select the code you want on the keyboard, and it's entered in the Address/Search bar.

Using Your Browsing History to Move to Websites

As you move about the Web, Safari tracks the sites you visit and builds a history list (unless you enable the Do Not Track option as explained in the Go Further sidebar "Safari Settings" at the end of this chapter). You can use your browsing history list to return to sites you've visited.

1 Tap Bookmarks.

2 Tap History.

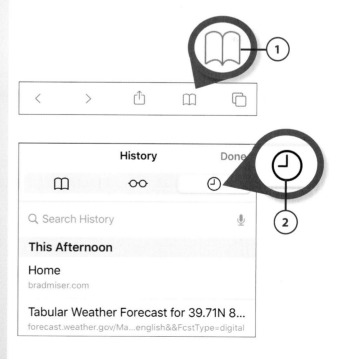

③ Swipe up and down the page to browse all the sites you've visited. The more recent sites appear at the top of the screen; the further you move down the screen, the further back in time you go. Earlier sites are collected in folders for various times, such as This Morning, or Monday Afternoon.

④ Tap the site you want to visit. The site opens and you can use the information in the section "Viewing Websites" to view the web page.

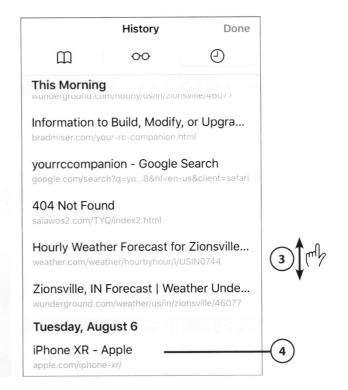

Erasing the Past

To clear your browsing history, tap Clear at the bottom of the History screen. At the prompt, tap the timeframe that you want to clear; the options are The last hour, Today, Today and yesterday, or All time. Your browsing history for the period of time you selected is erased. (Don't you wish it were this easy to erase the past in real life?)

Viewing Websites

Even though your iPhone is a small device, you'll be amazed at how well it displays web pages, even if they are designed for larger screens.

(1) Use Safari to move to a web page as described in the previous tasks.

(2) To browse around a web page, swipe your finger right or left, or up or down.

(3) Zoom in by unpinching your fingers.

Where Did the URL Go?

When you first move to a URL, you see that URL in the Address/Search bar. After you work with a site, the Address/Search bar and the toolbar are hidden and the URL is replaced with the high-level domain name for the site (such as sitename.com, sitename.edu, and so on). To see the Address/Search bar and toolbar again, tap the top of the screen. To see the full URL again, tap in the Address/Search bar.

(4) Zoom out by pinching your fingers.

(5) Tap a link to move to the location to which it points. Links can come in many forms, including text (most text that is a link is in color and underlined) or graphics. The web page to which the link points opens and replaces the page currently being displayed.

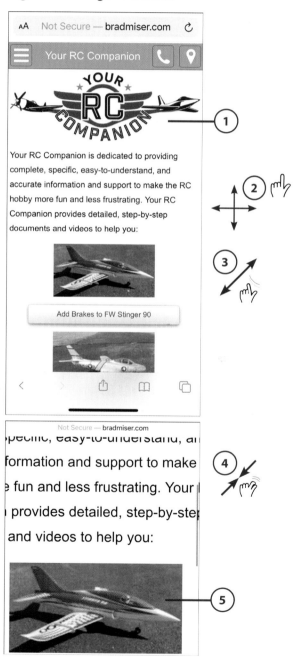
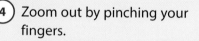

6. View the web page in landscape orientation by rotating the iPhone so that it is horizontal.

7. Scroll, zoom in, and zoom out on the page to read it, as described in steps 2–4.

8. Tap Refresh to refresh a page, which causes its content to be updated. (Note: While a page is loading, this is Stop [x]; tap it to stop the rest of the page from loading.)

9. To move to a previous page you've visited, tap back (<). (If the arrow is grayed out, it means you're at the beginning of the set of pages you've visited.)

10. To move to a subsequent page, tap forward (>). (If the arrow is grayed out, it means you're at the end of the set of pages you've visited.)

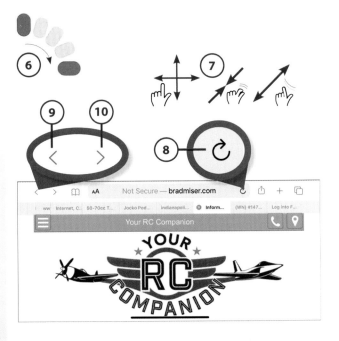

Disappearing Address/Search Bar

As you move around, the Address/Search bar at the top of the page and the toolbar at the bottom of the page are hidden automatically; to show them again, tap the top of the screen.

Working with Multiple Websites at the Same Time

When you move to a web page by using a bookmark, typing a URL, or tapping a link on the current web page, the new web page replaces the current one.

However, you can open and work with multiple web pages at the same time so that a new web page doesn't replace the current one.

When you work with multiple web pages, each open page appears in its own tab. You can use the Tab view to easily move to and manage your open web pages. You also can close open tabs or open new, empty tabs and then move to websites.

There are two ways to open a new web page in a new tab. One is to touch and hold on a link on the current web page; you can use the resulting Open command to open the new page. By default, Safari is set to open a new page using the Open in New Tab option. This causes the new page to open and move to the front so you see it instantly as the current page and its tab to move to the background.

The other option is Open in Background. When you open a link using this option, the new page opens and moves to the background; you don't see the new page until you use the Tab Manager to move to the new tab.

Open in Background

Because it's the default option, this section focuses on the Open in New Tab method. If you want to use the Open in Background option, refer to the Go Further sidebar "Safari Settings" at the end of this chapter to learn how to change this setting. After you do, when you touch and hold on a link, you see the Open in Background option instead of Open in New Tab.

Tapping Without Holding

When you tap, but don't hold down, a link on a web page, the web page to which the link points opens and replaces the current web page—no new tab is created. To create a new tab, you need to hold on the link so you can use the Open in New Tab command.

Opening New Pages in a New Tab

Using the Open in New Tab option, you can open new pages by doing the following:

(**1**) Touch and hold on the link you want to open in the background. The Quick Action menu appears.

(**2**) Tap Open in New Tab. The Tab view appears briefly, and a new tab opens and displays the page to which the link points. The web page from which you started moves into the background.

(**3**) Continue opening pages (not shown); see "Using Tab view to Manage Open Web Pages" to learn how to use the Tab view to manage your open pages.

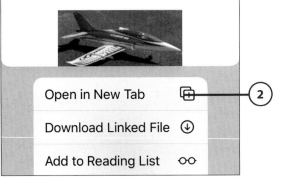

Using Tab view to Manage Open Web Pages

Safari's Tab view enables you to view and work with your open pages/tabs. Here's how:

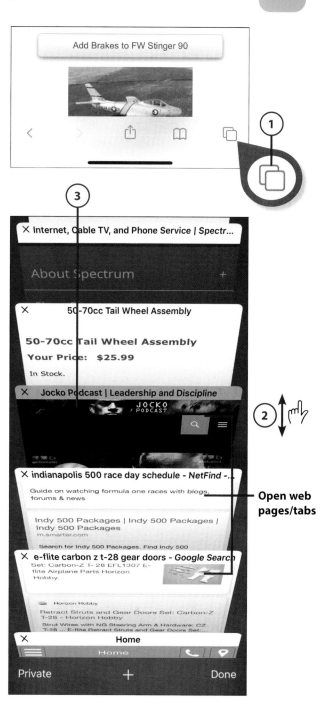

Open web pages/tabs

(1) Tap Tab view. Each open page appears on its own tab.

(2) Swipe up or down on the open tabs to browse them.

(3) Tap a tab/page to move into it. The page opens and fills the Safari window.

(4) Work with the web page.

(5) Tap Tab view.

Tabs Are Independent

Each tab is independent. So, when you're working with a tab and use back/forward to move among its pages, you are just moving among the pages open under that tab. Pages open in other tabs are not affected.

(6) Tap Close (x) to close a tab; alternatively swipe to the left on the tab you want to close.

(7) To open a new tab, tap Add (+) to create a new tab that shows your Favorites screen; navigate to a new page in that tab using the methods described in other tasks (tapping bookmarks or typing a URL).

(8) Tap Done to close the Tab view. The Tab view closes, and the page you were most recently viewing is shown.

Keep Private Things Private

If you aren't browsing in Private mode and tap Private at the bottom of the Tab view, Safari moves into Private mode. In Private mode, your activity on the Web isn't tracked so no history of sites or pages you've visited is created. To exit Private mode, tap Private again. If you are browsing in Private mode, tapping Private shows or hides the tabs in the Tab view.

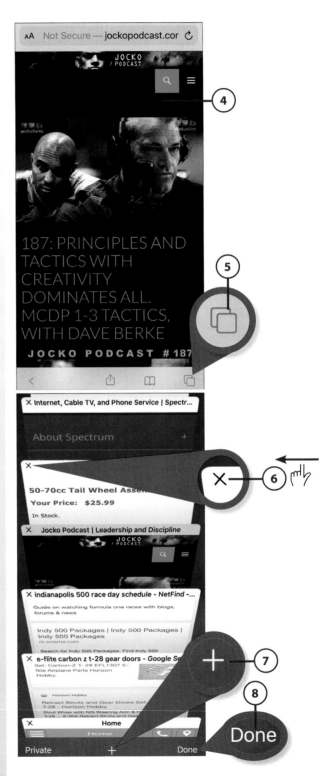

Searching the Web

Searching the Web might be the most useful thing you can do with your iPhone because you can quickly find information on just about anything.

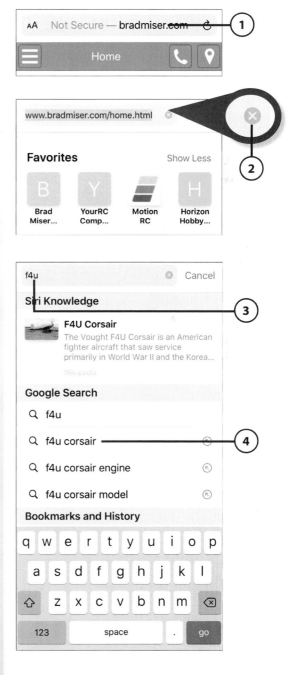

(1) Tap in the Address/Search bar (if you don't see this bar, tap at the top of the screen). The keyboard appears along with your Favorites.

(2) If there is any text in the Address/Search bar, tap Clear (x).

(3) Type your search word(s). As you type, Safari attempts to find a search that matches what you typed. The list of suggestions is organized in sections, which depend on what you are searching for and the search options you configured through Safari settings. One section, labeled with the search engine you're using (such as Google Search), contains the search results from that source. Other sections can include Siri Knowledge, Bookmarks and History, or Apps (from the App Store). At the bottom of the list is the On This Page section, which shows the terms that match your search on the page you are browsing.

(4) To perform the search using one of the suggestions provided, tap the suggestion you want to use. The search is performed and you can skip to step 6.

(5) If none of the suggestions are what you want, keep typing until you have entered the entire search term, and then tap go. The search engine you use performs the search and displays the results on the search results page.

(6) Use the search results page to view the results of your search. These pages work just like other web pages. You can zoom, scroll, and tap links to explore results.

These Tabs Are Made for Searchin'

Using the Open in New Tab command is particularly useful when you're searching. When you find something of interest, open it in a new tab. This leaves your search results open in the current tab so you can easily come back to them to explore something else.

Searching on a Web Page

To search for words or phrases on a web page you're viewing, perform these steps, except in step 4, tap the word or phrase for which you want to search in the On This Page section (you probably have to swipe up the screen to see this section). You return to the page you're browsing, and each occurrence of your search term on the page is highlighted.

Working with Bookmarks

In addition to moving bookmarks from a computer or iCloud onto your iPhone, you can save new bookmarks directly on your iPhone so you can easily return to the associated websites. Any bookmarks you create on your iPhone also become available on your other devices, such as an iPad. You can also organize bookmarks in folders on your iPhone to make them easier and faster to access.

Creating Bookmarks

When you want to make it easy to return to a website, create a bookmark with the following steps:

1. Move to a web page for which you want to save a bookmark.

2. Tap Share. The Share sheet appears.

3. Tap Add Bookmark (you might have to swipe up the screen to see this). The Add Bookmark screen appears, showing the title of the web page you're viewing, which is also the name of the bookmark initially; its URL; and the Location field, which shows where the bookmark will be stored when you create it.

Straight to Favorites

If you want to create a bookmark in your Favorites area, tap Add to Favorites and skip the rest of these steps. You can access your new bookmark by opening the Favorites screen.

4. Edit the bookmark's name as needed, or tap Clear (x) to erase the current name, and then type the new name of the bookmark. The titles of some web pages can be quite long, so it's a good idea to shorten them so the bookmark's name is easier to read on the iPhone's screen.

5. Tap the current folder shown under LOCATION to expand the section. You see all the folders of bookmarks on your phone. The folder that is currently selected is marked with a check mark.

6. Swipe up and down the screen to find the folder in which you want to place the new bookmark. You can choose any folder on the screen; folders are indented when they are contained within other folders.

7. Tap the folder in which to store the new bookmark. You return to the Add Bookmark screen, which shows the location you selected.

8. Tap Save. The bookmark is created and saved in the location you specified. You can use the bookmark to return to the website at any time.

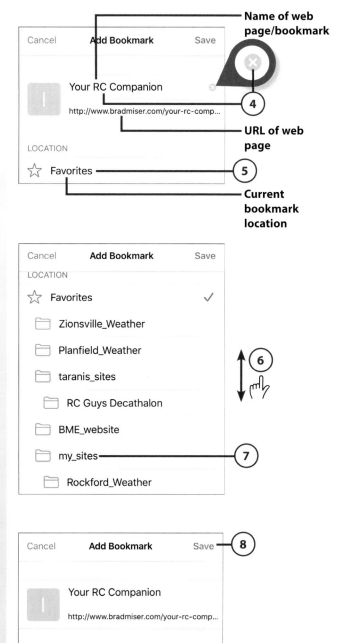

Name of web page/bookmark

URL of web page

Current bookmark location

New location

Creating Folders for Bookmarks

You can organize bookmarks into folders so they're easier to find and use. To create a new bookmark folder, perform the following steps:

(1) Move to the Bookmarks screen showing the location where you want to create the new folder. For example, creating folders on your Favorites page makes them very easy to access.

(2) Tap Edit.

(3) Tap New Folder.

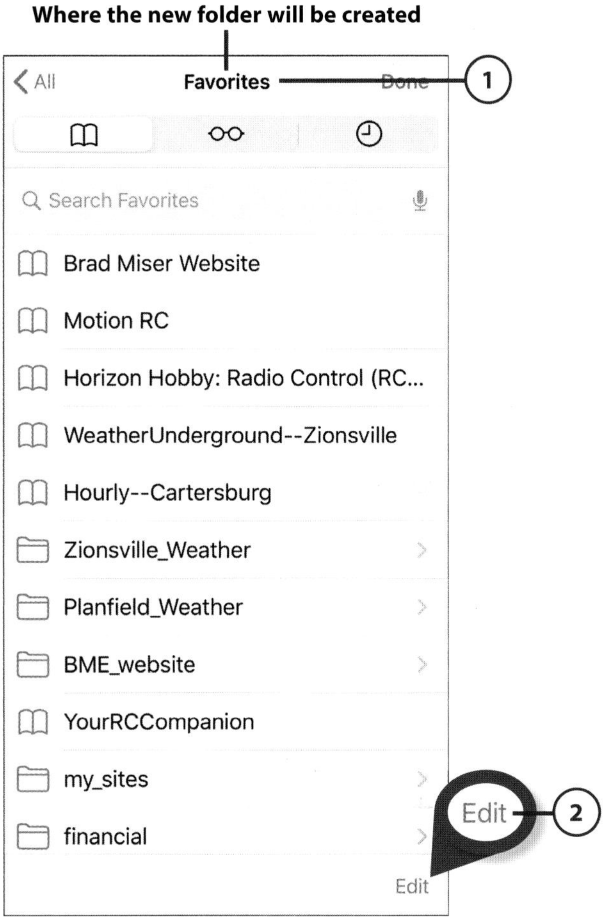

Where the new folder will be created

‹ All   Favorites   ~~Done~~ — (1)

Q Search Favorites

- 📖 Brad Miser Website
- 📖 Motion RC
- 📖 Horizon Hobby: Radio Control (RC...
- 📖 WeatherUnderground--Zionsville
- 📖 Hourly--Cartersburg
- 📁 Zionsville_Weather   ›
- 📁 Planfield_Weather   ›
- 📁 BME_website   ›
- 📖 YourRCCompanion
- 📁 my_sites   ›
- 📁 financial   ›

Edit — (2)

Edit

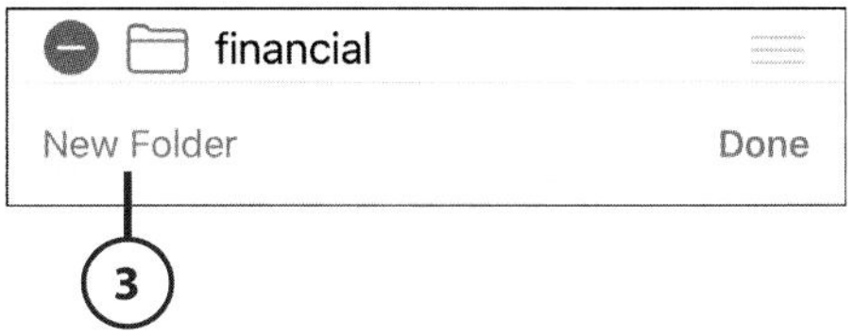

➖ 📁 financial

New Folder   Done

(3)

(4) Type the name of the new folder.

(5) Tap done. The folder is created at the top of the list you selected in step 1.

(6) Tap Done.

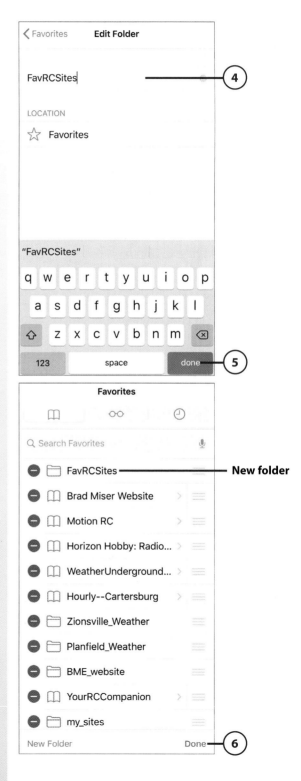

Deleting Bookmarks or Folders of Bookmarks

You can get rid of bookmarks or folders of bookmarks you don't want any more by deleting them:

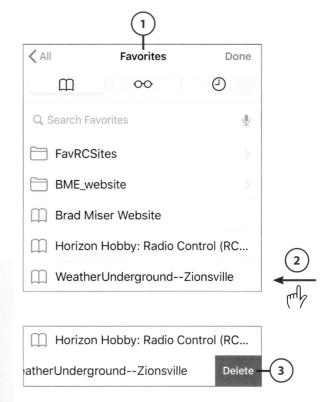

1. Move to the screen containing the folder or bookmark you want to delete.

2. Swipe to the left on the folder or bookmark you want to delete.

3. Tap Delete. The folder or bookmark is deleted. Note that when you delete a folder, all the bookmarks it contains are deleted, too.

Delete with a Swipe

You can combine steps 2 and 3 by swiping all the way to the left on the screen. The folder or bookmark is deleted immediately.

Signing Into Websites Automatically

Frequently, you'll need to enter your username and password to log into your account on websites you visit with your iPhone. Having to type this information every time is no fun. You can enable Safari to store and enter this information for you automatically. Your information is relatively secure because you still need a passcode (usually entered via Face ID or Touch ID) to unlock your iPhone. So, having Safari store your usernames and passwords is both relatively secure and very convenient.

Setting Safari to Remember Passwords

Your iPhone has to be configured to be able to save your passwords. To do so, perform the following steps:

(1) Open the Settings app and tap Passwords.

(2) Look at the screen (Face ID), touch the TouchID/Home button (Touch ID), or enter your passcode at the prompt. (Not shown on a figure.)

(3) Tap AutoFill Passwords.

(4) Set the AutoFill Passwords switch to on (green).

>>>*Go Further*

REMEMBERING USERNAMES AND PASSWORDS

When you enable Safari to save your passwords, you can access them by opening the Settings app and tapping Passwords. If you have enabled Face ID and are looking at the phone, you move to the Passwords screen; if you are using Touch ID or have to enter your passcode manually, use those methods to move to the Passwords screen. On the Passwords screen, you see a list of websites and usernames stored on your iPhone sorted by the URL of the website with which they are associated. You can quickly search for a website by typing information about it in the Search bar.

You can tap a URL to see the username and password for that site and a list of all websites using that information (looking up a password here can be handy if you've forgotten it). You can edit this information, or in some cases, you can tap Change Password on Website to visit the website to change the password associated with your account.

To remove a URL and its associated username and password from the list, move to the Passwords screen and swipe to the left on it and tap Delete. The next time you move to that site, you will need to manually enter your username and password again.

Saving Your User Account for a Website

When you visit a website on which you have an account that you sign into, Safari prompts you to store your login information. When saved, this information can be entered for you automatically or with a single tap.

(**1**) Move to a web page that requires you to log in to an account.

(**2**) Enter your account's username and password.

(**3**) Tap the icon to log in to your account, such as Continue, Sign In, Submit, Login, and such. You are prompted to save the login information.

(**4**) Tap Save Password to save the information. The next time you move to the login page, your username and password are entered for you automatically. Tap Never for This Website if you don't want the information to be saved and you don't want to be prompted again. Tap Not Now if you don't want the information saved but do want to be prompted again later to save it.

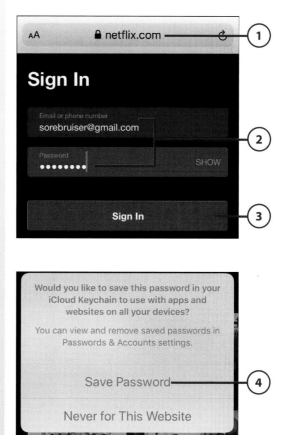

Using Saved User Accounts on Websites

(1) The next time you need to sign in to your account, tap your account information at the prompt. (You might need to confirm this action with Face ID or Touch ID.) Your user information is entered for you.

(2) Complete the sign-in process.

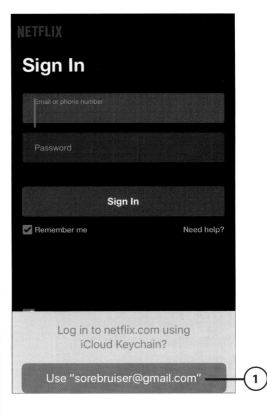

>>>Go Further
LETTING SAFARI CREATE PASSWORDS FOR YOU

Safari can create passwords for you. Go to a website that requires you to create a password, such as when you register for a new account. When you tap in a field that Safari recognizes as requiring a password, tap Suggest Password. Safari presents a password for you; most of these are not easy to remember, but that doesn't matter when you are using you iPhone because they are saved for you automatically. If you want to use the recommended password, tap Use Suggested Password; Safari enters the password in the password and verify password fields. When syncing is enabled, the password is stored on the synced devices, too, so you're able to sign in from those devices just as easily. If you don't use Safari and allow syncing on other devices, you'll need to manually enter the passwords Safari creates on those devices.

>>>Go Further
SAFARI SETTINGS

Like most apps, Safari offers settings you can use to adjust the way it works. You can likely use Safari with its default settings just fine; when the time comes that you want to tweak how Safari works for you, some of the more useful options are explained in the following table. To access these settings, tap the Settings icon on the Home screen, and then tap Safari.

Safari Settings

Settings Area	Location	Setting	Description
SEARCH	Search Engine	Search Engine	Enables you to choose your default search tool; the options are Google (default), Yahoo, Bing, or DuckDuckGo.
GENERAL	Autofill	Use Contact Info	Set the Use Contact Info switch to on (green) if you want to be able to automatically fill in your contact information (such as your address) on web forms.

Settings Area	Location	Setting	Description
GENERAL	Autofill	My Info	If you enable Use Contact Info, tap My Info and choose your card in the Contacts app. The information on this card will be used to automatically complete information on web forms.
GENERAL	Autofill	Credit Cards	Set the Credit Cards switch to on (green) to store and use credit cards on your iPhone so you can more easily enter their information to make purchases.
GENERAL	Autofill	Saved Credit Cards	Tap Saved Credit Cards and use Face ID or Touch ID or enter your passcode to view or change existing credit card information or to add new credit cards to Safari.
GENERAL	Favorites	Favorites	Use this option to choose the folder of bookmarks for sites that you use most frequently. The bookmarks in the folder you select appear at the top of the screen when you move into the Address/Search bar, making them fast and easy to use.
TABS	Open Links	Open Links	This tells Safari the option you want to see when you tap and hold a link on a current web page to open a new web page. The In New Tab option causes Safari to open and immediately take you to a new tab displaying the web page with which a link is associated. The In Background option causes Safari to open pages in the background for links you tap so you can view them later.

Settings Area	Location	Setting	Description
TABS	Close Tabs	Close Tabs	You can use this option to have Safari automatically close tabs you haven't viewed recently. For example, if you choose After One Week, Safari automatically closes tabs you haven't looked at for a week. This prevents an accumulation of tabs in Safari that you no longer need.
PRIVACY & SECURITY	N/A	Fraudulent Website Warning	If you don't want Safari to warn you when you visit websites that appear to be fraudulent, set the Fraudulent Website Warning switch to off (white). You should leave this switch enabled (green).
N/A	N/A	Clear History and Website Data	When you tap this command and confirm it by tapping Clear History and Data at the prompt, Safari removes the websites you have visited from your history list. The list starts over, so the next site you visit is added to your history list again—unless you have enabled private browsing. It also removes all cookies and other website data that have been stored on your iPhone.

Take photos and video

Take photos and
video and quickly
email or text
them to others

Taking Photos and Video with Your iPhone

The iPhone's cameras and Camera app capture high-quality photos and video. Because you'll likely have your iPhone with you at all times, it's handy to take photos with it whenever and wherever you are. And you can shoot video just as easily.

Whether you've taken photos and video on your iPhone or added them from another source, the Photos app enables you to edit, view, organize, and share your photos. (To learn how to use the Photos app, see Chapter 16, "Viewing and Editing Photos and Video with the Photos App.")

Getting Started

Each generation of iPhone has had different and more sophisticated photo and video capabilities and features than the previous versions. All current versions sport high-quality cameras; in fact, there's at least one camera on each side of the iPhone. The cameras on the back of the phone take photos of what you're looking at (the back-facing cameras), whereas the camera on the front side takes photos of what the screen is facing (the front-facing camera, which is usually for taking selfies or having FaceTime conversations, located on the face of the phone).

All current generations of iPhone have a flash; can zoom; take burst, panoramic, and time-lapse photos; and have other features you expect from a high-quality digital camera. They also can take Live Photos, which capture a small amount of video (about 3 seconds) along with the photo; when you view a Live Photo, you can hold on the screen to see its video.

The iPhone 7 and later models have automatic image stabilization, higher resolution, and other enhancements so you can take even higher quality photos and video with both the back-facing and front-facing cameras.

From there, the differences between the models become greater as additional cameras have been added to the backside of the phone. The following list describes the major photographic features of various groups of iPhone models that can run iOS 14:

- **SE, 6s, 6s Plus, 7, 8, and X$_R$**—These models have one camera on the back. While they lack some of the features of more advanced models, they are more than capable of taking excellent photos and video in many different situations.

- **7 Plus, 8 Plus, X, Xs, and Xs Max**—These models have two cameras on the back. With these models, you can use Telephoto zoom providing greater magnification than models with a single camera. These models also enable you to use Portrait mode in which the subject is in very sharp focus and the background is a soft blur.

- **11, 12, and 12 mini**—These models also have two cameras but don't have Telephoto zoom. Instead, they support the Ultra Wide mode, which is analogous to using a wide-angle lens. They also have Night mode, which enables them to take photos in dark conditions without the use of the flash. These models also support Portrait mode. The 12 and 12 mini support the newest version of HDR (High Dynamic Range, which is explained later in this chapter).

- **11 Pro, 11 Pro Max, 12 Pro, and 12 Pro Max**—These models have three cameras on the back and support many options including Telephoto zoom, Ultra Wide mode, Night mode, and Portrait mode. The 12 Pro and 12 Pro Max also have a Light Detection and Ranging (LiDAR) sensor that improves photos in low-light conditions and faster autofocus; these two models also support the current version of HDR.

Sections in this chapter correspond to each of these groups. At the beginning of each of these sections, you see which iPhone models the section applies to; if you don't have a model in the listed group, you can skip over that section.

Be aware that there are differences even among the models in the same group, so you might see some minor differences between what you see on your phone and the figures in this chapter. However, you can still follow the steps that apply for the group associated with your iPhone.

Regardless of the model, the iPhone's photo and video capabilities have been increasingly tied into iCloud. For example, you can store your entire photo library under your iCloud account; this offers many benefits, including backing up all your photos and making it easy to access your photos from any iCloud-enabled device. Therefore, I've assumed you are using iCloud and have configured it to work with photos as described in Chapter 4, "Setting Up and Using an Apple ID, iCloud, and Other Online Accounts." Like differences in iPhone camera capabilities, if you don't use iCloud with your photos, some of the information in this chapter doesn't apply to you and what you see on your phone might look different than what you see in this chapter.

Lastly, the Camera app's settings can change how it works. See the Go Further sidebar, "Camera Settings," at the end of this chapter for an overview of the more important settings that you might want to configure.

Using the Camera App to Take Photos and Video with Your iPhone

You use the Camera app to take photos and video with your iPhone. This app has a number of controls and features. Some features are easy to spot, whereas others aren't so obvious. By the end of this chapter, you'll know how to use these features to take great photos and video with your iPhone.

The general process for capturing photos or video follows:

1. Choose the type of photo or video you want to capture.

2. Set the options for the type of photo or video you selected.

3. Take the photos or video.

4. View and edit the photos or video you captured using the Photos app.

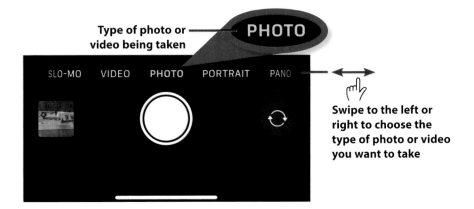

Type of photo or video being taken — PHOTO

Swipe to the left or right to choose the type of photo or video you want to take

The information you need to accomplish steps 1 through 3 of this process is provided in tables and tasks throughout this chapter. The details for step 4 are provided in Chapter 16.

The first step in taking photos or video is to choose the type of photo or video that you want to capture. You do this by swiping to the left or right on the selection bar just above the large Shutter icon at the bottom of the Camera app's screen, as shown in the previous figure. The option shown in yellow at the center of the screen just above the Shutter icon is the current type of photo or video you're capturing. The options available in the Camera app are explained in the following table. (If your iPhone isn't capable of taking a type of photo or video listed, you won't see it on your screen.)

Types of Photo and Video iPhones Can Capture

Type of Photo or Video	Description
TIME-LAPSE	Captures a video with compressed time so that the time displayed in the video occurs much more rapidly than "real time." This is what is often used to show a process that takes a long time, such as a plant growing, in just a few seconds.
SLO-MO	Takes slow motion video so that you can slow down something that happens quickly.
VIDEO	Captures video at a real-time speed. The steps to take video are provided in the task "Taking Video," later in this chapter.

Type of Photo or Video	Description
PHOTO	Captures static photos or Live Photos. Step-by-step instructions showing how to use this option are provided in the task "Taking Photos with Standard Zoom," later in this chapter. Using an iPhone's telephoto zoom capabilities is described in "Taking Photos with Telephoto Zoom." Taking photos in Ultra Wide mode is covered in "Taking Wide Angle Photos." Using Night mode is covered in "Taking Photos with Night Mode."
PORTRAIT	Takes portrait photos of people using a sharp focus in the foreground on the subject and a soft blur in the background. Additionally, there are a number of lighting options that you can apply to the portrait photos. On the 12 Pro, 12 Pro Max, Xs, Xs Max, 11 Pro, and 11 Pro Max, you can also adjust photo depth. On some models, you can also choose between 1x and 2x magnification for Portrait photos The steps to take portrait photos are in "Taking Portrait Photos."
SQUARE	Takes "square-shaped" photos in which the height and width are the same. This mode isn't provided on the 12, 12 mini, 12 Pro, 12 Pro Max, 11, 11 Pro, or 11 Pro Max because these models enable you to choose among several different proportions when you are using the PHOTO tool. (The iPhone SE doesn't offer it either.)
PANO	Takes panoramic photos that enable you to capture very wide images. An example of capturing a panoramic photo is provided in the task "Taking Panoramic Photos," later in this chapter.

After you choose the type of photo or video you want to take, there are options you can select to adjust how that photo or video is captured (the options available to you depend on the type of photo or video you're taking and the specific model of iPhone you're using). When you select options, the icons you see on the screen change to reflect your selection. For example, when you choose a self-timed photo, the Self-timer icon changes to show the time delay you've selected. Not all options are available at the same time. For example, you can't set the flash and HDR to go on at the same time because you can't take HDR images with the flash. If you use a model that has Smart HDR, you don't turn it off or on for specific photos; it works automatically if it's enabled.

The following table describes the icons and tools available on the Camera app's screen. Remember that the specific icons and tools you see depend on the type of photo or video you're capturing and the model of iPhone you're using. These icons can also appear in different places depending on the model you're using. For example, on an iPhone 12 Pro, they're shown on a toolbar just above the Shutter button, whereas on an iPhone Xs, they appear at the top of the screen. (In the table, iPhone 11 or later models refers to the iPhone 12, 12 Pro, 12 Pro Max, 11, 11 Pro, and 11 Pro Max models.)

Photo and Video Options and Icons

Icon	Description
	Show/Hide Toolbar—On iPhone 11 or later models, these arrows appear at the center of the top of the screen; tap one to show (arrow pointing up) or hide the toolbar (arrow pointing down). When shown, the toolbar appears above the Shutter icon.
	Toolbar—On iPhone 11 or later models, this toolbar enables you to quickly select options including (from left to right) Flash, Night mode, Live Photos, Proportion, Exposure, Self-timer, and Filter (swipe to the left on some models to see this).
	Flash—When you tap this icon, you see a menu with the flash options, which are Auto (the app uses the flash when required), On (flash is always used), or Off (flash is never used). Tap the option you want to use and the menu closes. When the flash is set to on, the icon is yellow. On iPhone 11 or later models, tap this at the top of the screen to turn the flash on or off. Tap it on the toolbar at the bottom of the screen to set it to Flash Auto, Off, or On.
	Flash Being Used—When this icon appears on the screen, it indicates the flash will be used when taking a photo or video. On iPhone 11 and later models, the flash icon at the top of the screen turns yellow to indicate the flash is on; it is white and has a slash when the flash is off.
	High Dynamic Range (HDR)—Tap this to set the HDR options (only models that don't support Automatic or Smart HDR). (You learn more about HDR in the Go Further sidebar, "More on Taking Photos and Video," later in this chapter.) The options are Auto (Flash Auto on iPhone 11 and later models), On, or Off. When the flash is set to on, HDR is disabled, and you see a line through the HDR icon because you can't use the flash with HDR images.

Icon	Description
	Live Photo on—When this feature is enabled, you take Live Photos (see the "Live Photos" note following this table) and the Live Photos icon is yellow. To turn Live Photos off, tap this icon. On iPhone 11 and later models, you can tap the Live Photo icon on the toolbar to choose Live Auto to have Live Photo in automatic mode as well as turn it on or off.
	Live Photo off—When disabled, you take static photos and the Live Photos icon is white. To turn Live Photos on, tap this icon.
	Self-timer—When you tap this icon, a menu appears on which you can choose a 3- or 10-second delay for photos. When you choose a delay, the icon is replaced with one showing the delay you set. When you tap the Shutter icon, the timer starts and counts down the interval you selected before capturing the image. On iPhone 11 and later models, the Self-timer icon appears on the toolbar. Tap it to choose a delay. When a delay is active, you see the amount of delay at the top of the screen.
	Filter—When you tap this icon, which is on the toolbar on iPhone 11 and later models (swipe to the left on the toolbar to see it) and at the top of the screen on other models, a filter selection bar appears above the Shutter icon. Swipe on this bar to select from the available filters. As you move among the filters, you see the effect of the current filter on the image you're viewing. The name of the current filter appears above the white box that is above the Shutter icon. For example, you can apply the NOIR filter to give the photo a cool Noir-movie look. You can also apply filters after you take a photo; typically, this is easier so the details of applying filters are provided in the task "Applying Filters to Photos" in Chapter 16.

Icon	Description
	Filter applied—When the Filter icon is in color, you know a filter is currently applied (on iPhone 11 and later models, this appears at the top of the screen and on the toolbar). You also see the filter highlighted on the selection bar. When you capture a photo using the filter, the filter preview is marked with a white dot (except for iPhone 11 and later models). Tap the Filter icon to close the filter selection bar. To remove a filter, tap the Filter icon, select the Original filter, and tap the Filter icon.
	Change Camera—When you tap this icon, you toggle between the back-facing and front-facing cameras (the front-facing camera is typically used for selfies). It looks a bit different on iPhone 11 and later models than on other models.
	Shutter—This icon changes based on the type of photo or video you are taking. For example, when you're taking a photo, this is a white circle as shown. When you take a video, it becomes red. It looks a bit different for other types as well, such as Time-Lapse. Regardless of what the icon looks like, its function is the same. Tap it to start the process, such as to take a photo or start capturing video. If applicable, tap it again to stop the process, such as stopping video capture. To take burst photos, you touch and hold it to capture the burst.
00:00:06 / 00:00:10	**Timer**—When you capture video, the timer shows the elapsed time of the video you're capturing. The red dot on the left side of the time indicates you're currently capturing video. On iPhone 11 and later models, the timer is highlighted in red while you're recording.
	Focus/exposure box—When you frame an image, the camera uses a specific part of the image to set the focus, exposure, and other attributes. The yellow box that appears on the screen indicates the focus/exposure area. You can manually set the location of this box by tapping on the part of the image that you want the app to use to set the image's attributes. The box moves to the area on which you tapped and sets the attributes of the image based on that area.

Icon	Description
	Exposure slider—When you tap in an image you're framing, the sun icon appears next to the focus/exposure box. If you tap this icon, you see the exposure slider. Drag the sun up to increase the exposure or down to decrease it. The image changes as you move the slider so you can see its effect immediately.
AE/AF LOCK	**AE/AF Lock**—When you tap an image to set the location of the focus/exposure box and keep your finger on the screen for a second or so, the focus and exposure become locked based on the area you selected. This icon indicates that the exposure and focus are locked so you can move the camera without changing the focus or exposure that is used when you capture the image. Tap the screen to release the lock and refocus on another area.
	Faces found—When your iPhone detects faces, it puts this box around them and identifies the area as a face. These are especially important when you take Portrait photos because they indicate where the image will be sharply focused. You can also use faces to organize photos by applying names to the faces in your photos.
	Zoom slider—You can unpinch on an image to zoom in or pinch on an image to zoom out. When you do this on models that don't support Telephoto zoom, the Zoom slider appears on the screen. The Zoom slider indicates the relative level of zoom you're applying. You can also drag the slider toward the – to zoom out or drag it toward the + to zoom in to change the level of zoom you are using.
	Zoom Level icon—On models that support Telephoto zoom, this indicates the level of zoom currently applied to the image, such as 1x or 2x. If you pinch or unpinch on an image to change the zoom level, you always see the current level in this icon. On the iPhone 12 Pro, 12 Pro Max, 11 Pro, and 11 Pro Max, you see three icons. Tap .5 for wide-angle shots, 1x for normal view, or 2 to magnify the image by 2x. The current level is highlighted in yellow. When you zoom using the slider (see next item), the current magnification level is shown in the center of the icon and highlighted in yellow.

Icon	Description
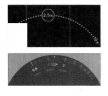	**Digital Zoom slider**—On models that support Telephoto zoom, this appears when you touch and hold on the Zoom Level icon. Dragging on the slider along the curve increases or decreases the level of magnification. On the iPhone 12 Pro, 12 Pro Max, 11 Pro, and 11 Pro Max, the slider looks like a dial, but works in the same way. Swipe to the left to decrease magnification or to the right to increase it. The current level is highlighted in yellow on the dial. After you lift your finger from the slider, it disappears and you see the current magnification level on the Zoom Level icon.
	Portrait Lighting Effect—Indicates the current lighting effect applied to an image when you're using the Portrait mode.
	Depth Level icon and slider—When you're using Portrait mode, you can set the depth level by tapping this icon (not available on all models). When active, it turns yellow and the Depth Level slider appears. Swipe to left and right on this slider to change the depth level. When you have manually set the depth level, you see the setting in the icon.
	Light Level icon and slider—When you're using Portrait mode on an iPhone 12 Pro, 12 Pro Max, 11 Pro, and 11 Pro Max, you can set the light level by tapping this icon. When active, it turns yellow and the Light Level slider appears. Swipe to left and right on this slider to change the light level.
	**Night Mode**—This icon only appears in low-light level conditions when Night mode is available (only on iPhone 11 and later models). Night mode activates automatically at a very low light level, or you can turn it on by tapping the icon when it's white (top of the screen or on the toolbar). When active, the icon is yellow, and you can set the amount of time used to take the photo by swiping to the left or right on the Night mode gauge that appears just above the Shutter icon.

Icon	Description
	Proportion—Tap the Proportion icon, which shows the current proportion, and then tap the proportion for the photo you want to take on the proportion bar (iPhone 11 and later models only). The proportion bar closes, the icon changes to show the proportion you selected, and you see the image in the proportion you selected.

And Now a Few Words on Live Photos

iPhones can capture Live Photos. A Live Photo is a static image, but it also has a few of what Apple calls "moments" of video around the static image that you take. To capture a Live Photo, you set the Live function to on and take the photo as you normally would. On iPhone 11 and later models, you can select the Live Auto option (in addition to turning Live Photo off or on), which takes a Live Photo automatically based on specific situations, such as if the camera detects motion or sound.

When you're viewing a Live Photo (these photos have the LIVE icon on them), tap and hold on the photo to see the motion and hear the sound associated with the video portion of that photo. When you aren't tapping and holding on a Live Photo, it looks like any other photo you've taken.

Like other types of photos, you can share Live Photos with others. If the recipients are using an iPhone model 6s or newer, they can view the motion part of the Live Photo too. If the recipient is using an older model or some other type of device (such as an Android phone), the recipient sees only the static image (what you see when you're looking at a Live Photo without pressing on it).

Taking Photos

Taking photos with your iPhone is fun and easy, but that doesn't mean you're limited in your photographic achievements. As you read in the previous section, the iPhone's photographic tools are extensive, so you'll be amazed at the photos you can take. Following are tasks dedicated to some very useful types of photos. After you understand how to take these types of photos, you're able to take advantage of all the photo types your iPhone supports.

Taking Photos

Models: All

You can use the Camera app to capture photos by performing the following steps:

1 On the Home screen, tap Camera.

2 To capture a horizontal photo, rotate your iPhone so that it's horizontal; of course, you can use either orientation to take photos just as you can with any other camera.

3 Swipe up or down (right or left if the phone is vertical) on the selection bar until PHOTO is in the center and in yellow.

4 If you want to change the camera you're using, tap Change Camera. When you change the camera, the image briefly freezes, and then the view changes to the other camera. The front-facing camera (the one facing you when you look at the screen) has fewer features than the back-facing cameras have. These steps show taking a photo with the back-facing cameras.

5 Set the Flash, HDR, Live, Self-timer, and other options you want to use for the photo; see the previous table for an explanation of these options. (You won't see options that don't apply to your model—for example, the HDR option doesn't appear on models that have Smart HDR.)

iPhone 11 and Later Models

On iPhone 11 and later models, you can tap the icons at the top of the screen to open or close the tool-bar above the Shutter icon to configure options. For example, tap the Flash icon at the top of the screen to turn it off or on, or you can open the toolbar to turn the flash on or off or put it in Flash Auto mode.

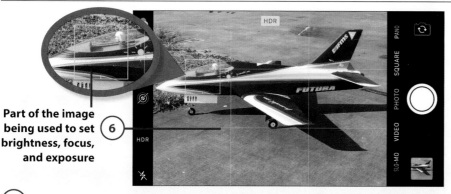

Part of the image being used to set brightness, focus, and exposure

(6) Frame the image by moving and adjusting the iPhone's distance and angle to the object you're photographing; if you have the Grid turned on, you can use its lines to help you frame the image the way you want it. When you stop moving the phone, the Camera app indicates the part of the image that is used to set focus, brightness, and exposure with the yellow box. If this is the most important part of the image, you're good to go. If not, you can set this point manually by tapping where you want the focus to be (see step 9).

(7) Zoom in by unpinching on the image. The camera zooms in on the subject. If you're using a model that doesn't have telephoto capabilities, the Zoom slider appears. If you're using a model with telephoto capabilities, you don't see this slider; instead you see the current zoom level in the Zoom level icon (not shown in the figure but shown on figures in the next section).

(8) Unpinch on the image (all models) or drag the slider (nontelephoto models only) toward the + to zoom in or pinch on the image or drag the slider toward the – to zoom out to change the level of zoom until it's what you want to use.

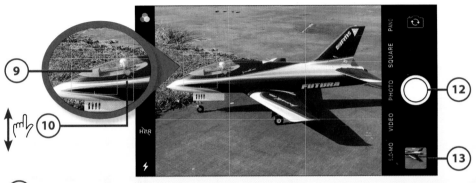

9 Tap the screen to manually set the area of the image to be used for setting the focus and exposure. The yellow focus box appears where you tapped.

10 To change the exposure, swipe up on the sun icon to increase the brightness or down to decrease it.

11 Continue making adjustments in the framing of the image, the zoom, focus point, and brightness until it's the image you want to take.

12 Tap the Shutter icon on the screen, either Volume button on the side of the iPhone, or press the center button on the EarPods. The Camera app captures the photo, and the shutter closes briefly while the photo is recorded. When the shutter opens again, you're ready to take the next photo.

13 You can repeat these steps to keep taking photos, or, if you want to view the photos you've taken, tap the thumbnail to see the photo you most recently captured.

14 Use the photo-viewing tools to view the photo (see Chapter 16 for the details).

15 If you don't want to keep the photo, tap Trash, and then tap Delete Photo.

16 Edit the photo by tapping Edit and using the resulting editing tools to make changes to the picture (see Chapter 16 for the details).

17 Tap back (<). You move back into the Camera app and can take more photos.

Taking Photos with Telephoto Zoom

Models: 7 Plus, 8 Plus, X, Xs, Xs Max, 11 Pro, 11 Pro Max, 12 Pro, 12 Pro Max

To take photos with iPhones that support Optical and Telephoto zoom, perform the following steps:

How Can You Tell?

You can tell if your phone has telephoto capabilities by unpinching the screen. If you see different zoom levels in the Zoom Level icon as you zoom in or out, you have a model that supports Telephoto zoom and can perform these steps. If the Zoom slider appears instead, use the information in "Taking Photos" to take photos.

1. Select the PHOTO mode, set up the image you want to capture, choose the options (such as flash or Live), frame the image, and set the exposure as described in the previous steps.

2. To zoom in at 2x using the optical zoom, tap the Zoom Level icon; on an iPhone 11 Pro, 11 Pro Max, 12 Pro, or 12 Pro Max, tap 2 on the Zoom Level icon. The magnification level changes to 2x using the iPhone's Optical zoom.

Just a Little Pinch Will Do You

The standard pinch and unpinch gestures to zoom or unzoom work, too. When you use a pinch or unpinch motion to zoom, the amount of magnification currently applied is shown in the Zoom Level icon. Using the pinch and unpinch motion to zoom is less precise than the method shown in these steps, but can be a bit faster.

3 Touch and hold on the Zoom Level icon. The Digital Zoom slider appears. On iPhone 11 Pro, 11 Pro Max, 12 Pro, or 12 Pro Max models, this slider looks like a dial, but works in the same way as the slider that you see in the figures.

4 Drag the Digital Zoom slider to the left to increase the level of magnification or to the right to decrease it. As you drag, the amount of magnification is shown in the Zoom Level icon, and, of course, you see the magnified image on the screen. When you've set the magnification level and lift your finger from the screen, the slider disappears.

Telephoto Zooming Applies Everywhere—Almost

The zooming features shown in these steps apply to all modes except Portrait. (On iPhone 11 Pro, 11 Pro Max, 12 Pro, and 12 Pro Max you can zoom to 2x.) However, there are different maximum levels of zoom in the various modes. For example, when using the VIDEO mode, you're limited to 6x, whereas in the SLO-MO mode, you're limited to 3x. In PANO mode, you only have the 1x and 2x options. Experiment with the zoom in the modes you use to see what zoom capabilities they have.

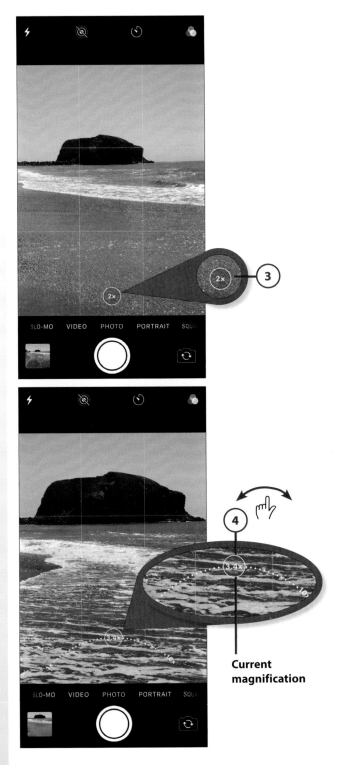

Current magnification

(5) Tap the Shutter icon to take the photo or video.

Quick Reset

To quickly return the magnification level to 1x, tap the Zoom Level icon. On iPhone 11 Pro, 11 Pro Max, 12 Pro, or 12 Pro Max, tap the center of the Zoom Level icon to set the magnification level to 1x.

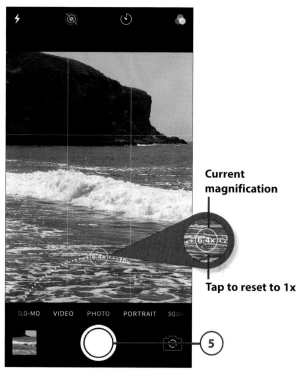

Current magnification

Tap to reset to 1x

5

Taking Portrait Photos

Models: 7 Plus, 8 Plus, X, Xs, Xs Max, 11, 11 Pro, 11 Pro Max 12, 12 mini, 12 Pro, 12 Pro Max
The Portrait mode captures the subject in sharp focus and blurs the background. You can also apply various lighting effects and set the depth level. Use these steps to take portrait photos:

Current lighting

(1) Swipe on the Selection bar until you reach PORTRAIT mode.

(2) Make sure the subjects are inside the yellow frames; the frames indicate where the image will be sharply focused. Once the subjects are recognized and captured inside the frame, the frame automatically stays on the subjects even if they move. If you're too close to the subjects, you see a message indicating you should move farther away.

(3) Touch and hold on the Lighting icon. The Portrait Lighting Effect slider appears.

Portrait Photos

When you take a portrait photo, you can use many of the Camera app's features, such as the self-timer. These work just like they do for other types of photos. You can't use some other features, such as HDR, when you're taking portrait photos.

Current lighting

(4) Swipe the Portrait Lighting Effect slider up or down (left or right if you are holding the phone vertically) to change the lighting effect applied to the image. In some cases, such as STAGE LIGHT, you see a focus area that shows where the light will be focused; you can use this to set up the image you want to capture.

(5) To capture the image with the current depth, tap the Shutter icon; to change the depth level, proceed to the next step.

(6) To manually set the depth level, tap the Depth Level icon. The Depth Level slider appears.

Zoom, Zoom

When using Portrait mode on some models, including the 11 Pro, 11 Pro Max, 12 Pro, and 12 Pro Max tap 2x to increase the magnification to 2x. Tap 2x to return the magnification to 1x.

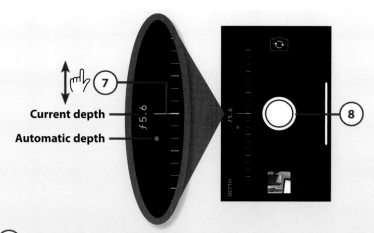

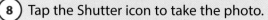

(7) Swipe the Depth Level slider down (or left) to increase the depth or up (or right) to decrease it.

Light Level

When using Portrait mode on some models, including the 11 Pro, 11 Pro Max, 12 Pro, and 12 Pro Max you can also change the light level when you're using the Portrait mode. Tap the Light Level icon and then use the slider to set the light level. See the table, "Photo and Video Options and Icons," earlier in this chapter to see what the icon and slider look like.

(8) Tap the Shutter icon to take the photo.

Taking Panoramic Photos

Models: All

The Camera app can take panoramic photos, which is ideal for landscape photography, by capturing a series of images as you pan the camera across a scene, and then "stitching" those images together into one panoramic image. To take a panoramic photo, perform the following steps:

(1) Open the Camera app.

2. Swipe on the selection bar until PANO is selected. On the screen, you see a bar representing the entire image that contains a smaller box representing the current part of the image that will be captured.

3. If you're using a model that has telephoto capabilities, you can zoom the image to 2x by tapping the Zoom Level icon (except 11 Pro, 11 Pro Max, 12 Pro, or 12 Pro Max). On iPhone 11 Pro, 11 Pro Max, 12 Pro, or 12 Pro Max models, tap 2 to set the magnification to 2x or .5 to use the Ultra Wide mode.

Current position in the image

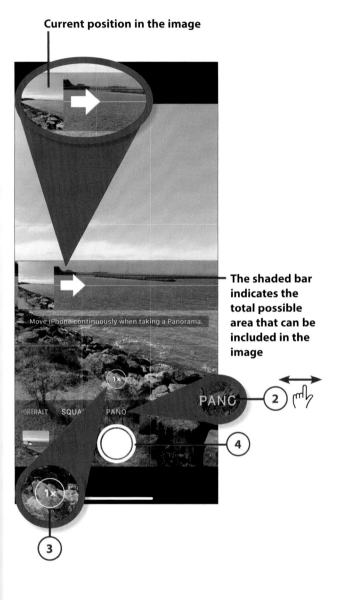

Move iPhone continuously when taking a Panorama.

The shaded bar indicates the total possible area that can be included in the image

(4) Tap the Shutter icon. The app begins capturing the image.

(5) Slowly sweep the iPhone to the right while keeping the arrow centered on the line on the screen. If you move the phone too fast, you see a message on the screen telling you to slow down. If the arrow goes too far above or below the line, you see a message telling you to move the phone to better align the arrow with the line. The better you keep the tip of the arrow aligned with the line, the more consistent the centerline of the resulting image will be.

(6) When you've moved to the "end" of the image you're capturing or the limit of what you can capture in the photo, tap the Shutter icon. You move back to the starting point and the panoramic photo is created. You can tap the panoramic image's thumbnail to view, delete, or edit it just as you can with other types of photos.

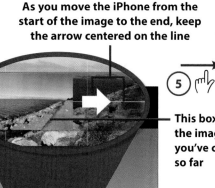

As you move the iPhone from the start of the image to the end, keep the arrow centered on the line

This box shows the image you've captured so far

Taking Wide Angle Photos

Models: 11, 11 Pro, 11 Pro Max , 12, 12 mini, 12 Pro, 12 Pro Max

The Ultra Wide mode is similar to using a wide-angle lens on a traditional camera. Here's how to use this option:

(1) Open the Camera app.

(2) Select the type of photo you want to take and configure its options.

(3) Tap .5. The magnification is decreased to .5x, which shows more of the image you're capturing. If that's the level you want to use, skip to step 6.

(4) To adjust the level of magnification, touch and hold on 0.5x. The Zoom Level slider or dial appears.

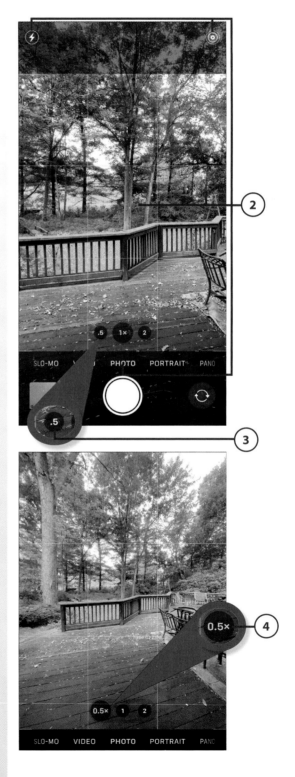

(5) Swipe to the right to increase the magnification or to the left to decrease it.

(6) When the photo is what you want, tap the Shutter icon.

Taking Photos with Night Mode

Models: 11, 11 Pro, 11 Pro Max , 12, 12 mini, 12 Pro, 12 Pro Max

The Night mode captures photos under low lighting conditions without using the flash. The resulting photos have better color and other properties that tend to be damaged by the bright light of the flash. You can change the amount of time that is used; the longer the time, the better quality of photo you capture, but you also have to hold the camera still longer.

Night mode is active automatically when the phone's sensors determine the light is low. When active, you can use the automatic setting or adjust it. Here's how to use Night mode:

(1) Open the Camera app.

2 Select the type of photo you want to take and configure its options; if it is very dark, you might not be able to see the image super clearly, but the Camera app attempts to present a usable image. If the conditions warrant, Night mode becomes active, and you see its icon in yellow and the current setting (for example, 1s) is indicated. If light levels are low, Night mode becomes available, but it isn't active. In this case, you see the Night mode icon in white at the top of the screen and on the toolbar.

3 If Night mode is active and you want to use the current setting, tap the Shutter icon and skip to step 7. To manually activate or adjust Night mode, move to step 4.

4 Tap Night mode at the top of the screen or on the toolbar. The Night mode slider appears.

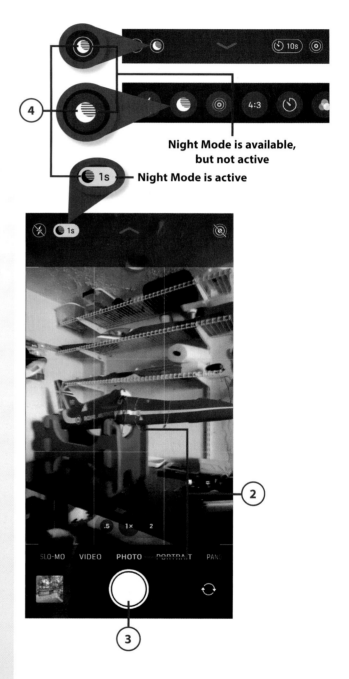

Night Mode is available, but not active

Night Mode is active

5) Swipe on the slider to the left to increase the effect or to the right to decrease it. The greater the time, the better the image will be, but you will also have to hold the phone steady for longer.

6) When you have set the amount of time, tap the Shutter icon to take the photo. The camera starts taking the photo. The image goes dark and slowly lightens.

7) Hold the phone steady until the process is complete; when it is, you return to the normal Camera screen and can preview the photo or take more photos as you normally can.

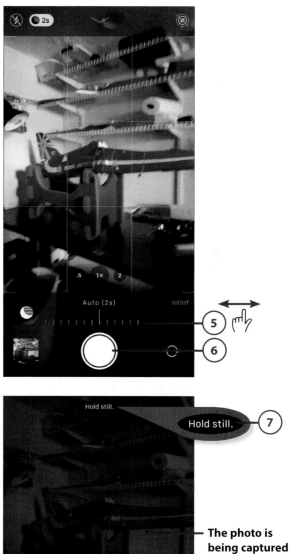

The photo is being captured

Taking Video

Models: All

You can capture video as easily as you can still images. Here's how.

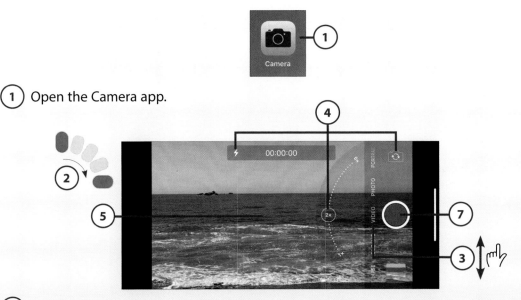

(1) Open the Camera app.

(2) To capture horizontal video, rotate the iPhone so that it's horizontal; of course, you can use either orientation to take video just as you can with any other video camera.

(3) Swipe on the selection bar until VIDEO is selected.

Taking Video on iPhone 11, 11 Pro, 11 Pro Max , 12, 12 mini, 12 Pro, and 12 Pro Max

You can take video on these models while you are in PHOTO mode by swiping to the right on the Shutter icon. The video process starts and you can control it as described in these steps.

---

(4) Choose the back-facing or front-facing camera, configure the flash, or zoom in, just like setting up a still image. (The Self-timer, Filter, and HDR modes are not available when taking video.)

(5) Tap on the screen where you want to focus.

(6) If needed, adjust the exposure by sliding the "sun" icon up or down just like a still photo (not shown on the figure).

(7) Tap the Shutter icon to start recording. You hear the start/stop recording tone, and the app starts capturing video; you see the timer on the screen showing how long you've been recording.

Length of video

⑧ Take still images while you take video by tapping the white Shutter icon. (If the Live Photos preference is enabled, the photos you take are Live Photos. If not, you take static images.)

⑨ Stop recording by tapping the red Shutter icon again. Also, like still images, you can then tap the video's thumbnail to preview it as well as any still images you took while taking the video. You can use the Photos app's video tools to view or edit the clip.

>>>Go Further

MORE ON TAKING PHOTOS AND VIDEO

The Camera app enables you to do all sorts of interesting and fun things with photos and video. Following are some additional pointers that help you make the most of this great app:

- **Set and forget**—You need to set the Flash, HDR, and most other options only when you want to change the current settings because these settings are retained even after you move out of the Camera app and back into it. The Camera Mode, Creative Controls (such as filters), and Live Photo behaviors are controlled by the Preserve Settings described in the Go Further sidebar at the end of this chapter.

- **HDR**—The High Dynamic Range (HDR) feature causes the iPhone to take three shots of each image, with each shot having a different exposure level. It then combines the three images into one higher-quality image. HDR works best for photos of subjects that aren't in motion and are in good lighting. (You can't use the iPhone's flash with HDR images.)

  Smart HDR is supported on Xs, Xs Max, 11, 11 Pro, 11 Pro Max, 12, 12 mini, 12 Pro, and 12 Pro Max models. Smart HDR uses more images than standard HDR and is a more sophisticated application of the technique to produce even higher quality images (for example, multiple exposures of the images are also used).

When you're using an 11, 11 Pro, 11 Pro Max, 12, 12 mini, 12 Pro, or 12 Pro Max HDR is either on and happens automatically, or it is off, which means it isn't used at all. This is controlled by a setting you can read about in the Go Further sidebar at the end of this chapter. By default, Smart HDR is turned on.

On other models, you choose whether to use HDR when you take photos. (In these cases, HDR photos display the HDR icon in the upper-left corner when you view them.)

When the Keep Normal Photo switch (only on models that don't use Smart HDR) in the Camera Settings is on (green), you see two versions of each HDR photo in the Photos app: One is the HDR version, and the other is the normal version. If you prefer the HDR versions, set the Keep Normal Photo switch to off (white) so that your photos don't use as much space on your iPhone, and you don't have twice as many photos to deal with.

- **Location**—The first time you use the Camera app, you're prompted to decide whether you allow it to use Location Services. If you allow the Camera app to use Location Services, the app uses the iPhone's GPS to tag the location where photos and video were captured. Some apps can use this information, such as the Photos app on your iPhone, to locate your photos on maps, find photos by their locations, and identify where photos were taken.

  To configure Location Services, open the Settings app and tap Privacy. Tap Location Services. On the Location Services screen, you can globally enable or disable Location Services. When Location Services are enabled, you can also configure them for specific apps. For example, to configure how the Camera app uses Location Services, tap Camera and then tap the type of Location Services you want the Camera app to use. The options are Never, Ask Next Time, or While Using the App.

- **Sensitivity**—The iPhone's camera is sensitive to movement, so if your hand moves while you're taking a photo, it might be blurry; current iPhone models have image stabilization that mitigates this to some degree. Sometimes, part of the image will be in focus and part of it isn't, so be sure to check the view before you capture a photo. This is especially true when you zoom in. If you're getting blurry photos, the problem is probably your hand moving while you're taking them. Of course, because it's digital, you can take as many photos as you need to get it right; delete the rejects as you take them, and use the Photos app to periodically review and delete photos you don't want to keep (see Chapter 16), so you don't have to waste storage room or clutter up your photo library with photos you don't want to keep.

- **Burst photos**—When you touch and hold on the Shutter icon while taking photos or quickly swipe to the left (don't apply pressure) on iPhone 11, 11 Pro, 11 Pro Max, 12, 12 mini, 12 Pro, 12 Pro Max, a series of images is captured rapidly, and you see a counter showing the number being taken. When you release the Shutter icon, a burst photo is created; the burst photo contains all of the images you captured but appears as a single image in the Photos app. You can review the images in the burst and choose to keep only the images you want to save. Burst photos are best suited to capturing action.

- **Self-timer**—When you set the Self-timer option, you choose either a 3- or 10-second delay between when you tap the Shutter icon and when the image is captured. Like the other settings, the Self-timer is persistent, so you need to turn it off again when you want to stop using it.

- **Self-timer and burst**—If you set the timer, and then tap and hold on the Shutter icon for a second or so, a burst of 10 photos is captured when the timer expires.

- **Slow-motion video**—You can also take slow-motion video. Choose SLO-MO on the selection bar. Set up the shot and take the video as you do with normal speed video. When you play it back, the video plays in slow motion except for the very beginning and ending.

- **Time-lapse video**—When you choose the TIME-LAPSE option, you can set the focus and exposure level and choose the camera you want to use. You record the video just like "real-time" video. When you play it back, the video plays back rapidly so you seemingly compress time.

- **Screenshots**—You can take screenshots of your iPhone's screen by pressing and holding the Side button and upper Volume button (iPhones without a Home button) or by pressing and holding the Touch ID/Home and Side buttons at the same time (iPhones with a Home button). The screen flashes white and the shutter sound plays to indicate the screen has been captured. You see a thumbnail of the screen capture. You can tap this to open the screen capture to edit it; when you're finished editing it, tap Done to close the preview (tap Save to Photos to save it or Delete Screenshot if you don't want to save it). If you don't tap the thumbnail, after a couple of seconds, it disappears (you can swipe it off the screen to get rid of it immediately).

  If you didn't preview the screenshot, it is saved automatically. If that happened or you chose to save the preview in the editor, the resulting image is stored in the Screenshots album in the Photos app. You can view the screen captures you take, email them, send them via Messages, or do other tasks as you can with photos you take with the iPhone's camera.

Taking Photos and Video from the Lock Screen

Because it's likely to be with you constantly, your iPhone is a great camera of opportunity. You can use its Quick Access feature to quickly take photos when your iPhone is asleep/locked. Here's how:

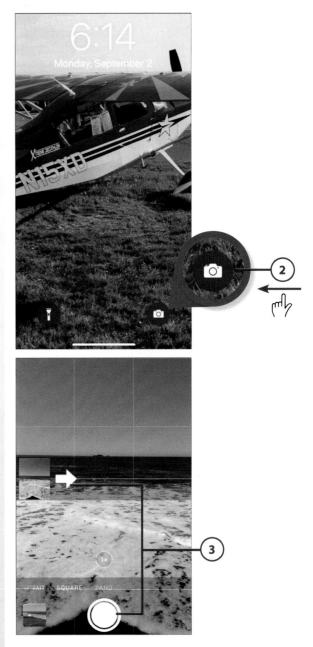

(1) When the iPhone is locked, press the Side button, tap the screen, touch the Touch ID/Home button, or lift your phone up (if you have a model that supports the Raise to Wake feature, and you have it enabled). The Lock screen appears.

(2) Swipe to the left (all models) or press the Camera icon (iPhones without a Home button). The Camera app opens.

(3) Use the Camera app to take the photo or video as described in the previous tasks. You can only view the most recent photos or videos you captured from within the Camera app when your iPhone is locked; you have to unlock the phone to work with the rest of your photos.

Yet Another Way

You can also open the Camera app from the Control Center. To do this, open the Control Center (on iPhones without a Home button, swipe down from the upper-right corner, or on models with a Home button, swipe up from the bottom of the screen) and then tap the Camera icon. (If you don't see the Camera icon, you can add it to the Control Center using the information in Chapter 6, "Making Your iPhone Work for You.")

Taking Photos and Video from the Home Screen

The Quick Access menu offers a selection of photos and video commands that you can choose right from a Home screen.

(1) Touch and hold on Camera until the Quick Actions menu opens.

(2) Tap the type of photo or video you want to take. The Camera app opens and is set up for the type you selected.

(3) Use the Camera app to capture the photo or video (not shown in the figures).

Take Selfie

Record Video

Take Portrait

Take Portrait Selfie

Edit Home Screen

Remove App

>>>Go Further

SCANNING QR CODES

QR Code — MEET YOUR RI
SCAN NOW
LEARN MORE
ONRAMP ONRAMP.EHI.CC

QR (Quick Response) codes provide information about or enable you to take action on objects to which they are attached or associated. A QR scanner reads these codes and presents the information they contain. For example, rental cars sometimes have a QR code sticker on a window; when you scan this code, you get information about the car. Or displays in a museum might have a QR code that you can scan to get more information.

QR codes also enable you to take action, such as scanning the QR code for a Wi-Fi network and then joining it, or scanning the code for an email address and then creating an email.

The Camera app can scan QR codes (when enabled to do so as explained in the next section). To do so, choose the PHOTO mode and focus on the QR code. The code is recognized and scanned. After the code is scanned, you're prompted to take action on it—for example, to move to the related website.

>>>Go Further
CAMERA SETTINGS

You can probably use the Camera app as it is just fine. However, you might want to tweak the way it works. The following table describes some of the more useful options in the Settings app that you can access by tapping Settings on the Home screen, and then tapping Camera. You should also configure settings for the Photos app because they impact how your photos are displayed and stored (see Chapter 16 for more information).

The settings available for the Camera app are dependent on the type of phone. On your iPhone, you might see fewer, more, or different settings. However, this table explains the most important settings you are most likely to want to change.

Camera Settings

Section	Setting	Description
Formats	CAMERA CAPTURE	Choose High Efficiency if you want your photos to be captured in the HEIF/HEVC format so they use less storage space. Not all devices and apps can use this format, but any that are related to the iPhone (such as iPads and Macs) should be able to. This format results in smaller file sizes so you can store more photos on your iPhone. If you use photos on other types of devices or if you want to make sure your photos are compatible with as many devices and apps as possible, choose Most Compatible instead.

Section	Setting	Description
Record Video/ Record Slo-mo	Resolution and Frame Rate	Use these options to determine how video is recorded. The options available depend on the model of iPhone you have. You can choose from among different combinations of resolution and frame rate. Higher resolution and frame rates mean better-quality video but also larger files. For example, 4K at 60 fps is very high quality, but the resulting video files are large. iPhone 12, 12 mini, 12 Pro, and 12 Pro Max offer HDR Video to further improve image quality.
N/A	Record Stereo Sound	When enabled (green), your iPhone records audio in stereo. If you disable this, it records the audio in mono. Stereo provides better quality but also produces larger file sizes.
Preserve Settings	Camera Mode	When this switch is on (green), the Camera app retains the mode you most recently used, such as VIDEO or PANO. When off (white), the camera is reset to the PHOTO mode each time you move into the Camera app.
Preserve Settings	Creative Controls	When this switch is on (green), the Camera app retains the filters, light, or depth settings you most recently used, such as the DRAMATIC filter. When off (white), the camera is reset to the default settings for each of these adjustments each time you move into the Camera app.
Preserve Settings	Exposure Adjustment	When this switch is on (green), the Camera app retains the Exposure adjustment you most recently made.
Preserve Settings	Live Photo	When this switch is on (green), the Camera app retains the Live Photo setting you used most recently, such as Off. When off (white), Live Photo is turned on automatically each time you move into the Camera app.
N/A	Use Volume Up for Burst	When enabled, you can press and hold the Up Volume button to capture Burst photos.
N/A	Scan QR Codes	When this switch is on (green), you can use the Camera app to scan QR codes that provide information about the object to which they're attached or with which they're associated. See the Go Further sidebar, "Scanning QR Codes," for some examples.

Section	Setting	Description
COMPOSITION	Grid	When this switch is on (green), you see a grid on the screen when you're taking photos with the Camera app. This grid can help you align the subject of your photos in the image you're capturing.
COMPOSITION	Mirror Front Camera	When this switch is on (green), the image you see when taking selfies is a mirror image of what the camera is pointing at. This looks more natural because we are used to seeing our reflections in a mirror. When this switch is off (white) selfie images might look "backward" to you.
COMPOSITION	View Outside the Frame	When this switch is on (green), you see areas outside the frame that will be captured when you take the photo. This is useful because you can make sure you can see everything you want to capture inside the framed image.
PHOTO CAPTURE	Scene Detection	When enabled, the app attempts to improve photos in various scenes using the iPhone's image recognition capability.
PHOTO CAPTURE	Prioritize Faster Shooting	When this switch is on (green), image quality is adjusted to provide for faster image capture. This enables you to take photos faster, though the resulting images might be of lower quality.
PHOTO CAPTURE	Lens Correction	When enabled, the app attempts to lessen distortion caused by the front or ultra wide cameras.
N/A	Smart HDR (X or later models)	When enabled, Smart HDR captures a series of images and blends them into one image. The intention is to produce higher quality images. Note that if Smart HDR is disabled, you won't be able to take HDR images at all.
HDR (Pre-X models)	Auto HDR	When enabled, the Camera app automatically uses the HDR option to capture photos. When this switch is off (white), you can use the HDR icon on the screen to turn HDR on or off for specific photos.
HDR (Pre-X models)	Keep Normal Photo	When enabled, the Camera app saves both the normal and HDR versions of the photo. This means you have two of every photo you take using HDR. This can be useful if you prefer the normal version in some cases, but also adds a lot of duplicate images to your photo library.

Organize, view, edit, and
share photos and videos

Use Messages, Mail,
and other apps to
share your photos
and video

In this chapter, you explore all the photo- and video-viewing and editing functionality that your iPhone has to offer. Topics include the following:

→ Getting started
→ Finding and viewing photos
→ Editing and improving photos
→ Working with photos
→ Finding and viewing videos

16

Viewing and Editing Photos and Video with the Photos App

Chapter 15, "Taking Photos and Video with Your iPhone," explains how to take photos and video using the iPhone's Camera app. The photos and video you take with your iPhone's cameras are stored in your photo library, which you access using the Photos app. In this chapter, you learn how to use the Photos app to view, organize, edit, and share those photos and videos.

Getting Started

The Photos app provides many useful tools to view and edit photos and video stored on your iPhone, whether you used the iPhone's camera to capture them or you downloaded them from another source, such as images attached to email messages.

As you take photos, capture video, and save photos in email or messages onto your iPhone, you can quickly build up a large photo library. Fortunately, the Photos app automatically organizes your photos and video so that you can find specific photos or video you want to view, edit, or share quite easily. You can also manually organize photos in albums with just a few steps.

After you find photos and video in which you're interested, you can view them in the app manually, or you can use groups that the Photos app builds for you automatically; these groups contain collections of photos and video based on location, dates, people, and other factors.

You can also edit photos and videos to fix mistakes, make improvements, or express yourself artistically.

You'll probably want to share photos and videos with others, and the Photos app makes that a snap, too. You can do that directly from within Photos or using other apps on your iPhone, such as Messages or Mail.

Finding and Viewing Photos

To work with your photos in the Photos app, you first find the photos that you're interested in. You use the tabs along the bottom of the Photos screen to choose different ways to find the photos you want to view and with which you want to work (such as editing them).

After you find photos, you can view them in a number of ways, including individually, in groups that Photos creates for you automatically, or in groups that you create for yourself.

Finding Photos Using the Photos Tab

The Photos tab automatically organizes photos based on the time and location associated with them (this information is embedded in the photos you take with the iPhone's camera, assuming you haven't disabled Location Services for it). The top level is Years, which shows your photos grouped by the year in which they were taken. From there, you can move down into photos by month, and then day. You can use the All Photos option to browse all of your photos.

To find photos based on their time, follow these steps:

(**1**) On the Home screen, tap Photos.

(**2**) Tap Library and then tap Years. You see photos collected by the year in which they were taken.

(**3**) Swipe up and down the screen to browse all the years.

(**4**) Unpinch on the year whose photos you want to view. You see that year's photos organized by month.

Tabs Work, Too

You can jump from the Years view to the Months view by tapping the Months tab. Likewise, you jump directly to photos collected by their dates by tapping the Days tab. However, the unpinch method often makes it easier to get to the specific group of photos you want to view.

Tap to change to years, months, days, or all your photos

(5) Swipe up and down the screen to browse the months in the year you selected; if the app identified photos taken over a timeframe, you see groups by those timeframes too.

(6) Unpinch on the timeframe whose photos you want to view. You see those photos organized by their capture date; if there are multiple photos associated with a date, you see those collected under that date using thumbnails of different sizes.

(7) Swipe up and down the screen to browse the photos collected by date.

(8) To go back to a previous level, such as moving from Days to Months, pinch on the screen.

Photos taken on the same date

(**9**) Continue browsing until you find photos you want to view.

(**10**) Tap a photo to view it.

(**11**) You're ready to view the photo in detail as described in the task "Viewing Photos" later in this chapter.

View Them All

Tap All Photos to browse all your photos. Swipe up or down to move back or forward in time (respectively). As you browse, the title at the top of the screen changes to indicate the timeframe and location of the photos you're browsing.

All Photos Collections

When you're viewing the All Photos tab, pinch to show photos collected into groups based on larger timeframes, such as by month. If you keep unpinching, you'll eventually see photos grouped by year. Unpinch on a group to drill down into shorter timeframes, such as from year to month.

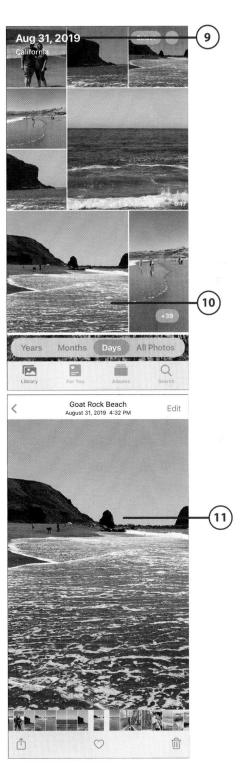

>>>*Go Further*

THE FOR YOU TAB

When you tap the For You tab, you see different groupings based on your activity. For example, when you share photos using iCloud, you see a Shared Album Activity section; tap that to see photos that you're sharing or that are being shared with you. You also see the Memories section; memories are collections of photos the Photos app creates for you automatically based on a variety of factors that can include location, timeframe, or people. You also see Featured Photos, which include photos that the app has highlighted for you.

Tap any groups you see to move into those groups to view the photos they contain; these work similarly to other options you learn about in this section. Tap back (<) to move back up to the For You tab (you might need to tap it several times depending on where you are).

Finding Photos Using the Albums Tab

Albums in the Photos app are analogous to albums in which you organize and present physical photos; except albums in the Photos app are much more flexible and easier to create and organize.

There are four basic types of albums within the Photos app:

- **Albums the Photos App Creates**—The Photo app creates and organizes a number of albums for you automatically. These include albums based on people and places along with albums based on types of photos such as Selfies, Live Photos, and Portraits.

- **Albums You Create**—You can create albums to keep your photos organized in many different ways; the same photos can be in multiple albums. You learn how to create and manage albums in "Working with Photos" later in this chapter.

- **Shared Albums**—Shared Albums are collections of photos you're sharing with others or that others are sharing with you.

- **Albums Other Apps Create**—Some other apps (such as Instagram) that you use for photos might create their own albums within the Photos app.

To use albums to find photos, perform the following steps:

1. Continuing in the Photos app, tap Albums. You see albums organized by type including My Albums, Shared Albums, People & Places, Media Types, and Other Albums.

2. Swipe up and down the screen to browse the albums (such as albums you have created) or automatic albums (for example, Selfies).

3. Swipe left or right on groups of albums, such as My Albums, to browse them.

4. To see all the albums in a group, tap See All and then browse the resulting group.

5. Tap the album you want to view.

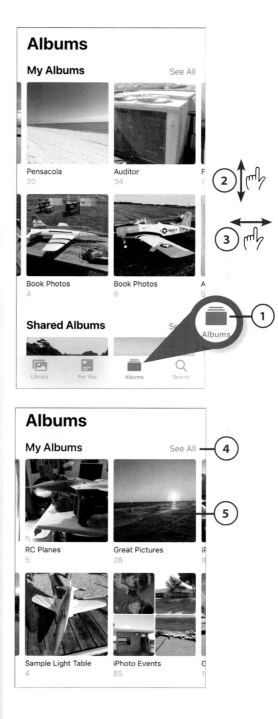

(6) Swipe up and down to browse the photos in the album.

(7) Tap the photo you want to view.

Go Back

As you move into albums and to specific photos, you see back (<) in the upper-left corner of the screen. Tap this to move back to a previous screen; for example, to move from a photo you are viewing to the album it is contained in.

(8) You're ready to view the photo in detail as described in the task "Viewing Photos" later in this chapter.

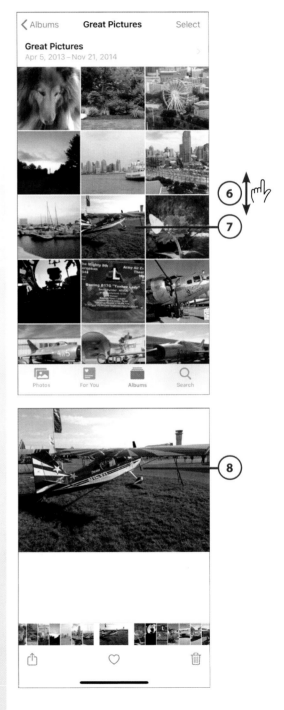

Finding Photos by Searching

Browsing photos can be a fun way to find photos, but at times you might want to get to specific photos more efficiently. The Search tool enables you to quickly find photos based on their time, date, location, content, and other factors.

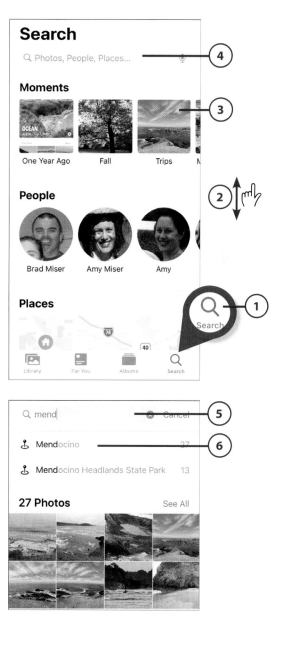

(1) Continuing in the Photos app, tap Search. At the top of the screen, you see the Search bar that you can use to perform a specific search. Below that are potential searches you might want to use; these are based on a variety of criteria, such as moments, people, or places.

(2) Swipe up and down the searches on the screen. For example, Places finds photos based on where they were taken.

(3) To use a search, tap it and skip to step 6; if you don't want to use a current search, move to step 4.

(4) Tap in the Search bar.

(5) Type your search term. This can be any information associated with your photos. As you type, collections of photos that match your search criteria are listed under the Search bar. The more specific you make your search term, the smaller the set of photos that will be found. Suggested searches appear immediately under the Search bar.

(6) Tap a suggested search to perform it. The search results refresh based on the search you selected.

7 Tap search (this is optional because you can work with the results as they appear when you do a search).

8 Swipe up and down the screen to browse all the results. At the top, you see the individual photos that were found (tap See All to see all of them). Under that, you see collections of photos based on various criteria, such as Moments, Places, and People. (You learn more about the collections Photos creates for you in "Viewing Photos in Collections" later in this chapter.)

9 Tap the results you want to explore.

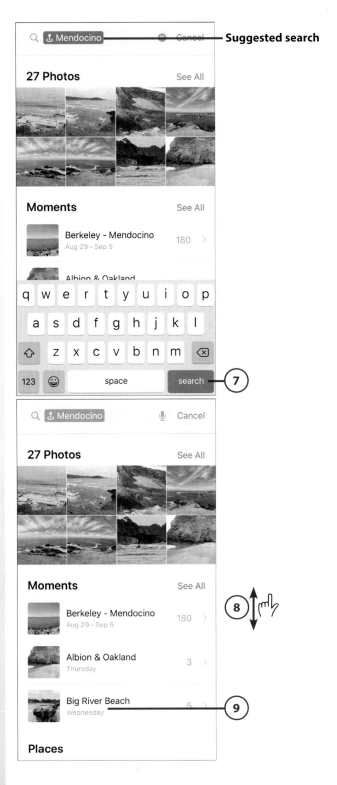

Suggested search

10 To see all the photos in the results, tap Show More.

11 Tap a photo to view it.

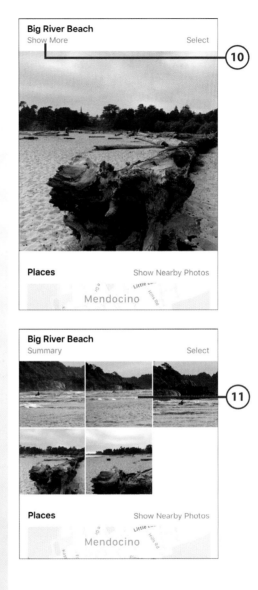

(12) View the photo (covered in the next section).

(13) Tap back (<) to return to the group of photos you were viewing.

(14) Tap Search (<) to return to the search results.

(15) Tap a different result to explore its photos.

(16) To change the search, tap in the Search bar and change the current search term (you can delete the current term by tapping Delete [x]).

(17) Tap Cancel to exit the search.

Viewing Photos

The Photos app enables you to view your photos individually. Here's how:

(1) Using the skills you learned in the previous tasks, open the group of photos that you want to view.

Orientation Doesn't Matter

Zooming, unzooming, and browsing photos works in the same way whether you hold your iPhone horizontally or vertically.

(2) Swipe up and down to browse all the photos in the group.

(3) Tap the photo you want to view. The photo display screen appears.

(4) If it is a Live Photo, touch and hold on the photo to see its motion.

My, Isn't That Special

When there's something special about a photo, you see an icon above or on the image. In this figure, you see the LIVE icon indicating it is a Live photo. You might also see Burst for Burst photos, HDR for a photo captured with HDR, and so on. Keep an eye out for these icons as you explore your photos.

(5) To see the photo without the app's toolbars, tap the screen. The toolbars are hidden.

(6) Rotate the phone horizontally.

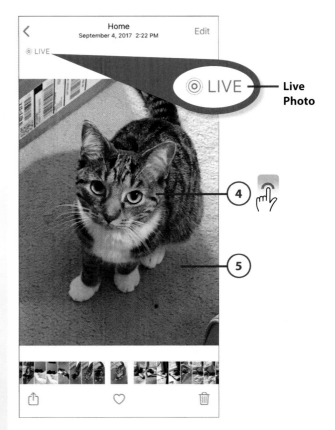

Live Photo

(7) Unpinch on the photo to zoom in.

(8) When you're zoomed in, drag around the image to view different parts of the zoomed image.

Lose the Zoom

Double-tap on a zoomed photo to unzoom all the way.

 (7)

(9) Pinch on the photo to zoom out.

No Zooming Please

You can't swipe to move to the next or previous photo when you're zooming in on a photo, so make sure you are zoomed out all the way before performing step 10. If not, you move the photo around instead.

(10) Swipe to the left to view the next photo in the group.

(11) Swipe to the right to view the previous photo in the group.

Auto-Play

As you swipe through photos, any that have motion associated with them, such as Live Photos or videos, play automatically. This behavior is enabled or disabled by a setting; refer to the Go Further sidebar "Photos Settings" at the end of the chapter for details.

(12) When you're done viewing photos in the group, tap the screen to show the toolbars again.

(8)   (9)   (10)

(12)

(11)

13 Swipe to the left or right on the thumbnails at the bottom of the screen to view all the photos in the current group. As you swipe, the photo you're viewing changes to be the one in the larger thumbnail at the center of the screen.

14 Tap a photo to view it.

15 Swipe up on the photo to get information about it and to see photos related to it. If the app can apply effects to the photo, you see the Effects section.

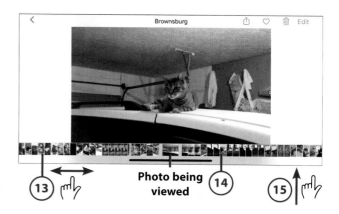

16 Swipe to the left and right to see the available effects. Each thumbnail shows the image with the effect applied.

17 Tap an effect to apply it to the image. When you return to the image, you see it with the effect you selected.

18 Swipe up to get to the Location section.

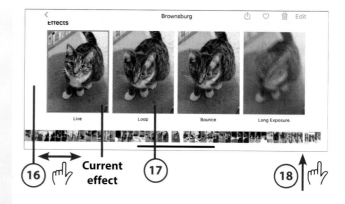

19 Tap Show Nearby Photos to see photos that were captured near the location associated with the photos you're viewing.

20 Tap the photo on the map to move to the map view (using the map view is covered in the "Viewing Photos by Location" section later in this chapter).

21 When you are done viewing the details, tap back (<). You move back to the group of photos you were most recently viewing.

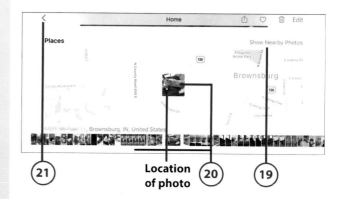

Viewing Photos in Collections

As you take and gather photos, the Photos app automatically creates collections of photos in many different ways. Here are some examples:

- **Memories**— Memories are collections of photos for you to view in a number of ways. You can view them in a slideshow, individually, by selecting places, or by selecting people. This feature provides lots of options and at times can be a very interesting way to view photos because you might be surprised by some of the photos included in a particular memory. Memories pop up throughout the Photos app.

  The Photos app creates memories for you dynamically, meaning they change over time as the photos in your library change. You can save memories that you want to keep as they are. Otherwise, the memories you see change as you take more photos or edit photos you have. This keeps memories a fresh and interesting way to view your photos.

- **Locations**—Unless you've disabled Location Services for the Photos app, the app associates photos with the locations in which they were taken. This is a really useful way to collect photos because you're often interested in finding and viewing photos based on where you were, such as when you traveled somewhere for vacation. You can use the Map tool to see where photos are located and quickly view photos in that location.

  You see the Show Nearby Photos command in different places; tap this to see photos that were captured near the location associated with the photos you're currently viewing.

- **Moments**—Moments are collections of photos based on location, timeframe, people, and other factors. You see moments in different locations as you use the app.

- **People**—Through the iPhone's facial recognition technology, it can identify photos featuring specific people and group those for you. You can associate those photos with a name and the Photos app attempts to tag future photos with that person with her name. You can view photos of that person simply by opening the group associated with her name.

- **Other**—The Photos app uses other techniques to group photos, too. For example, it can group photos based on the type of subject, such as planes or animals. Like the other collections, you see these groups throughout the Photos app.

While there are lots of ways Photos collects photos and videos for you, you can work with these groups in similar ways. The following tasks show you how to work with photos in memories and by location. Using these techniques, you'll be able to work with other types of collections just as easily.

Viewing Photos in Memories

To view the photos in a memory, perform the following steps:

(1) Tap the For You icon on the Dock at the bottom of the screen. The Memories section appears on the For You screen.

(2) Swipe to the left or right to browse memories there or tap See All to see all your memories.

(3) Tap the memory you want to view.

Moments Versus Memories

Moments and memories work in the same way. For example, when you click a moment in the search results, it opens, and you can view its contents as shown in these steps.

(4) To view the photos in a slideshow, tap Play. The slideshow begins to play in full screen. If some of the photos in the slideshow aren't currently stored on your iPhone, there might be a pause while they're downloaded (you see the Downloading status on the opening screen while this is done).

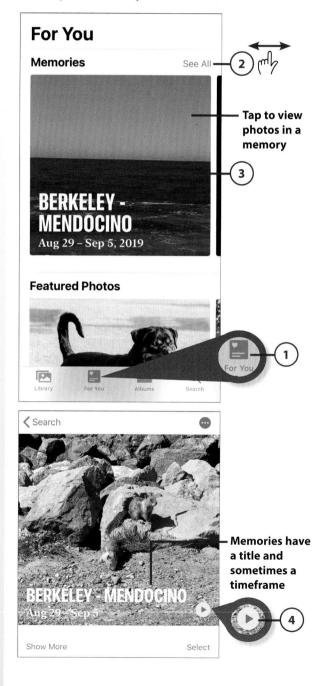

(5) To view the slideshow in landscape orientation, rotate the iPhone. As the slideshow plays, you see effects applied to the photos; videos included in the memory also play.

(6) Tap the screen to reveal the slideshow controls.

(7) To pause the slideshow, tap Pause. (When paused, this becomes the Play icon you can tap to resume the slideshow.)

(8) Swipe to the right or left on the theme bar to change the slideshow's theme. As you change the theme you might see changes on the screen, such as a different font for the title, and hear different music while the slideshow plays.

(9) Swipe to the right or left on the duration bar to change the slideshow's length. As you make changes, you see the slideshow's current length just above the theme bar.

(10) Swipe to the left or right on the thumbnails at the bottom of the screen to move back or forward, respectively.

(11) To restart the slideshow with the new settings, tap Play (not shown on the figures).

(12) Tap back (<) to return to the memory.

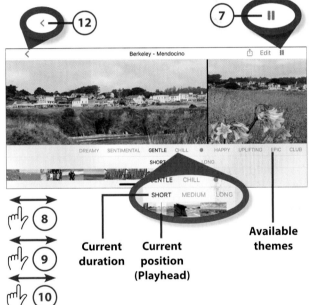

Current duration

Current position (Playhead)

Available themes

⑬ Swipe up and down the photos section to see all the photos the memory contains.

Show All, Tell All

If the memory has lots of photos, you might see only a summary view of them on the Photos section. Tap Show More to see all the photos. Tap Summary to return to the summary view.

⑭ To see all of the memory's photos, tap Show More.

⑮ Tap a photo to view it.

⑯ Use the techniques you learned in "Viewing Photos" earlier in the chapter to work with the photos in the memory.

⑰ Tap back (<) to return to the memory.

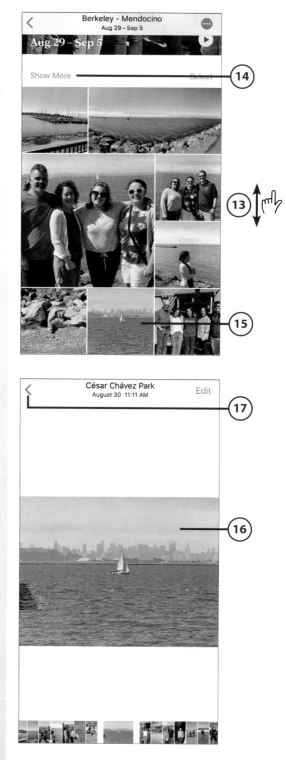

(18) Swipe up the screen to see other collections related to the memory.

(19) Tap a collection to view its photos. (You learn about using locations to view photos in the next section.)

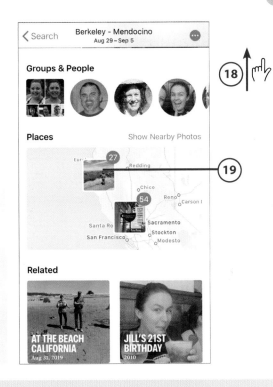

>>>Go Further
MAKING THE MOST OF YOUR MEMORIES

Here are a couple of ways to manage your memories:

- Memories can change over time. If you want to save a memory as it is, view it and tap Options (...) at the top of the screen. Then tap Add to Favorite Memories to add a memory you are viewing to the Favorite Memories album. Open the menu for a memory you've designated as a favorite and tap Remove from Favorite Memories to remove the memory from the Favorite Memories album.

  To see your favorite memories, tap For You and tap See All in the Memories section. Tap Favorites in the upper-right corner of the screen. You see only memories you've tagged as a favorite. To see them all again, tap See All.

- If you want to remove a memory, open it, tap Options (...), tap Delete Memory, and then confirm you want to delete a memory by tapping Delete Memory again; the memory is deleted. When you delete a memory, only the memory is deleted; the photos that were in that memory remain in your photo library. (However, if you delete a photo from within a memory, that photo is deleted from your photo library, too.)

Viewing Photos by Location

As you learned earlier, locations are associated with the photos and video that you take with your iPhone's cameras. The Photos app can use this information to display photos by location.

As you move around in the Photos app, you see the Places section in various locations, such as in a memory. Within the Places section, you can use a map to find and view photos as follows:

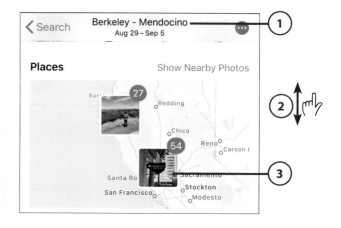

(1) Open a memory or other collection of photos.

(2) Swipe up or down the screen until you see the Places section.

(3) Tap a place. The map expands to fill the screen and you see more locations associated with photos in the memory.

Nearby Photos

Tap Show Nearby Photos to show other photos that were taken near the locations that you're viewing on the map but that aren't currently included in the memory. Tap Hide Nearby Photos to hide those photos again.

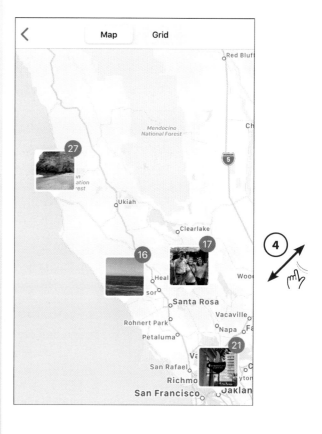

(4) Unpinch your fingers on the screen to zoom in to reveal more detailed locations.

5 Swipe around the screen to move around the map.

6 Tap a location with photos to see the photos associated with it.

7 Swipe up and down the screen to browse the photos taken at the location.

8 Tap a photo to view it.

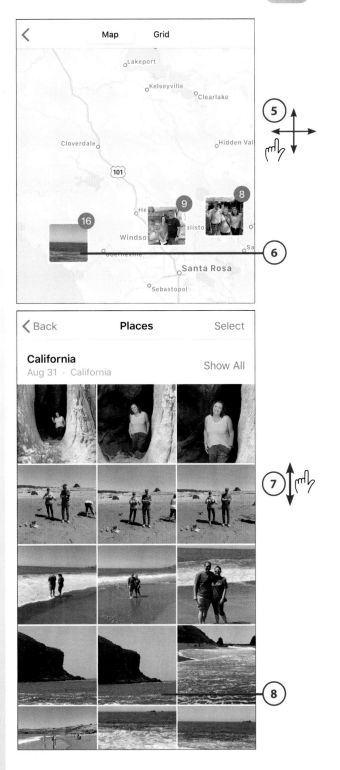

(9) Use the techniques covered in "Viewing Photos" earlier in the chapter to work with the photos you view.

(10) Tap back (<) to return to the place.

(11) Tap Back (<) to return to the map. (You might have to tap Back an additional time to return to the map depending on whether the location you viewed has multiple moments associated with it.)

(12) Tap other locations to view their photos.

(13) Tap back (<) to return to your previous location.

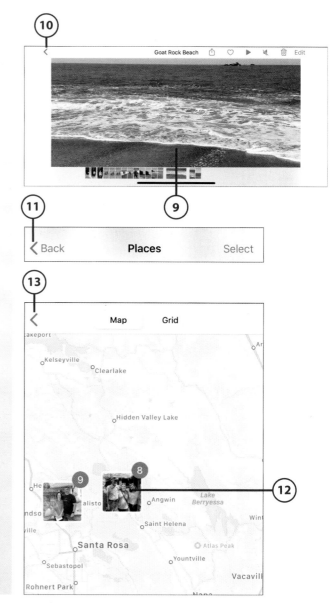

Editing and Improving Photos

Even though the iPhone has great photo-taking capabilities, not all the photos you take are perfect from the start. Fortunately, you can use the Photos app to improve your photos. The editing tools available to you include the following:

- **Automatic Adjustments**—This tool attempts to automatically adjust the colors and other properties of photos to make them better. You can change how much of the automatic adjustments are applied to an image.

- **Other Adjustments**—You can adjust the exposure, brilliance, highlights, and other properties of your photos.

- **Filters**—You can apply different filters to your photos for artistic or other purposes.

- **Straighten, Rotate, and Crop**—You can rotate your photos to change their orientation and crop out the parts of photos you don't want to keep.

- **Red-eye**—This one helps you remove that certain demon-possessed look from the eyes of people in your photos.

Making Automatic Adjustments to Photos

To improve the quality of a photo, use the Enhance tool.

(1) View the image you want to adjust.

(2) Tap Edit.

3 Tap Adjust.

4 Tap AUTO. The image is enhanced and the AUTO icon is highlighted to show the AUTO adjustment is applied. If you are happy with the adjusted photo, skip to step 6.

Undo

To remove the AUTO adjustment, tap the AUTO icon. The icon becomes unhighlighted and the adjustment is removed from the photo. You can also tap Cancel and then confirm you don't want to save the changes to leave the image unchanged.

5 If you want to further adjust the photo, swipe on the gauge below the AUTO icon to the left or right to change the amount of adjustment applied to the image. As you swipe, you see the changes in the photo as you make them. The dot indicates the amount of adjustment that was applied when you tapped the AUTO icon.

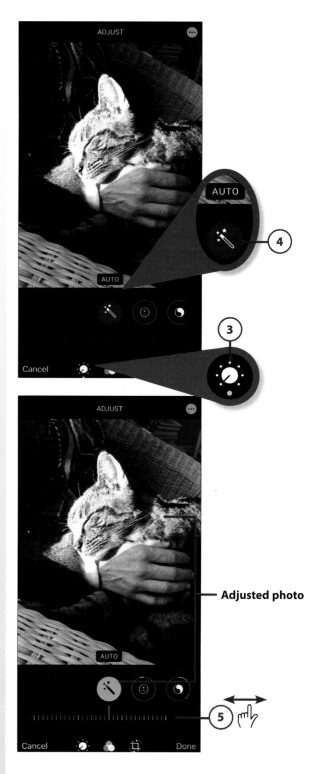

Adjusted photo

6 Tap Done to save the adjusted image.

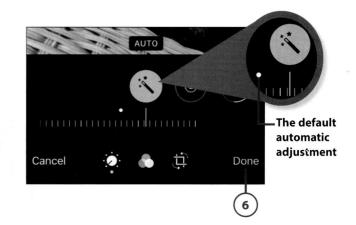

—The default automatic adjustment

6

Making Advanced Adjustments to Photos

There are many adjustment tools available to you in the Photos app. The steps to use these are similar to using the AUTO tool you learned about in the previous section:

1 View the image you want to adjust.

2 Tap Edit.

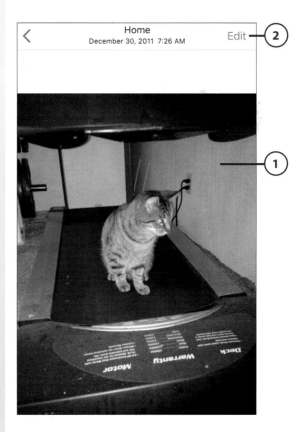

(3) Tap Adjust.

(4) Swipe to the left or right on the tool bar until you see the property of the photo you want to change, such as Exposure. As you swipe on the bar, the name of the current tool is shown.

(5) Swipe to the left and right on the gauge to change the amount of that property applied to the photo. As you make changes, you see the results in the image. You also see the amount of change you are making in the tool icon's circle; the dot indicates the original amount of that property in the image.

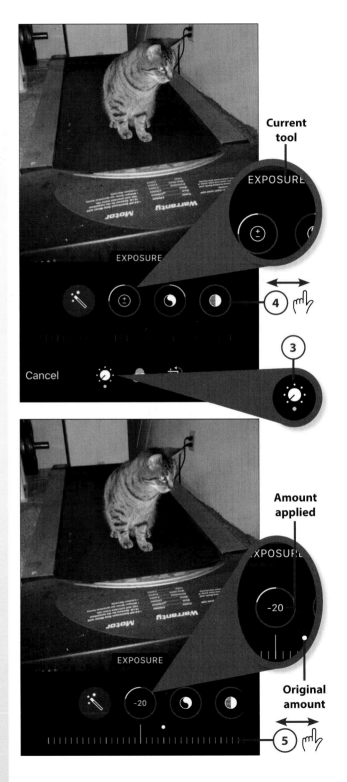

(6) Select other tools to make more adjustments to the image.

(7) Use the resulting gauge to change the amount of the adjustment you selected in step 6 that is applied to the image.

(8) When you're done making changes, tap Done to save the adjusted image.

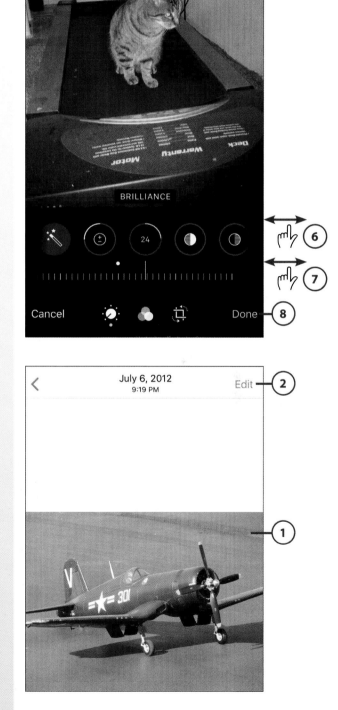

Applying Filters to Photos

To apply filters to photos, do the following:

(1) View the image to which you want to apply filters.

(2) Tap Edit.

(3) Tap the Filter icon. The palette of filters appears. If you haven't applied a filter, you see the current filter as Original, which is highlighted with a white box.

(4) Swipe to the left or right on the filter palette to browse all of the filters.

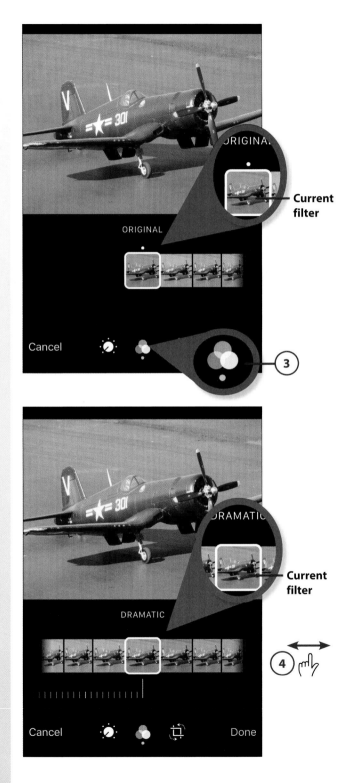

Current filter

ORIGINAL

Cancel

3

Current filter

DRAMATIC

Cancel   Done

4

5 Stop on the filter you want to apply so it's highlighted. The filter is applied to the image and you see a preview of the image as it will be with the filter; the filter currently applied is highlighted with a white box.

6 Swipe to the left or right on the gauge below the filter palette to change the intensity of the filter. You see the current intensity value within the white box.

7 When you are happy with the filter, tap Done. The photo with the filter applied is saved.

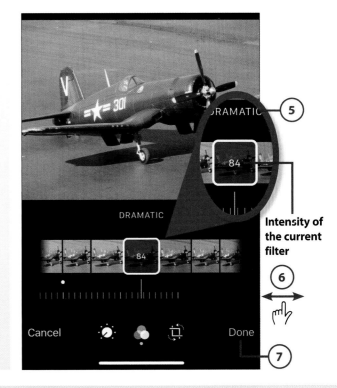

Intensity of the current filter

>>>Go Further

UNDOING WHAT YOU'VE DONE

To restore a photo to its unedited state, tap Revert, which appears when you edit a photo that you previously edited and saved. At the prompt, tap Revert to Original, and the photo is restored to its "like new" condition.

The original version of photos is saved in your library so you can use the Revert function to go back to the photo as it was originally taken or added to the library, even if you've edited it several times.

However, you can't have the edited version and the original version displaying in your library at the same time. If you want to be able to have both an edited and original version (or multiple edited versions of the same photo), make a copy of the original before you edit it.

To make a copy of a photo, view the photo, tap Share, and tap Duplicate. You can do this as many times as you want. Each copy becomes a new photo. If you edit one copy, the original remains available in your library for viewing or editing differently.

Straightening, Rotating, and Cropping Photos

To change the alignment, position, and part of the image shown, perform the following steps:

1. View the image you want to change.

2. Tap Edit.

3. Tap Straighten/Crop. The tools you use to change the orientation and part of the image displayed appear. A white box appears around the image; this shows what part of the image you will keep if you crop it.

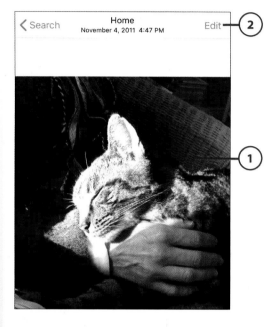

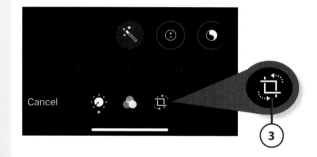

(4) To flip the image along its vertical axis, tap Flip.

(5) To rotate the image in 90-degree increments, tap Rotate. Each time you tap this icon, the image rotates 90 degrees in the clockwise direction.

(6) To rotate the image within the white box, swipe on the toolbar until Straighten is in the center of the screen.

(7) Swipe to the left or right on the gauge to rotate the image; within the icon, you see the current amount of rotation. The grid appears to help you see how the image aligns with vertical and horizontal lines.

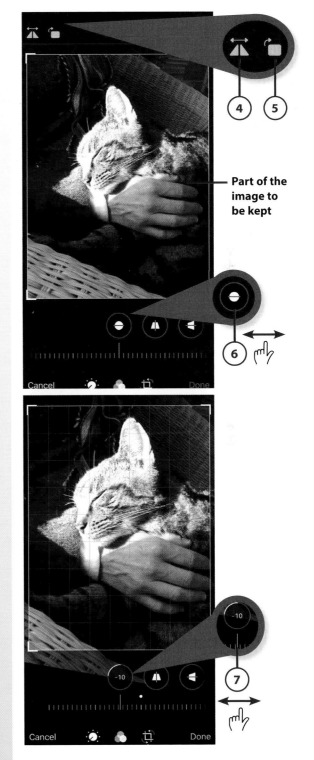

Part of the image to be kept

8 To adjust the image in the vertical direction, swipe on the tool bar until Vertical is in the center of the screen and then swipe to the left and right to adjust the image.

9 Use a similar process to adjust the image in the horizontal direction.

10 To crop the image without respect to keeping a specific proportion, drag the edges of the box around the part of the image you want to keep so that part is inside the box; if you want to crop the image to a specific proportion, jump to step 12.

11 Drag on the image to move it around inside the crop box; when the image is what you want it to be, skip to step 15.

12 Tap Constrain.

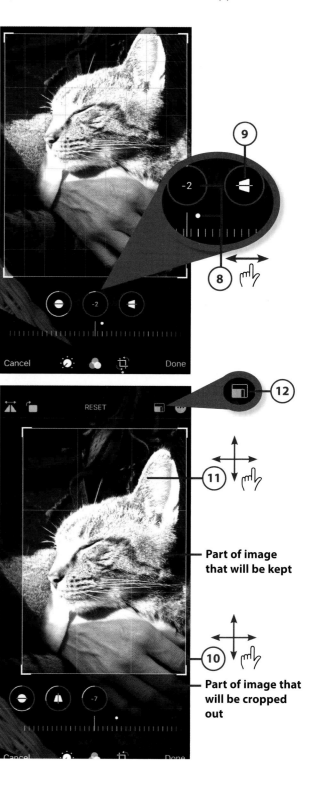

Part of image that will be kept

Part of image that will be cropped out

(13) Swipe to the left and right on the toolbar to see the options, and then tap the proportion you want to use; the crop box assumes the current proportion. You can use this to configure the image for how you intend to display it. For example, if you want to display it on a 16:9 TV, you might want to constrain the cropping to that proportion so the image matches the display device.

(14) Resize the box and position the image within it until the image is what you want it to be; as you drag the edges of the box, it remains in the proportion you selected in step 13.

(15) When the image is cropped and positioned as you want it to be, tap Done. The edited image is saved.

Auto Straighten and Crop

The Photos app can automatically straighten and crop some photos, usually photos that contain people. When this option is available, it's applied automatically when you tap Straighten/Crop. In addition to the changes in the photo (such as it being cropped or the subjects repositioned in the frame), you see AUTO highlighted in yellow at the top of the screen. To remove the automatic changes, tap AUTO. The automatic changes are removed from the image, and AUTO is not highlighted any more. Tap AUTO to reapply the automatic changes. If you make any manual changes to the photo (for example, recrop it), the automatic changes are removed.

Removing Red-Eye from Photos

When you edit a photo with people in it, the Red-eye tool becomes available (if no faces are recognized, this tool is hidden). To remove red-eye, perform the following steps:

 (1) View an image with people that have red-eye.

(2) Tap Edit.

(3) Tap Red-eye.

(4) Zoom in on the eyes from which you want to remove red-eye; as you zoom in, drag the photo to keep the eyes you want to fix on the screen.

5 Tap each eye containing red-eye. The red in the eyes you tap is removed.

6 Repeat steps 4 and 5 until you've removed all the red-eye.

7 Tap Done to save your changes.

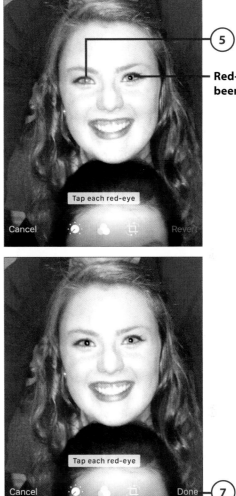

Red-eye has been corrected

Working with Photos

You can do a lot of things with the photos on your phone, including the following:

- Emailing one or more photos to one or more people (see the next task).

- Sending a photo via a text message (see Chapter 11, "Sending, Receiving, and Managing Texts and iMessages").

- Sharing photos via AirDrop (see Chapter 17, "Working with Other Useful iPhone Apps and Features").

- Posting your photos on your Facebook wall or timeline or on Instagram.
- Assigning photos to contacts (see Chapter 8, "Managing Contacts").
- Using photos as wallpaper (see Chapter 7, "Customizing How Your iPhone Looks").
- Sharing photos via tweets.
- Printing photos from your printer (see Chapter 3, "Using Your iPhone's Core Features").
- Deleting photos (covered later in this chapter).
- Organizing photos in albums (also covered later in this chapter).

Copy 'Em

If you select one or more photos and tap the Copy icon, the images you selected are copied to the iPhone's clipboard. You can then move into another app and paste them in.

Individual Versus Groups

Some actions are available only when you're working with an individual photo. For example, you might be able to send only a single photo via some apps, whereas you can email multiple photos at the same time. Any commands that aren't applicable to the photos that are selected won't appear on the screen.

Sharing Photos via Email

You can email photos via iPhone's Mail app starting from the Photos app.

(1) View the source containing one or more images that you want to share.

(2) Tap Select.

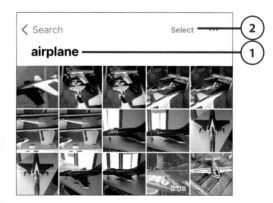

3 Browse the photos to find the ones you want to share.

4 Select the photos you want to send by tapping them. When you tap a photo, it is marked with a check mark to show you that it is selected.

5 Tap Share.

Too Many?

If the photos you have selected are too much for email, the Mail icon won't appear. You need to select fewer photos to attach to the email message.

6 Tap Mail. A new email message is created, and the photos are added as attachments. If the photos need to be downloaded to your phone, this might take a few moments.

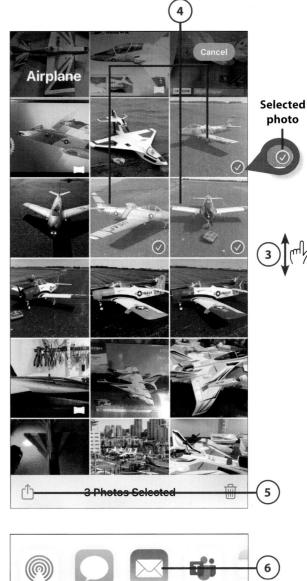

Selected photo

(7) Use the email tools to address the email, add a subject, type the body, and send it. (See Chapter 10, "Sending, Receiving, and Managing Email," for detailed information about using your iPhone's email tools.)

(8) Tap the size of the images you want to send. Choosing a smaller size makes the files smaller and reduces the quality of the photos. You should generally try to keep the size of emails to 5MB or smaller to ensure the message makes it to the recipient. (Some email servers block larger messages.) After you send the email, you move back to the photos you were browsing.

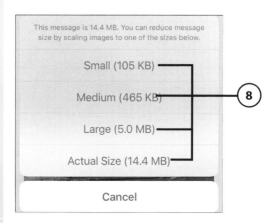

Creating New Albums

You can create photo albums and store photos in them to keep your photos organized.

To create a new album, perform these steps:

(1) Move to the Albums screen by tapping Albums on the toolbar.

(2) Tap Add (+). (If you don't see this, tap back [<] until you do.)

(3) To create a personal album, tap New Album or to share an album, tap New Shared Album. These steps show a personal album.

(4) Type the name of the new album.

(5) Tap Save. You're prompted to select photos to add to the new album.

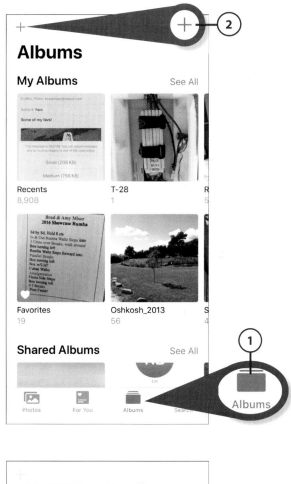

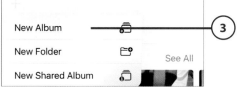

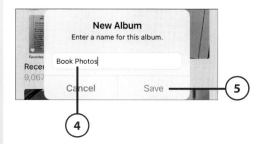

(6) Move to the source of the photos you want to add to the new album.

(7) Swipe up and down to browse the source and tap the photos you want to add to the album. They're marked with a check mark to show that they're selected. The number of photos selected is shown at the top of the screen.

(8) Tap Done. The photos are added to the new album and you move back to the Albums screen. The new album is shown on the list, and you can work with it just like the other albums you see.

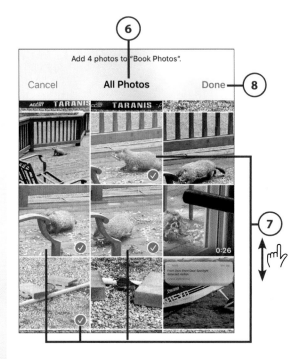

Playing Favorites

You can tap the Favorite icon (the heart) to mark any photo or video as a favorite. The heart fills in with blue to show you that the image you're viewing is a favorite. Favorites are automatically collected in the Favorites album, so this is an easy way to collect photos and videos you want to be able to easily find again without having to create a new album or even put them in an album. You can unmark a photo or video as a favorite by tapping its Favorite icon again.

New album containing the photos you selected

Folder o' Photos

In step 3, you see there is an option to create a folder. A folder can be used to group albums together. To do this, create a folder. Open it and tap Add (+). Tap New Album to create an album within the folder or New Folder to create a nested folder. You can add and work with photos stored within albums that are inside folders just as you can any other album. Folders are useful when you have lots of albums and want to keep them organized. For example, you might want to create folders of albums for specific activities by year.

Adding Photos to Existing Albums

To add photos to an existing album, follow these steps:

(1) Move to the source containing the photos you want to add to an album.

(2) Tap Select.

(3) Tap the photos you want to add to the album.

(4) Tap Share.

(5) Swipe up the sheet.

(6) Tap Add to Album.

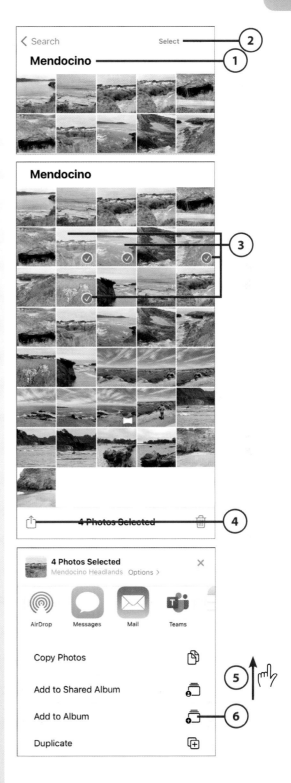

(7) Swipe up and down the list to find the album to which you want to add the photos.

(8) Tap the album; the selected photos are added to the album. (If an album is grayed out and you can't tap it, that album was not created on the iPhone, so you can't change its contents.)

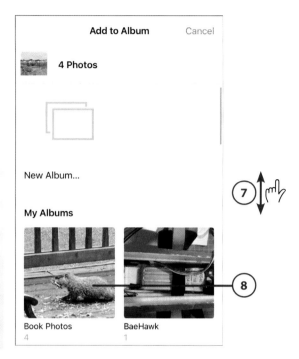

>>>Go Further
MORE ALBUM FUN

You can create a new album from photos you've already selected. Perform steps 1 through 5 and tap New Album instead of tapping an existing album as in step 7. Name the new album and save it. It's created with the photos you selected already in it.

You can change the order in which albums are listed on the My Albums screen. Move to the My Albums screen by moving to the Albums page and tapping See All in the My Albums section. Then tap Edit. Drag albums up or down the screen by their Unlock icons (red circle with a –) to reposition them. To delete an album that you created in the Photos app, tap its Unlock icon (–) and then tap Delete Album. (This deletes only the album; the photos in your library remain there.) When you're done making changes to your albums, tap Done.

To remove a photo from an album, view the photo from within the album, tap the Trash icon, and then tap Remove from Album. Photos you remove from an album remain in your photo library; they're removed only from the album. (If you tap Delete instead, the photo is deleted from your photo library.)

Deleting Photos

You can delete photos and videos that you don't want to keep on your iPhone. If you use iCloud to store them, deleting photos from your phone also deletes them from your photo library and from all the other devices using your library. So, make sure you really don't want photos any more before you delete them.

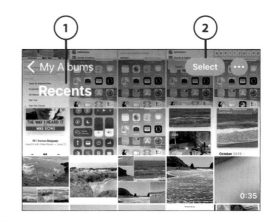

(**1**) Open the source containing photos you want to delete.

(**2**) Tap Select.

(**3**) Tap the photos you want to delete. Each item you select is marked with a check mark.

(**4**) Tap Trash.

Deleted Means Deleted

Be aware that when you delete a photo from your iPhone, it's also deleted from your iCloud Library and all the devices sharing that library—not just from your iPhone. (After it has been deleted from the Recently Deleted album of course.)

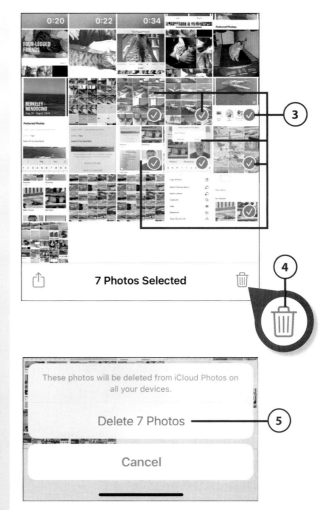

(**5**) Tap Delete *X* Photos, where *X* is the number of photos you selected. The photos you selected are moved to the Recently Deleted album, where they remain until they're automatically deleted (the amount of time they remain there depends on your storage space and other factors).

Deleting Individual Photos

You can delete individual photos that you're viewing by tapping the Trash icon, and then tapping Delete Photo.

>>>Go Further
RECOVERING DELETED PHOTOS

As you learned earlier, photos you delete are moved to the Recently Deleted album; to see this folder, swipe up on the Albums tab until you get to the bottom of the screen. You can recover photos you've deleted by opening this album. You see the photos you've deleted; each is marked with the time remaining until it's permanently deleted. To restore photos in this album, tap Select, tap the photos you want to recover, and tap Recover. Tap Recover X Photos, where X is the number of photos you selected (if you select only one, this is labeled as Recover Photo). The photos you selected are returned to the location from which you deleted them.

Finding and Viewing Videos

As explained in Chapter 15, you can capture video with your iPhone. Once captured, you can view videos on your iPhone, edit them, and share them.

Watching videos you've captured with your iPhone is simple.

(1) Move to the Albums screen.

(2) Tap the Videos album (swipe up the screen until you see the Media Types section to see it). Video clips display their running time at the bottom of their thumbnails. (Videos can also be stored in other albums, in collections, and in memories. This Videos album just collects videos no matter where else they are stored.)

Auto-Play

Videos automatically play when they appear in collections of photos, and at other times so don't be surprised if they start playing without you doing anything to start them.

3 Swipe up and down the screen to browse your videos.

4 Tap the video you want to watch.

5 Rotate the phone to change its orientation if necessary.

6 If the video doesn't start playing automatically, tap Play (not shown on the figure). The video plays. After a few moments, the toolbars disappear automatically.

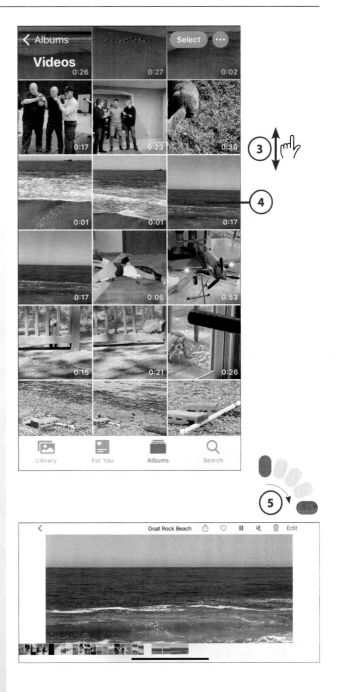

(7) Tap the video. The toolbars reappear.

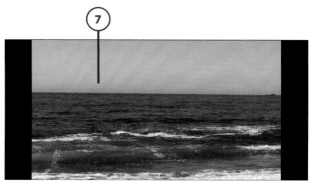

Deleting Video

To remove a video from your iPhone (and all other devices that access your library), select it, tap the Trash icon, and then tap Delete Video at the prompt.

(8) Pause the video by tapping Pause.

(9) Jump to a specific point in a video by swiping to the left or right on the thumbnails at the bottom of the screen. When you swipe to the left, you move ahead in the video; when you swipe to the right, you move back in the video.

Current frame

Watching Slow-Motion and Time-Lapse Video

Watching slow-motion video is just like watching regular speed video except after a few frames, the video slows down until a few frames before the end, at which point it speeds up again. Watching time-lapse is similar except the video plays faster instead of slower than real time.

>>>Go Further
EDITING VIDEO

You can trim a video to remove unwanted parts. View the video you want to edit. Tap Edit. If the video isn't stored on your phone, it's downloaded. When that process is complete, you can edit it.

Drag the left trim marker to where you want the edited clip to start; the trim marker is the left-facing arrow at the left end of the timeline. Drag the right trim marker to where you want the edited clip to end.

When the part of the video you want to keep is enclosed in the yellow box, tap Done. The shortened video is saved.

To return the video to full length again, tap Edit, and tap Revert, and then tap Revert to Original. The clip returns to be what you originally captured.

>>>Go Further
PHOTOS SETTINGS

You can probably use the Photos app just as it is, but there are some settings you might want to change. The most important setting is the one that determines whether your photos are stored on the cloud using your iCloud account. The following table explains the most useful Photos settings that you can access using the Settings app:

Photos Settings

Setting	Description
iCloud Photos	When enabled, your photos and videos are stored in your iCloud account on the cloud so that they are backed up and you can access them from multiple devices. See Chapter 4, "Setting Up and Using an Apple ID, iCloud, and Other Online Accounts," for information about using iCloud with your photos and video.

Setting	Description
Optimize iPhone Storage	If you select this option, only versions of your photos that are optimized for the iPhone are stored on your phone; this saves space so that you can keep more photos and videos on your iPhone. Full resolution photos are uploaded to the cloud.
Download and Keep Originals	This option downloads full resolution versions of your photos and videos on your iPhone. They consume more space than optimized versions.
Cellular Data	When the Cellular Data switch is enabled (green), photos are copied to and from your iCloud account when you are using a cellular Internet connection. If you have a limited data plan, you might want to set this to off (white) so photos are copied only when you are using a Wi-Fi network.
Cellular Data, Unlimited Updates	When enabled, updates to your photos are made constantly, which uses more data. With this disabled, updates are made periodically, which lowers the data use.
Auto-Play Videos and Live Photos	When this switch is on (green), when you view videos and Live Photos in various locations, they play automatically. When disabled, you have to manually play them.
View Full HDR	When this switch is enabled (green), you always see the maximum quality of photos as created with the HDR feature.
Show Holiday Events	With this switch enabled (green), the Photos app attempts to collect photos taken on the holidays in your country. When disabled (white), the Photos app ignores holidays when it groups photos into memories for you.

Capture notes containing text, attachments, lists, and more

Find your way

Use apps to make travel easier and less stressful

Listen to your favorite tunes

Store passes, membership cards, and payment information for quick and easy use

Get the news you want when you want it

Listen to or watch podcasts

In this chapter, you learn about some other really useful apps and iPhone functionality. Topics include the following:

→ Getting started
→ Touring other cool iPhone apps
→ Configuring and using Emergency Calling
→ Using Bluetooth to connect to other devices
→ Using AirDrop to share and receive content
→ Working with the Wallet app and Apple Pay

Working with Other Useful iPhone Apps and Features

An iPhone can truly become *your* iPhone over time as you use the apps and features that you find to be useful for traveling, communicating, and virtually every other facet of modern life. The great thing about iPhone apps is that there are so many to choose from, and because most of them are free or have a low cost, you can try lots of apps. Over time, you'll develop a group of core apps that you use constantly and a few that you use just occasionally. The rest you can just delete or move out of the way.

Getting Started

Previous chapters cover many of the iPhone's most useful apps, such as Phone, Mail, Messages, Photos, Safari, and Calendar. This chapter provides an overview of a number of other apps.

Don't be hesitant to try apps. Apple maintains extremely tight control over the apps that make it into the App Store, so there's little chance that an app you download and install can put you or your information at any risk. It usually takes only a few minutes of using an app to determine whether it will be useful to you, so you're not risking much of your time either.

In the event of an emergency, your iPhone can be a useful tool to quickly summon help and communicate with others at the same time. Later in this chapter, you learn how to configure and use the very important Emergency Calling feature.

You also learn how you can connect your iPhone to other devices in several ways. You can use Bluetooth to connect to speakers, headphones, fitness trackers, keyboards, and more. You can connect to other iPhones, iPads, and Macs using AirDrop, so you can quickly and easily share content across those devices.

With the Wallet app, you can store various types of cards, such as store loyalty cards, or passes—most significantly boarding passes—so you can access them with a couple of taps. You'll also see how Apple Pay enables you to safely and easily pay for purchases in online stores or in the "real world." You can store credit or debit cards in the Wallet app and then use them to pay for purchases with a wave of your phone or tap of a button.

Touring Other Cool iPhone Apps

There are many thousands of apps available for the iPhone. Some of these are pre-installed on your iPhone; others you download from the App Store. No matter how they get installed on your iPhone, these apps can be really useful in so many ways.

Touring Other Cool Apple Apps

The following table provides an overview of a number of other Apple apps (most of which are already installed when you get your iPhone) that you might find useful or entertaining:

Icon	App	Description
	Books	Access and organize books and PDFs so that you always have something to read. The app enables you to change how the books appear on the screen, such as making the font larger or smaller. Books syncs across your devices so that you can read something on your iPhone and then pick it up later at the same spot on an iPad.

Icon	App	Description
	Calculator	In portrait orientation, the Calculator is the equivalent of one you would get at the local dollar store. Rotate the iPhone to access scientific calculator options. You can get to the Calculator quickly by opening the Control Center; then, tap Calculator. You can also tap Calculator on a Home screen to launch the app.
	Clock	Get the current time in multiple locations around the world, set alarms, set a consistent time to sleep every day, use a stopwatch, or count down time with a timer. You can launch the Clock app from the Control Center or a Home screen.
	Compass (Extras folder)	Transform your iPhone into a compass. Using this app, you can see your current location on the analog-looking compass showing your exact position shown in latitude and longitude.
	Files	When you enable the iCloud Drive under your iCloud account, use the Files app to access the documents stored there.
	Find My	Use the Devices tab of this app to locate other iPhones, iPads, or Macs that have enabled Find My iPhone using the same iCloud account. (Learn about Find My iPhone in Chapter 18, "Maintaining and Protecting Your iPhone and Solving Problems.") On the People tab, you can see the current location of "friends" who are part of your Family Sharing group or who you add to the app manually. You can also share your location with others. On the Me tab, you can configure if and how your location information is available to others.
	Flashlight	Use your iPhone's flash as a flashlight. Open the Control Center and tap Flashlight to turn the light on; tap it again to turn it off. Press and hold on the icon to pop up the brightness slider; drag the slider up to increase the brightness or down to decrease it. On iPhones without a Home button, you can use the Flashlight by briefly pressing on Flashlight on the Lock screen. This is a very simple app, but it's also extremely useful.

Icon	App	Description
	Health	You can use this app to track and manage a wide variety of health-related information, from basic statistics for you, such as age, to the results of exercise you perform. Many health-related apps, such as exercise trackers, can report data into the Health app so it becomes the one-stop-shop for all your health information. One of the most useful features is the ability to set up a Medical ID that contains vital information about you, such as conditions you have and medications you take, that you can use for reference and that others can use in emergency situations.
	Home	After you configure it to work with your home automation devices, use the Home app to control those devices.
	iTunes Store	Browse or search for music, TV shows, movies, and other content. When you find something you want to listen to or watch, you can easily download it to your iPhone. You access the iTunes Store using your Apple ID; any purchases you make are completed using the purchase information associated with that account. Content you download to your iPhone is available in the associated app; for example, when you download music, you listen to that music using the Music app.
	Maps	This app enables you to search for and find just about any location on Earth using addresses or general information, such as "Restaurants near me." When you find a location you want to travel to, you can select it and tap Directions to generate turn-by-turn voice instructions that guide you each step of the way by car or by walking. You can also find public transportation routes and even a ride via a ride-sharing app. If you run into trouble along the way or take a wrong turn, the app automatically reroutes you to make sure you arrive where you want to go. When you're on the way, see estimated arrival time (and share it with others) and other information about your trip. You can swipe up from the bottom of the screen to see other options, such as locating someplace to get coffee; you can choose one of these locations to navigate to it.

Icon	App	Description
Measure	Measure	Use the Measure app to measure distances and surface areas using the iPhone's camera. You can tap Level to measure angles as you do with a physical level.
Music	Music	Use this app to listen to music wherever you are. You find music to listen to in several sources. If you subscribe to Apple Music, you can listen to any music in its vast library (you can download music to your phone so you don't have to be connected to the Internet to listen to it). You can also purchase music using the iTunes Store app and listen to it in the Music app. There are also free radio stations you can listen to. When you select a source of music, you can use the app's controls to play it. These are available within the app, on the Control Center, and on the Lock screen.
News	News	Read news from a variety of sources. You can search news and save items of interest. You can also choose your favorite news sources so that you can focus more on what you care about and avoid what you don't have any interest in.
Notes	Notes	Capture different kinds of information, including text notes, lists, photos, and maps. You can also draw in the app and attach different types of objects to notes, and then use the Attachment Manager to view them. Using online accounts with the Notes app, such as iCloud, means that your notes are available on all your devices. If you want to share a note, tap the Share icon and choose how you want to share it.
Podcasts	Podcasts	This app enables you to subscribe and listen to podcasts on a variety of topics. Episodes of podcasts to which you're subscribed download to your iPhone automatically. You can configure how many episodes are saved on your phone globally and for individual podcasts. There are podcasts on any topic you can imagine, so it's very likely you can find podcasts that are of interest to you. You can search for specific podcasts or browse for them based on category and topic.

Icon	App	Description
	Reminders	Create reminders for just about anything you can imagine, such as to-do and shopping lists. You can configure alarms for reminders that trigger notifications based on date and time or by location. You can set up repeating reminders, too. If that isn't enough, you can also create multiple lists of reminders, such as one for clubs and one for personal items. Like other apps, you can store your reminders on the cloud so that you can access them from multiple devices.
	Stocks	Use this app to track stocks in which you're interested. You can add any index, stock, or mutual fund as long as you know the symbol for it, and you can even use the app to find a symbol if you don't know it. You can see current performance and view historical performance for various time periods. Rotate the iPhone to see a more in-depth view when you're examining a specific investment.
	Tips (Extras folder)	Use this app to get a few pointers about working with your iPhone. Of course, because you have this book, you're way ahead of the game.
	Translate	This amazing app translates from one language to another; the app speaks and shows you the translated sentences or phrases. It currently supports about a dozen different languages with more being added regularly. To use this app, select the language you want to translate by tapping the language in the upper-left corner of the screen. Select the language you want to translate to by tapping the language shown in the upper-right corner of the screen. Tap the Microphone and speak what you want to translate. The translation is immediately spoken in the selected language. Tap Play to play it again. You also see both versions on the screen. You can tap words or phrases in the translated version to open a dictionary. As if all that isn't enough, you can store favorite phrases so that you can easily repeat them at any time. Go to the Translate screen in the Settings app to enable On-device mode so that you can use this app even when your phone isn't connected to the Internet.

Icon	App	Description
	Apple TV	Use this app to watch TV shows and movies. You can sign in to a cable provider to watch TV content, and you can subscribe to video services, such as Apple TV+ or HBO Now. Movies or TV shows you purchase or rent using the iTunes Store app are added to this app so you can watch them.
	Voice Memos (Extras folder)	Record audio notes using the Voice Memos app. Play them back, and, through syncing, move them onto your computer, or share them by tapping the Share icon. You can record through the iPhone's microphone or via the mic on the earbud headset.
	Watch	Use this app to pair and configure your iPhone with an Apple Watch. After it is paired with your iPhone, you can configure how your Apple Watch accesses content from your iPhone's apps such as Mail, Calendar, Messages, Phone, Contacts, Photos, Music, and Reminders.
	Weather	See current weather conditions and a high-level forecast for any number of locations. You can use the default locations, and if you tap the List icon in the lower-right corner of the screen, you can add, remove, and organize the locations you want to track. Swipe through the screens to see each location's forecast.

No Joy

If you don't see one of the apps listed in this table on your iPhone, you can download it from the App Store. The instructions to download and install apps are in Chapter 5, "Customizing Your iPhone with Apps and Widgets."

Touring Other Cool Apps You Can Download

The following table provides "mini-reviews" of some apps you might find useful, but that aren't developed by Apple and thus aren't installed on your iPhone by default. Fortunately, it's easy to download and install apps from the App Store onto your iPhone (refer to Chapter 5).

Other Useful Apps Not Installed by Default

Icon	App	Description
AARP Now	AARP Now	You can watch the daily AARP Minute, read relevant articles, and find local events. You can also take advantage of valuable AARP benefits and access your AARP account.
Southwest	Airline apps	All the major airlines have apps you can use to make reservations or check flight status. The best of these apps also enable you to check in for flights and access an electronic boarding pass through the Wallet app.
AroundMe	AroundMe	Use the app to locate "things" that are around you. You can find hospitals, restaurants, gas stations, hotels, and much more. This app is really useful when you are in a new area because you can quickly find and get to places of interest.
GRUBHUB Grubhub	Food ordering and delivery apps	You can order and get food delivered through a number of apps, such as BeyondMenu, DoorDash, GrubHub, and UberEATS. You can use these apps to search for nearby restaurants. You can choose the type of food you want and then view menus and place orders for delivery in many cases. These apps are especially useful when you travel, but you should also try them at home to see what local restaurants participate in these services.
lyft Lyft	Ride sharing apps	These apps, prime examples of which are Uber and Lyft, enable you to request a ride from your current location to wherever you want to go. When you submit a request and a driver accepts it, you see exactly when your ride will arrive. You can use the app to communicate with your driver in case you need to clarify the pickup location. You also get fare information, and the app handles all payments for you, so, when you arrive at your destination, you can just get out of the car and go. These services generally provide better service, are easier to use, and are less expensive than taxis. You can also request a ride with a ride-sharing service through the Maps app. When you do, it uses a ride-sharing app that you have installed on your iPhone.

Icon	App	Description
	TripIt	This is a great app if you travel frequently because it consolidates your travel plans in one place. When you book a trip, you email your itinerary to the TripIt address. The details are extracted and a TripIt itinerary is created. The app then tracks the status of flights and other information and it keeps you informed of any changes. For example, if a flight gets delayed, you see a notification and your itinerary is automatically updated.
	Yelp	This is another app you can use to find places of interest to you, including shopping, restaurants, gas stations, and more. You can also read reviews posted by other Yelp users and get deals at participating organizations by checking-in.

Configuring and Using Emergency Calling

Using the Emergency SOS feature, you can quickly place a call to local emergency services and also send notifications to people to let them know an emergency has occurred. To be able to use this feature, you need to configure it as explained in the next section. Once configured, you can call for help and notify others as explained in "Using Emergency Calling."

Configuring Emergency Calling

To configure emergency calling, perform the following steps:

(1) Open the Settings app.

(2) Tap Emergency SOS.

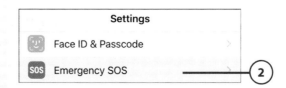

(3) Note how you can activate the Emergency Call slider by pressing and holding the Side Button and either Volume buttons down.

(4) Set the Call with Side Button switch to on (green) if you want to be able to activate a call by quickly pressing the Side button five times.

Different Presses for Different Phones

If you have an older phone, it might use a different set of presses to activate Emergency Calling. When you open the Emergency SOS Settings screen, you read the specific presses your phone uses in the Call with Side Button area. If your model doesn't have a side button (SE), this is labeled with the type of button your model has. In all cases, read this section to make sure you understand the exact presses needed for your phone.

(5) Set the Auto Call switch to on (green). If you don't do this, the emergency call won't be automatically placed.

(6) Tap Edit Emergency Contacts in Health (if you have configured a Medical ID) or Set up Emergency Contacts in Health (if you haven't configured a Medical ID) to identify who you want to be notified when you place an emergency call. You move into the Health app.

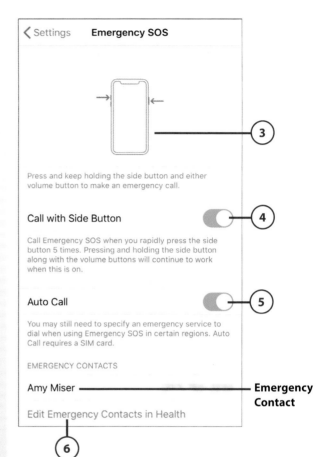

‹ Settings   **Emergency SOS**

Press and keep holding the side button and either volume button to make an emergency call.

Call with Side Button

Call Emergency SOS when you rapidly press the side button 5 times. Pressing and holding the side button along with the volume buttons will continue to work when this is on.

Auto Call

You may still need to specify an emergency service to dial when using Emergency SOS in certain regions. Auto Call requires a SIM card.

EMERGENCY CONTACTS

Amy Miser —————————————— **Emergency Contact**

Edit Emergency Contacts in Health

We Need Some Basic Information First

If you tap Set up Emergency Contacts in Health in step 6, you're prompted to create your Medical ID. Tap Create Medical ID and fill out the information by tapping it and entering your data. You don't need to include very much for a basic Medical ID. Your name is added automatically. You can add more information, such as your birthday and medications you take, but that's not required for the emergency contact feature. When you're done adding information, tap Next and then tap Done. After your Medical ID is created, you can pick up at step 7 to complete the emergency contact information.

⑦ Tap Edit.

⑧ Tap add emergency contact.

⑨ Use your contact list to select the emergency contact.

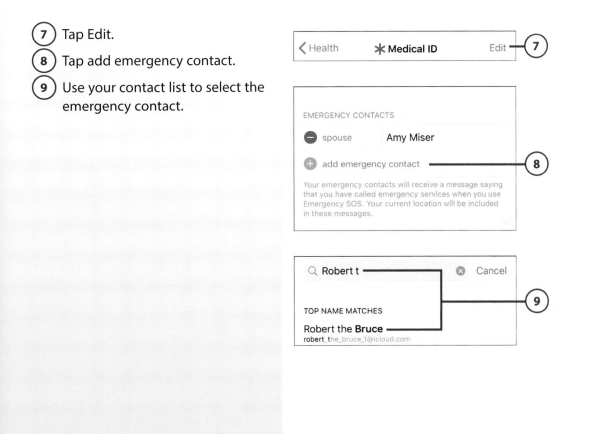

(10) Tap the contact's relationship to you.

(11) Repeat steps 8 through 10 to add more contacts.

(12) Tap Done.

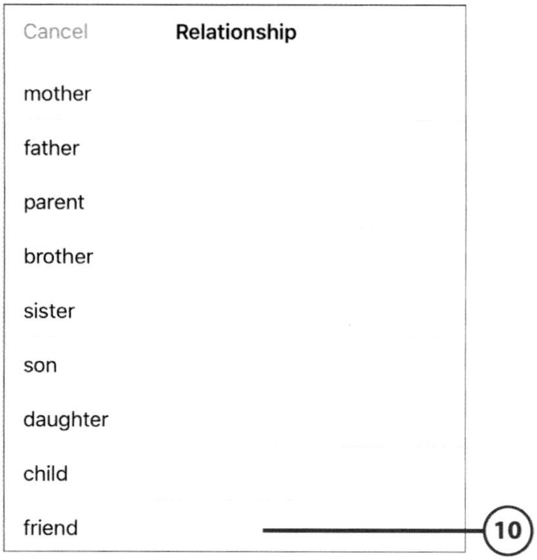

Cancel	**Relationship**
mother	
father	
parent	
brother	
sister	
son	
daughter	
child	
friend	————(10)

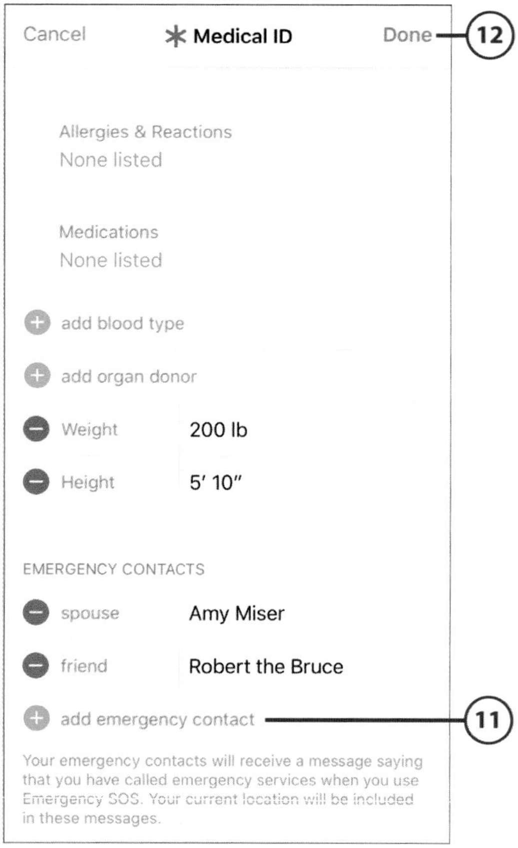

Cancel	✳ **Medical ID**	Done ——(12)

Allergies & Reactions
None listed

Medications
None listed

⊕ add blood type

⊕ add organ donor

⊖ Weight   200 lb

⊖ Height   5' 10"

EMERGENCY CONTACTS

⊖ spouse   Amy Miser

⊖ friend   Robert the Bruce

⊕ add emergency contact ————(11)

Your emergency contacts will receive a message saying that you have called emergency services when you use Emergency SOS. Your current location will be included in these messages.

(13) Tap Health.

(14) Tap Settings.

(15) Tap Emergency SOS.

(16) Set the Countdown Sound switch to on (green) so you hear a warning sound before the call is placed. This can help prevent accidental emergency calls because you hear the Countdown Sound before the call is made.

Emergency Contact Number

The number you should call for emergency services depends on your specific location. These steps cover someone in the United States or other locations with a national emergency calling system. If where you are doesn't have this type of system, search for how to use Emergency SOS in your area on Apple's support site located at www.apple.com/support.

⟨ Health ———— ✱ Medical ID ———— Edit (13)

(14)

⟨ Settings   **Health**

Settings

😊 Face ID & Passcode ⟩

SOS Emergency SOS ————————— (15)

Auto Call ⬤

You may still need to specify an emergency service to dial when using Emergency SOS in certain regions. Auto Call requires a SIM card.

EMERGENCY CONTACTS

Amy Miser ———— ————————————

Robert the Bruce ———— ————————————
Emergency Contacts

Edit Emergency Contacts in Health

You can add and edit emergency contacts for Emergency SOS in your Medical ID in the Health app. About Emergency SOS & Privacy

Countdown Sound ⬤ — (16)

Play a warning sound while Emergency SOS is counting down to call emergency services.

Using Emergency Calling

With Emergency SOS set up as described in the previous section, there are a couple of ways you can activate emergency calling.

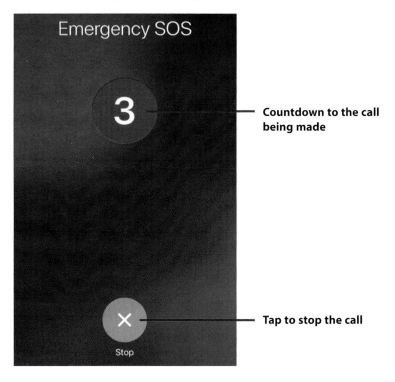

Countdown to the call being made

Tap to stop the call

Press the button sequence to activate Emergency SOS that you noted in step 4 of the previous task. For example, on most current models of iPhone, you press the Side button five times. The Emergency SOS feature activates, starts a countdown from three, and plays a very loud alert sound.

When the countdown expires, you're connected to your local emergency services just as if you had placed the call manually. Notifications are also sent to your emergency contacts to inform them you have made an emergency call.

If you didn't intend to place the call, tap Stop (x) before the timer finishes its countdown.

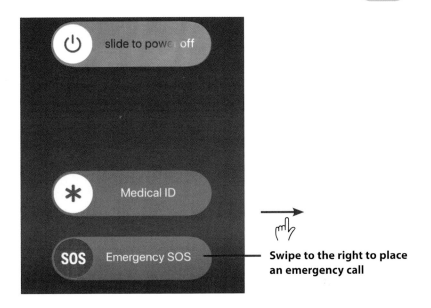

Swipe to the right to place an emergency call

You can also activate Emergency SOS by pressing and holding the Side and either Volume buttons down until you see sliders on the screen. Swipe to the right on the Emergency SOS slider. The call for emergency services is immediately placed and notifications are sent to your emergency contacts.

No Going Back

The emergency call is placed immediately after you swipe on the Emergency SOS slider. Even if you stop the emergency call before it's answered, you might be contacted (via text or voice call) by the emergency response organization to determine if you have an emergency. Be very careful about using this feature so you don't accidentally place an emergency call.

Using Bluetooth to Connect to Other Devices

The iPhone includes built-in Bluetooth support so you can use this wireless technology to connect to other Bluetooth-capable devices. The most likely devices to connect to your iPhone in this way are Bluetooth headphones (such as Apple AirPods), speakers, headsets, or car audio/entertainment/information systems. You also connect to an Apple Watch via Bluetooth.

To connect Bluetooth devices, you pair them. Pairing enables two Bluetooth devices to communicate with each other. The one constant requirement is that the devices can communicate with each other via Bluetooth. For devices to find and identify each other so they can communicate, one or both must be discoverable, which means they broadcast a Bluetooth signal that other devices can detect and connect to.

There is also a "sometimes" requirement, which is a pairing code, passkey, or PIN. All those terms refer to the same thing, which is a series of numbers, letters, or both, entered on one or both devices being paired. Sometimes you enter this code on both devices, whereas for other devices you enter the first device's code on the second device. Many Bluetooth devices, such as headphones or speakers, don't require a pairing code at all.

When you have to pair devices, you're prompted to do so, and you have to complete the actions required by the prompts to communicate via Bluetooth. This might be just tapping Connect, or you might have to enter a passcode on one or both devices to connect them.

Connecting to Bluetooth Devices

This task demonstrates pairing an iPhone with Bluetooth headphones; you can pair it with other devices similarly.

1. Move to the Settings screen. The current status of Bluetooth on your iPhone is shown.

2. Tap Bluetooth.

	Settings	
✈	Airplane Mode	
📶	Wi-Fi	ATTH7n7RMS >
✳	Bluetooth	On
📡	Cellular	>
📷	Notifications	>
🔊	Sounds & Haptics	>
🌙	Do Not Disturb	>

Bluetooth is on

(3) If Bluetooth isn't on (green), tap the Bluetooth switch to turn it on. If it isn't running already, Bluetooth starts up. The iPhone immediately begins searching for Bluetooth devices. You also see the status Now Discoverable, which means other Bluetooth devices can discover your iPhone. In the MY DEVICES section, you see devices to which you've previously paired your iPhone; their current status is either Connected, meaning the device is currently communicating with your iPhone, or Not Connected, meaning the device is paired with your iPhone, but is not currently connected to it. In the OTHER DEVICES section, you see the devices that are discoverable to your iPhone but that are not paired with it.

(4) If the device you want to use appears on the MY DEVICES list with Connected as its status, skip the rest of these steps because your iPhone is already connected to the device. (Not shown in a figure.)

< Settings   Bluetooth

Bluetooth   ◯ — 3

Now discoverable as "Brad Miser's iPhone".

MY DEVICES

BH-M9A   Not Connected ⓘ — **Device that has been paired with the iPhone previously but isn't currently connected to it**

Brad's Apple Watch   Connected ⓘ

TIBS01BK   Not Connected ⓘ

UConnect   Not Connected ⓘ

OTHER DEVICES

WAVESOUND 2 ——

Device that isn't paired with the iPhone   **Device connected to the iPhone**

5 If the device you want to use isn't shown in the OTHER DEVICES section, put it into Discoverable mode. (Not shown in a figure; see the instructions provided with the device. Often, this involves holding one or more buttons down while you turn the device on.) When it is discoverable, it appears in the OTHER DEVICES section.

6 Tap the device to which you want to connect. If the device isn't currently paired with your iPhone, you might need to provide a passkey. If a passkey is required, you see a prompt to enter it on the device with which you are pairing; perform step 7. If no passkey is required, skip to step 8.

7 If a pairing code, passkey, or PIN is required, input it on the device (not shown in a figure), such as typing the passkey on a keyboard if you are pairing your iPhone with a Bluetooth keyboard.

8 If required, tap Connect (not shown in a figure)—most devices connect as soon as you select them or enter the passkey, and you won't need to do this. You see the device to which the iPhone is connected in the MY DEVICES section of the Bluetooth screen, and its status is Connected, indicating that your iPhone can communicate with and use the device.

‹ Settings	**Bluetooth**	
Bluetooth		⬤
Now discoverable as "Brad Miser's iPhone".		
MY DEVICES		
BH-M9A	Not Connected	ⓘ
Brad's Apple Watch	Connected	ⓘ
TIBS01BK	Not Connected	ⓘ
UConnect	Not Connected	ⓘ
OTHER DEVICES		
WAVESOUND 2		

6

Using and Managing Bluetooth Devices

When your iPhone is paired with and connected to Bluetooth devices, you can use and manage them in a number of ways. Some examples follow:

Tap to choose a
device to listen to
music

Audio Player on
the Lock screen

- **Choose Bluetooth Devices**—When your iPhone is connected to a Bluetooth device, you can use that device with an associated app. For example, you can choose to listen to music on Bluetooth headphones. In the app you're using, tap the Output icon. This icon appears in various locations, such as the upper-right corner in the Audio Player on the Lock screen.

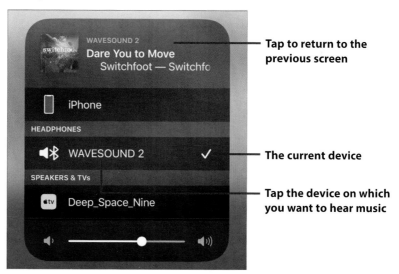

Tap to return to the
previous screen

The current device

Tap the device on which
you want to hear music

On the resulting menu, you see the Bluetooth and other devices available to use because they are currently communicating with your iPhone. The current device is marked with a check mark. Tap the device you want to use; the iPhone starts using that device, such as playing music on it. When you're done making a selection, tap outside the device list to return to the previous screen.

- **No Pairing Required**—If you've previously connected to a Bluetooth device, you don't need to go through the pairing process again. You can select it on the Bluetooth settings screen, the device selection menu (as shown in the previous figure), or via the Control Center. Your iPhone immediately begins using that device (as long as the device is powered up and within range of your iPhone).

Tap to turn Bluetooth on or off   Tap and hold to see more options

- **Manage Bluetooth from the Control Center**—You can quickly control Bluetooth and select devices to use from the Control Center. Open the Control Center and tap the Bluetooth icon to turn Bluetooth on or off; you see its current status by the color of the button (blue is on) and onscreen text. Touch and hold on the icon to expand the options you see.

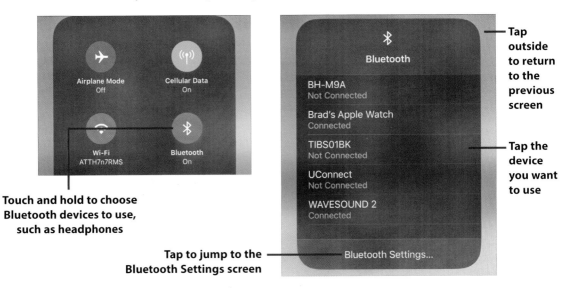

Touch and hold to choose Bluetooth devices to use, such as headphones

Tap to jump to the Bluetooth Settings screen

Tap outside to return to the previous screen

Tap the device you want to use

Touch and hold on the Bluetooth icon on the expanded options screen. The Devices menu appears and you can select devices just as you can on the Bluetooth Settings screen.

Tap outside the options box to close it; then tap on the top or bottom of the screen to return to the Control Center.

MY DEVICES		
BH-M9A	Not Connected	ⓘ
Brad's Apple Watch	Connected	ⓘ
TIBS01BK	Not Connected	ⓘ
UConnect	Not Connected	ⓘ
WAVESOUND 2	Connected	ⓘ

❮ Bluetooth   **WAVESOUND 2**

Disconnect

Forget This Device

ⓘ — **Tap to see options**   **Tap to disconnect**

- **Change Devices**—Your iPhone can be paired with multiple devices at the same time, and it remembers every device you pair it with. It can also work with multiple devices at the same time; for example, you can use Bluetooth to work with an Apple Watch and Bluetooth headphones simultaneously. However, your iPhone can only work with one of the same type of device at the same time. Suppose you have both Bluetooth headphones and a speaker. Both can be paired with your iPhone, but you can't be connected to both at the same time.

You can change devices you're using by selecting the device you want as described earlier. Your iPhone switches from one device to the next.

However, many Bluetooth devices can be connected to only one device at a time. For example, if you have an iPhone and an iPad, both can be paired with the same Bluetooth headphones, but only one can be connected to the headphones at a time.

When you want to change the device you're using, such as changing from your iPhone to your iPad, you need to disconnect the Bluetooth device from the iPhone so that you can connect to it with the iPad.

Open the Bluetooth Settings screen and tap Info (i) for the device from which you want to disconnect. On the resulting screen, tap Disconnect. Your iPhone disconnects from the device, and you can connect to it with another device, such as your iPad. It remains paired with your phone so you can use it again.

- **Forget Devices**—Like other connections you make, the iPhone remembers Bluetooth devices to which you've connected before and reconnects to them automatically, which is convenient—most of the time anyway. If you don't want your iPhone to keep connecting to a device, move to the Bluetooth Settings screen and tap Info (i) for the device. Tap Forget This Device and then tap Forget Device. The pairing is removed, and your iPhone stops connecting to the device automatically. Of course, you can always pair the device again. If you just want to stop using the device, but keep the pairing in place, tap Disconnect instead.

Using AirDrop to Share and Receive Content

You can use AirDrop to share content directly with people using a Mac running OS X Yosemite or later, or using a device running iOS 7 or later (including iPadOS). For example, if you capture a great photo on your iPhone, you can use AirDrop to instantly share that photo with iPhone, iPad, and Mac users near you.

AirDrop can use Wi-Fi or Bluetooth to share, but the nice thing about AirDrop is that it manages the details for you. You simply open the Share menu—which is available in most apps—tap AirDrop, and tap the people with whom you want to share.

When you activate AirDrop, you can select Everyone, which means you see anyone who has a Mac running OS X Yosemite or later, an iOS device running version iOS 7 or newer, or an iPad running any version of iPadOS and is on the same Wi-Fi network as you (or has a paired Bluetooth device); those people can see you, too. (You can decline any request to share with you via AirDrop, so even if you choose Everyone, you can still decide which content is moved onto your iPhone.) Or you can select Contacts Only, which means only people who are in your Contacts app are able to use AirDrop to communicate with you. In most cases, you should choose the Contacts Only option so you have more control over who uses AirDrop with you.

Is AirDrop Safe?

Anything you share with AirDrop is encrypted, so the chances of someone else being able to intercept and use what you share are quite low. Likewise, you don't have to worry about someone using AirDrop to access your information or to add information to your device without your permission. However, like any networking technology, there's always some chance—quite small in this case—that someone will figure out how to use this technology for nefarious purposes. The best thing you can do is to be wary of any requests you receive to share information and ensure they're from people you know and trust before you accept them.

Enabling AirDrop

To use AirDrop, you must enable it on your iPhone.

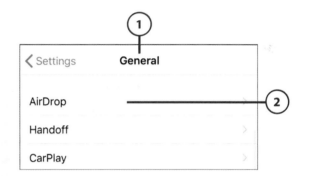

1. Open the General screen of the Settings app.

2. Tap AirDrop.

3. Tap Receiving Off if you don't want to receive any content via AirDrop (you can still share your content with others), tap Contacts Only to allow only people in your Contacts app to be able to request to share content with you, or Everyone to allow anyone using an AirDrop device in your area to request to share content with you.

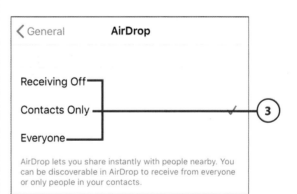

Another Way to Configure AirDrop

You can also configure AirDrop by opening the Control Center and pressing on the Airplane mode, Wi-Fi, Cellular Data, or Bluetooth icons. On the resulting panel, tap the AirDrop icon. Then choose the AirDrop status you want to set (these are the same as described in step 3).

Share and Share Alike?

You should choose the Receiving Off option when you aren't using AirDrop, especially if you usually use the Everyone option. When Receiving Off is selected, people in your area won't see you as an option when they use AirDrop. This prevents you from being contacted for AirDrop sharing when you don't want to be. When you want to receive content again, you can quickly reset AirDrop to Contacts Only or Everyone again.

Using AirDrop to Share Your Content

To use AirDrop to share your content, do the following:

(1) Open the content you want to share. This example shows sharing a photo using the Photos app (this app is covered in detail in Chapter 16, "Viewing and Editing Photos and Video with the Photos App"). The steps to share content from any other app are quite similar.

(2) Tap Share.

3 If the app you selected in step 1 supports it, you can swipe to preview other content and select it by tapping it. Each item you select is available for sharing. When marked with the check mark, the content will be shared.

4 Swipe to the left or right on the Recents bar to review AirDrop recipients you've communicated with previously (these are marked with the AirDrop icon); tap a recent recipient to send the content there and skip to the information after step 6.

5 If the recipient is not shown on the Recents bar, tap AirDrop. You see the potential recipients for your AirDrop sharing; these are grouped as People, Devices, or Other People (which shows available devices that you haven't shared with before or that aren't identifiable as a device).

6 Tap the people with whom or device with which you want to share the content.

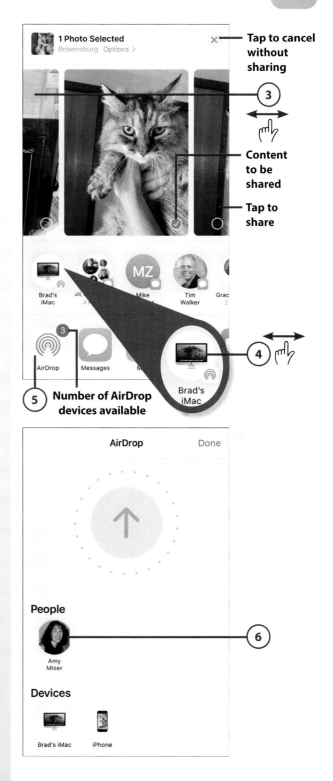

A sharing request is sent. Under the AirDrop icons you selected, the Waiting status is displayed. When a recipient accepts your content, the status changes to Sending and when delivered, the status becomes Sent. If a recipient rejects your content, the status changes to Declined.

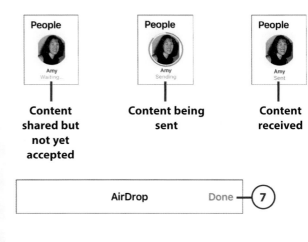

Content shared but not yet accepted

Content being sent

Content received

(7) When you're done sharing, tap Done. You return to the Share sheet and can share other content or tap Close (x) to close the Share sheet.

Using AirDrop to Work with Content Shared with You

When someone wants to share content with you with AirDrop, you receive an AirDrop sharing request. (If you use AirDrop to share content among your own devices such as sending a photo from your iPhone to a Mac, it might be accepted automatically.) Respond to sharing requests that you receive by doing the following:

(1) Make sure you know the person attempting to share with you.

(2) Make sure the content being shared with you is something you want. In this case, a photo is being shared.

(3) To accept the content on your iPhone, tap Accept. To reject it, tap Decline.

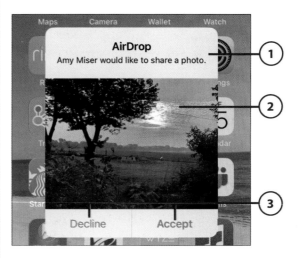

4 If you accepted the content, use the associated app to work with what you receive. For example, the Photos app provides tools to edit and share photos that are shared with you. In some cases, such as saving a contact shared with you, you need to tap Save to save the content on your iPhone or Cancel to not save it. (Other apps provide different controls depending on the type of content and the app it opens in.)

Working with the Wallet App and Apple Pay

The Wallet app manages all sorts of information that you need to access, from airline tickets to shopping and credit cards to movie tickets and rewards cards. Instead of using paper or plastic to conduct transactions, you can simply have your iPhone's screen scanned.

You can also use the Wallet app to access your Apple Pay information to make payments when you're in a physical location, such as a store or hotel (Apple Pay is covered in detail in "Working with Apple Pay" later in this chapter). You can also use Apple Pay when you make purchases online using some apps on your iPhone.

And, if that isn't enough, you can send payments directly to other people; this capability is highlighted in the Go Further sidebar at the end of this chapter, "Paying People with Apple Cash."

Working with the Wallet App

You can store a wide variety of information in your Wallet so that it's easily accessible. Examples include boarding passes, membership cards (to a gym, for example), store cards (such as Starbucks if you're an addict like I am), and loyalty or discount cards (like those loyalty cards that grocery stores or gas stations offer). The Wallet app eliminates the need to carry physical cards or paper for each of these; instead, your information is available to you digitally, and you can enter it as needed by scanning the iPhone's screen.

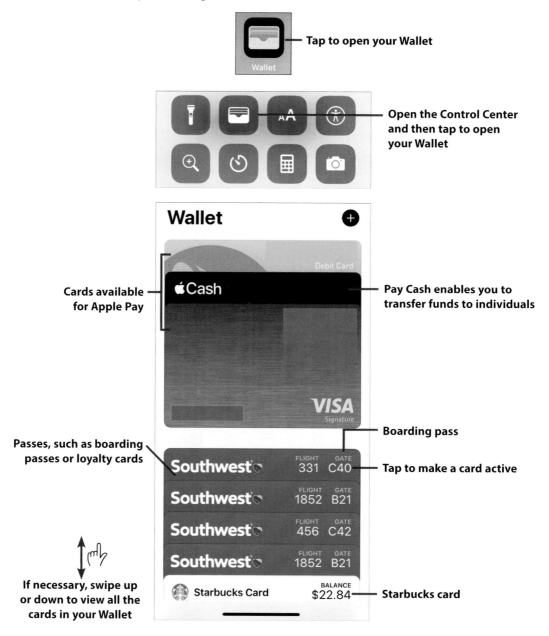

Tap to open your Wallet

Open the Control Center and then tap to open your Wallet

Cards available for Apple Pay

Pay Cash enables you to transfer funds to individuals

Boarding pass

Passes, such as boarding passes or loyalty cards

Tap to make a card active

If necessary, swipe up or down to view all the cards in your Wallet

Starbucks card

To open your Wallet, tap Wallet on a Home screen or first open the Control Center and then tap the Wallet icon. The Wallet app opens. At the top, you see the Apple Pay section that shows the cards you've enabled for Apple Pay and Apple Cash that you can use to send payment to others directly. In the Passes section, you see other cards, including boarding passes and loyalty cards.

As you add cards and boarding passes to your Wallet, they "stack up" at the bottom of the screen. Tap the stack to see the list of items available in your Wallet. The cards expand so that you can see at least the top of each card; if you have a lot of cards installed, swipe up or down on the screen to view all of them.

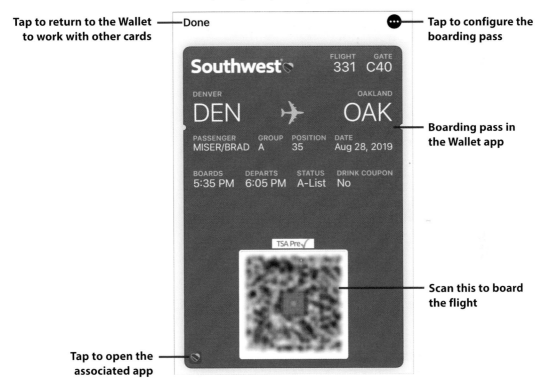

Tap to return to the Wallet to work with other cards — Done

Tap to configure the boarding pass

Boarding pass in the Wallet app

Scan this to board the flight

Tap to open the associated app

To use a card in your Wallet, tap it. It becomes the active card, and you can access its information. For example, you can quickly access a boarding pass by opening the Wallet app and tapping the pass you want to use. To board the plane, you scan the code on the boarding pass. Even better, boarding passes start appearing on the Lock screen a few hours before the take-off time. Press on the boarding pass on the Lock screen, and you jump directly to the boarding pass in the Wallet app (you don't even need to unlock your phone).

Some passes enable you to share them or information about them by tapping the Share icon and using the standard sharing tools that appear.

Others are connected to an app; you can tap the icon on the pass or card to move into the associated app.

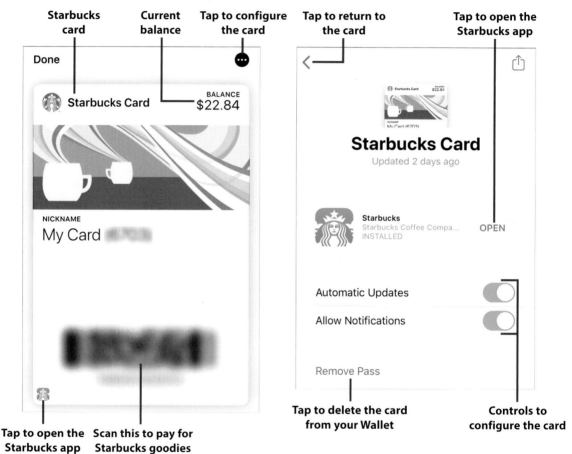

Starbucks card · Current balance · Tap to configure the card · Tap to return to the card · Tap to open the Starbucks app

Tap to open the Starbucks app · Scan this to pay for Starbucks goodies · Tap to delete the card from your Wallet · Controls to configure the card

When you tap a card's Options (…), you can use the controls on a card's Configure screen to adjust certain aspects of how the card works. If you set the Suggest on Lock Screen switch to on (green) for a card that works based on location, such as your favorite stores, the card automatically appears on your Lock screen at the appropriate times. And some cards can be automatically refilled from your bank account or credit/debit card so that they always have a positive balance.

To delete a card from your Wallet (such as a boarding pass when the flight is finished), tap Options (…), tap Remove Pass, and then confirm that you want to remove it. The pass or card is removed from your Wallet.

When you're done configuring a card, tap back (<). The Configure screen closes, and you return to the card you were configuring.

There are a couple of ways to add cards to your Wallet. The most frequent way add a card is by using the Add to Wallet command in the app associated with the card. In some cases, you might be able to scan the code on a card to add it.

Adding Passes or Cards to Your Wallet Using an App

If you're a frequent patron of a particular business that has an iPhone app, check to see if it also supports the Wallet app. For example, when you use the Starbucks app configured with your account, you can add a Starbucks card to your Wallet as follows:

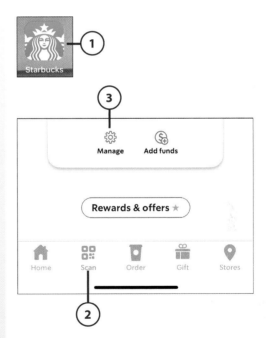

(1) Open the app for which you want to add a pass or card to your Wallet.

(2) Tap Scan. (These steps are specific to the Starbucks app, but you can add other cards to your Wallet in similar ways. Look for the command that enables you to configure how you pay for purchases associated with that card.)

Starbucks Card

You need to have a card configured in the Starbucks app to be able to add it to the Apple Wallet. Also, if you have more than one card, you need to individually add each card to the Wallet.

(3) Tap Manage; for other apps, look for the app's command to manage the card's information; this command can be labeled with different names in different apps, or it might be accessed with an icon or menu.

(4) Tap Add to Apple Wallet.

(5) Tap Add. The card or pass is
added to your Wallet and is
ready to use. (You might be
prompted to do some additional
configuration of the card, such as
to indicate favorite locations.)

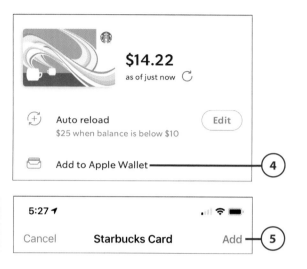

Finding Apps that Support the Wallet

You can search for apps that support the Wallet app by opening it, tapping Edit Passes
at the bottom of the Passes section, and tapping Find Apps for Wallet. This takes you
into the App Store app, and you see apps that support Wallet. You can then download
and install apps you want to use as described in Chapter 5.

Adding Cards Without an App

If there isn't an app associated with a card you want to add to your Wallet, you can
add it by swiping all the way up on the Wallet screen and tapping Edit Passes. Tap
Scan Code. Position the iPhone's camera over the card. If the card's information can be
scanned, it's added to your Wallet.

Cards Without Using the Wallet?

If a card you use has a bar code or QR code on it, but you aren't able to scan it for some reason, you can almost replicate what the Wallet app does by taking a photo of the bar code
or QR code. When you need to scan the code, such as to enter a gym, hold the photo of the
code up to the code reader. This isn't as convenient as using the Wallet app, but it can work.

This trick can also work for cards that don't have a bar or QR code too. Just take a
photo of the card and use that instead of the actual card. Some places will accept this,
whereas others insist on the old-fashioned paper or plastic card. (Certainly, don't try this
trick with official identification, such as a driver's license.)

Adding Apple Pay Cards Already Associated with Your Apple ID to Your Wallet

If you have a credit or debit card associated with your Apple ID, you can add it to your Wallet as follows:

① Open the Wallet app.

② Tap Add (+).

③ Tap Credit or Debit Card. You move to the Add Card screen.

④ Tap Continue.

⑤ Verify the card number shown is the one you want to make available for Apple Pay (a number is not shown in the figure, but you'll see dots and the last four digits for your card).

⑥ Type the card's three-digit security code (not shown in the figure).

⑦ Tap Next.

Wallet

Card Type

Choose the type of card to add to Apple Pay.

PAYMENT CARDS

Credit or Debit Card

 Pay

Add cards to Apple Pay to send money to friends and make secure purchases in stores, in apps, and on the web.

Card-related information, location, device settings, and device use patterns will be sent to Apple and may be shared together with account information with your card issuer or bank to set up Apple Pay. See how your data is managed...

Continue

‹ Back   Next

Add Card

Verify your card information on file with iTunes or App Store.

Card on File   VISA

Security Code

8 If prompted to do so, tap Agree. The card becomes available to Apple Pay. (If you don't agree, the card won't be added to your Wallet.)

Apple Pay Support

A credit or debit card must support Apple Pay for it to work with that card. The easiest way to figure out if your cards support it is to try to add a card to Apple Pay. If you can do so, the card is supported, and you can use it. If not, you can check with the credit or debit card company to see when support for Apple Pay will be added.

Adding Cards to Apple Pay by Scanning Their Codes

In some cases, you can add a card or pass to the Wallet by scanning its code. To do so, follow these steps:

1 Follow steps 1 through 4 in the previous section. You move to the Scan screen; a box appears on the screen.

2 Tap Add a Different Card.

3 Use the iPhone's camera to scan the bar code on the card by positioning the phone so that the white box encloses the bar code on the card (not shown on a figure). If the bar code is recognized and is available for the Wallet, you're prompted to enter the card's security code. If the bar code isn't recognized or doesn't support the Wallet, you see an error, and you need to find an associated app for the card to use it with the Wallet.

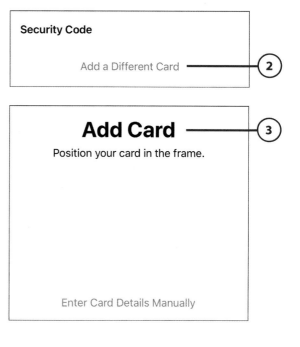

When you add a Chase Card to a Wallet, you agree to these Terms:

1. Adding Your Chase Card. You can add an eligible Chase Card to a Wallet by either following our instructions as they appear on a Chase proprietary platform (e.g., Chase Mobile® app or chase.com) or by following the instructions of the Wallet provider. Only Chase Cards that we determine are eligible can be added to the Wallet. If your Chase Card or underlying account is not in good standing, that Chase Card will

Disagree   Agree — **8**

Security Code

Add a Different Card — **2**

Add Card — **3**
Position your card in the frame.

Enter Card Details Manually

(4) Type the card's three-digit security code (not shown on the figure); this is required for credit or debit cards, but not all other types of cards.

(5) Tap Next.

(6) If prompted to do so, tap Agree (not shown on a figure). The card becomes available for Apple Pay. (If you don't agree, the card won't be added to your Wallet.)

< Back   Next — (5)

Card Details

Verify and complete your card information.

Expiration Date   03/22

Security Code ————— (4)

Another Way to Add an Apple Pay Card

If you don't want to scan a card, or the scan process won't work for some reason, move to the Scan screen as explained in steps 1 and 2. Then tap Enter Card Details Manually. This enables you to manually type your name, the card number, expiration date, and security code to configure the card for Apple Pay.

More Verification Required

In some cases, there might be an additional verification step after you tap Agree. When this is the case, the Complete Verification screen appears. Tap how you want to receive the verification code, such as via Email or Text Message, and then tap Next. The card configuration completes. You should receive a verification code. When you have the code, open the Settings app and tap Wallet & Apple Pay. Tap the card you need to verify, and then tap Enter Code. Enter the verification code you received; when you enter the correct code, the card is verified and becomes available for Apple Pay.

Working with Apple Pay

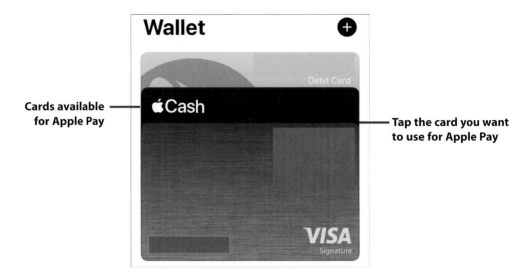

Cards available for Apple Pay

Tap the card you want to use for Apple Pay

With Apple Pay, you can store your credit and debit cards in the Wallet app, and they're instantly available to make purchases. Apple Pay enables you to pay for things more easily and securely than with physical credit or debit cards.

When you use Apple Pay in a physical store, the steps to use Apple Pay are slightly different between iPhones that don't have a Home button and those that do.

Press the Side button twice to pay

Card that will be used to pay

Apple Pay on an iPhone without a Home button is a great way to pay for transactions quickly and easily. You can just hold your iPhone near the scanner to activate Apple Pay, press the Side button twice, and look at your phone.

Pay by Default

If you have only one card in your Wallet, it is used automatically. If you have more than one card, you designate the default card, which is used automatically. If you want to use a card that isn't the default, you have to select it in the Wallet before you pay.

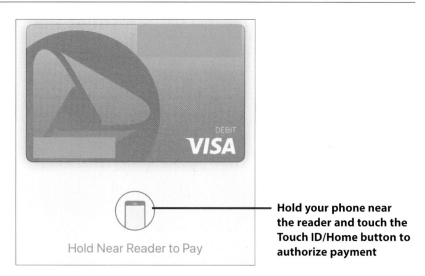

Hold Near Reader to Pay

Hold your phone near the reader and touch the Touch ID/Home button to authorize payment

If you're using Apple Pay on a iPhone with a Home button, you can simply hold your iPhone up to a contactless reader connected to the cash register and touch your finger on the Touch ID/Home button. The iPhone communicates the information required to complete the purchase.

Check out with Apple Pay

Check out with other payment options

Other payment options include
Debit/Credit card, Barclaycard Visa,
PayPal, Apple Store gift cards.

When you are making a transaction in an app that supports Apply Pay, tap this to pay

Apple Pay also simplifies purchases made in online stores. When you use an app or a website that supports Apple Pay, you can tap Check out with Apple Pay, Buy with Apple Pay, or a similar onscreen button to complete the purchase.

Apple Pay is actually more secure than using a credit or debit card because your card information is not passed to the device; instead, a unique code is passed that ties back to your card, but that can't be used again. And, you never present your card so the number is not visible to anyone, either visually or digitally.

Managing Apple Pay

Settings	**Wallet & Apple Pay**

Use the Wallet & Apple Pay Settings screen to manage Apple Pay

Apple Cash

Enable sending and receiving money in Messages on this iPhone.

PAYMENT CARDS

Apple Cash
Not Set Up

Rewards Visa
....

Debit Card
....

Add Card

Double-Click Side Button

Get cards and passes ready at any time by double-clicking the side button.

TRANSIT CARDS

Express Transit Card   None

To manage Apple Pay, open the Settings app and tap Wallet & Apple Pay. On the resulting screen, you can do the following:

- Enable and configure Apple Cash (see the Go Further sidebar, "Paying People with Apple Cash").

- Tap a card to configure it, such as to determine if notifications related to it are sent. You also see a list of recent transactions for the card.

- Add a new credit or debit card (tap Add Card and follow the prompts).

- Use the Double-Click Side Button switch (iPhones without a Home button) or Double-Click Home Button switch (iPhones with a Home button) to determine whether you can access the cards in your Wallet by pressing the Side button or the Touch ID/Home button twice. This makes using your Wallet even easier because you press the associated button twice and your cards appear; tap a card to use it and then use Face ID or Touch ID to complete the process.

• I don't recommend this option, but you can tap Express Transit Card and choose a card to use without Face ID, Touch ID, or your phone's passcode to be entered. With this option enabled, you can simply hold your phone next to a reader, and Apple Pay activates to provide a payment. This makes transactions easier but doesn't require some physical action on your part, which is less secure.

• Set your default Apple Pay card. The default card is used automatically; you have to manually select other cards in the Wallet app to use one of them instead.

• Update your shipping address, email addresses, and phone numbers.

• Set the Allow Payments on Mac switch to on (green) if you use a Mac computer and want to be able to use Apple Pay for online transactions.

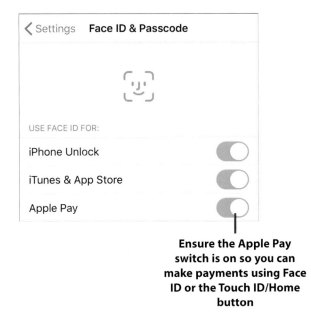

Ensure the Apple Pay switch is on so you can make payments using Face ID or the Touch ID/Home button

You should configure Apple Pay is set to use Face ID (iPhones without a Home button) or Touch ID (iPhones with a Home button) so paying for things is easier and faster. Open the Settings app, tap Face ID & Passcode (iPhones without a Home button) or Touch ID & Passcode (iPhones with a Home button), and then enter your passcode. Ensure that the Apple Pay switch is on (green).

>>>Go Further

PAYING PEOPLE WITH APPLE CASH

Apple Cash gives you the ability to transfer and receive money via the Messages app. For example, if you need to reimburse someone for your part of a restaurant tab, you can do so quickly and easily using Apple Cash; people to whom you send money with Apple Cash must have their devices configured to receive funds in Apple Pay. You use the Messages app to send money to a person very similarly to how you send messages.

Before you can use Apple Cash, you need to set it up and add money to your Apple Cash account. Open the Wallet & Apple Pay Settings screen and tap Apple Cash. After you tap Continue and agree to the terms, tap Done. Tap Apple Cash again and tap Add Money. Enter the amount of Apple Cash you want to have available and tap Add. Use the resulting Apple Pay prompt to allow the transaction; you can choose the card in your wallet from which the funds come if you don't want to use the card that is automatically selected. The amount you selected is added to your Apple Cash account, and you can use it to make payments.

You can use the Apple Cash screen to manage your account, such as choosing how you accept payments.

Once Apple Cash is configured, you can use the Apple Pay app within the Messages app to pay people with whom you are messaging or to request payment. When you pay someone, money is transferred from your Apple Cash account to that person.

Funds you receive via Apple Cash are added to your Apple Cash account. You can use those funds for Apple Pay purchases, or you can transfer them to a bank account using the Wallet & Apple Pay Settings screen (tap Apple Cash and then tap Transfer to Bank).

An iPhone is easy to maintain and isn't likely to give you much trouble

In this chapter, you learn how to keep an iPhone in top shape and what to do should problems happen. Topics include the following:

→ Maintaining an iPhone's software
→ Backing up your iPhone
→ Blocking unwanted calls, messages, or FaceTime requests
→ Finding and securing your iPhone with Find My iPhone
→ Maintaining an iPhone's power
→ Solving iPhone problems
→ Getting help with iPhone problems

Maintaining and Protecting Your iPhone and Solving Problems

iPhones usually work very well, and you are unlikely to have problems with one, especially if you keep its software current. When problems do occur, you can usually solve them with a few simple steps. If that fails, there's lots of help available for you on the Internet or you can visit the Genius bar at an Apple Store or a local retail store for your cellular provider.

Getting Started

The iOS system software that makes the iPhone work is currently on version 14. This software runs very reliably so you won't have to spend a lot of time dealing with problems.

When it comes to preventing and solving iPhone problems, this chapter teaches you three key things you need to know:

- Keeping your iPhone's software current (covered in the next section) minimizes the chances of you having problems with your phone and also ensures you have the most secure software available.

- Restarting your iPhone (see "Restarting Your iPhone") solves many problems you might encounter. Any time your iPhone starts behaving oddly or not working the way you expect it to, restart it. This simple task often gets your iPhone working normally again quickly and easily.

- Getting help (see "Getting Help with iPhone Problems") when you need it. Your best resource is an Apple Genius at an Apple Store, who can help you solve any hardware or software problem you might experience. You can also contact Apple for help via chat or phone through the Apple Support website.

Maintaining an iPhone's Software

Like software on computers, the software on your iPhone gets updated periodically to fix bugs, solve issues, and introduce new features. For best results, you should ensure the software you use is current to the latest releases.

There are two types of software you need to maintain: the iOS software that runs your iPhone and the apps you have installed on it. Fortunately, maintaining both types is simple.

Maintaining the iOS Software with the Settings App

You can check for updates to the iOS using the Settings app. If an update is found, you can download and install it directly on the iPhone.

Update Notification

You might receive a notification on the iPhone when an update to iOS is available; you also see a badge on the Settings app's icon. You can tap Update in the notification to move to the Update screen (and so won't need to perform steps 1–3).

(1) On the Home screen, tap Settings.

2 Tap General.

3 Tap Software Update. The app checks for an update. If one is available, you see information about it; installing the update is shown starting with step 8. If no update is needed, you see a message saying so.

4 To ensure updates are downloaded and installed automatically, tap Automatic Updates.

5 Set the Download iOS Updates switch to on (green). Updates are downloaded automatically when your iPhone is connected to a Wi-Fi network.

6 Set the Install iOS Updates switch to on (green); you receive a notification prior to the update being installed. The update is installed overnight after it has been downloaded. Your iPhone must be connected to a charger and to a Wi-Fi network for the automatic update process to perform.

7 Tap Back (<). If a download is ready to be installed as described in step 3, proceed to step 8 to install it; if not, skip the rest of these steps.

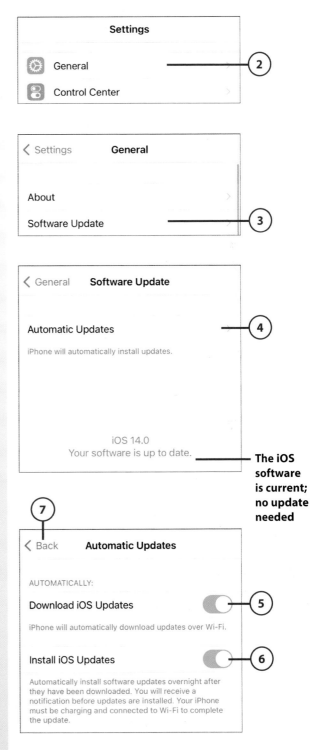

Settings
General — **2**
Control Center

< Settings   **General**

About

Software Update — **3**

< General   **Software Update**

Automatic Updates — **4**

iPhone will automatically install updates.

iOS 14.0
Your software is up to date. — **The iOS software is current; no update needed**

7

< Back   **Automatic Updates**

AUTOMATICALLY:

Download iOS Updates ⬤ — **5**

iPhone will automatically download updates over Wi-Fi.

Install iOS Updates ⬤ — **6**

Automatically install software updates overnight after they have been downloaded. You will receive a notification before updates are installed. Your iPhone must be charging and connected to Wi-Fi to complete the update.

8 If automatic update downloads are enabled, tap Install Now; if not, tap Download and Install. The update is downloaded if it hasn't been already, and the installation process begins. You might see a brief prompt asking if the update should be installed; you can tap Install or just ignore it, and the installation starts automatically. Depending on your iPhone's status, you might see different warnings, such as if you aren't connected to a power supply.

9 Enter your iPhone's passcode.

10 If required, accept Apple's Terms and Conditions (not shown in a figure). You have to accept these Terms and Conditions to be able to install the update.

11 After the update has been downloaded, tap Install Now (not shown in the figure). (If the Automatic option is active, the installation happens without this step.)

The software update installs on your iPhone and you see the progress on the screen. Although you can do something else while the install completes, as part of the update process, the iPhone might restart so you might not have much time. When the iPhone restarts, the updated iOS software is ready for you to use. If a restart isn't required, you see that the iPhone's software has been updated to the current version.

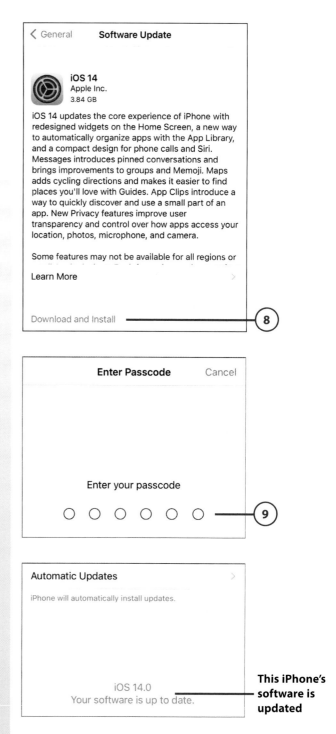

< General   **Software Update**

iOS 14
Apple Inc.
3.84 GB

iOS 14 updates the core experience of iPhone with redesigned widgets on the Home Screen, a new way to automatically organize apps with the App Library, and a compact design for phone calls and Siri. Messages introduces pinned conversations and brings improvements to groups and Memoji. Maps adds cycling directions and makes it easier to find places you'll love with Guides. App Clips introduce a way to quickly discover and use a small part of an app. New Privacy features improve user transparency and control over how apps access your location, photos, microphone, and camera.

Some features may not be available for all regions or

Learn More   >

Download and Install ———— **8**

Enter Passcode   Cancel

Enter your passcode

○ ○ ○ ○ ○ ○ ———— **9**

Automatic Updates   >

iPhone will automatically install updates.

iOS 14.0 ———— **This iPhone's
Your software is up to date.   software is
updated**

Maintaining iPhone Apps

Fortunately, you can configure your iPhone so that it automatically installs updates. regularly updated to fix bugs or add features and enhancements. Fortunately, you can configure your iPhone so that it automatically installs updates.

(1) Open the Settings app, tap your Apple ID, and then tap App Store.

(2) Ensure the App Updates switch is on (green).

(3) If you don't want updates to download when you're using your cellular data, such as if you have a limited amount of data per month, set the Automatic Downloads switch to off (white).

(4) If Automatic Downloads is enabled (green), tap App Downloads.

(5) Choose how you want updates to be downloaded from the options you see. For example, if you're okay with relatively small updates being downloaded, choose Ask If Over 200 MB. Any updates smaller than that are downloaded automatically. You have to allow any larger updates to download.

(1)

‹ Settings   **App Store**

AUTOMATIC DOWNLOADS

Apps

App Updates   **(2)**

Automatically download new purchases (including free) made on other devices.

CELLULAR DATA

Automatic Downloads   **(3)**

App Downloads   Ask If Over 200 MB   **(4)**

Only allow apps under 200 MB to download automatically using cellular data.

‹ App Store   **App Downloads**

Always Allow

Ask If Over 200 MB   **(5)**

Always Ask

Only allow apps under 200 MB to download automatically using cellular data.

>>>Go Further
I GOTTA KNOW

You can get information about how your apps have been updated using the App Store app. Open the App Store app and tap on your account icon located in the upper-right corner of the screen. Swipe up on the Account screen to see the list of updated apps in the UPDATED RECENTLY section. For each app, you see its name and a description of the update; tap more to read all of the descriptions. Tap OPEN to launch an updated app.

Backing Up Your iPhone

Like all things data related, it's important to back up your iPhone's data. If you ever need to restore your iPhone to solve a problem or replace your iPhone with a new one, having a current backup can get you back to where you want to be as quickly and easily as possible.

Ideally, you'll configure your iPhone to automatically back up to iCloud whenever your iPhone is connected to a Wi-Fi network that provides an Internet connection.

To learn how to configure your iPhone to back up to your iCloud account, see "Configuring Your iCloud Back Up" in Chapter 4, "Setting Up and Using an Apple ID, iCloud, and Other Online Accounts." With this enabled, your iPhone is backed up automatically. You can manually back up at any time as described in the following task.

Manually Backing Up Your iPhone to iCloud

To manually back up your iPhone to your iCloud account, do the following:

(1) Tap Settings.

2. Tap your Apple ID account.

3. Tap iCloud.

4. Swipe up the screen.

5. Tap iCloud Backup.

6. Tap Back Up Now. Your iPhone's data is backed up to iCloud. You can see the progress on the screen; however, you can continue using your iPhone because the process works in the background. You just need to start the manual back-up process, and then do whatever you want while it completes.

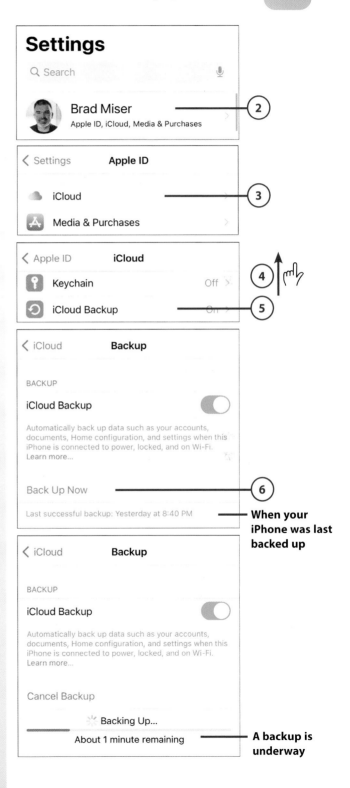

Settings

Q Search

Brad Miser — 2
Apple ID, iCloud, Media & Purchases

< Settings   **Apple ID**

☁ iCloud — 3

🅰 Media & Purchases

< Apple ID   **iCloud**

🔑 Keychain   Off >   4 ☝

🔄 iCloud Backup   On >   5

< iCloud   **Backup**

BACKUP

iCloud Backup   ⬤

Automatically back up data such as your accounts, documents, Home configuration, and settings when this iPhone is connected to power, locked, and on Wi-Fi. Learn more...

Back Up Now — 6

Last successful backup: Yesterday at 8:40 PM — **When your iPhone was last backed up**

< iCloud   **Backup**

BACKUP

iCloud Backup   ⬤

Automatically back up data such as your accounts, documents, Home configuration, and settings when this iPhone is connected to power, locked, and on Wi-Fi. Learn more...

Cancel Backup

☼ Backing Up...

About 1 minute remaining — **A backup is underway**

Blocking Unwanted Calls, Messages, or FaceTime Requests

It's inevitable that you'll encounter calls, messages, or FaceTime requests from people or organizations you don't know and those you do know but don't want to communicate with. You can use the iPhone's blocking feature to block calls, messages, and FaceTime requests from specific phone numbers or email addresses.

If you want to block calls, messages, or FaceTime requests from someone, do the following:

1. Open the app through which the person you want to block is trying to contact you; you can use Phone, Messages, or FaceTime. These steps show the Phone app, but blocking people in other apps is similar.

2. Tap Recents.

3. Tap Info (i) for the person you want to block.

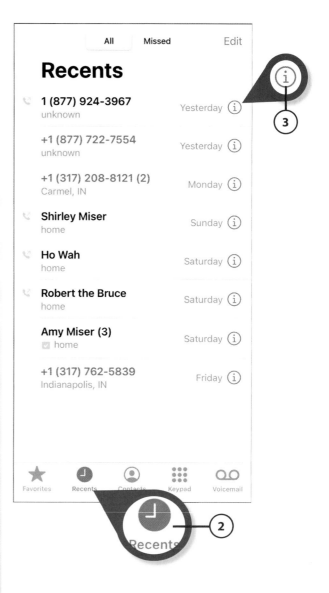

(4) Swipe up the screen.

(5) Tap Block this Caller.

(6) Tap Block Contact. The contact becomes blocked, and you no longer receive calls, messages, or FaceTime requests from them. When the person who is blocked tries to contact you, the attempt fails, but the person doesn't know that the reason it failed is that you blocked him.

Unblock

To remove the block on a contact, open the Settings app, and tap either Phone, Messages, or FaceTime to get to the associated Settings screen. Tap Blocked Contacts. On this screen, you see all of the numbers, email addresses, or names you have blocked. Swipe to the left on a name, number, or address that you want to unblock and then tap Unblock.

I Used To Hear From Him

If someone who you have been in contact with suddenly stops contacting you for no apparent reason, check to make sure that person isn't accidentally blocked. If she is, unblock her, and normal communication resumes.

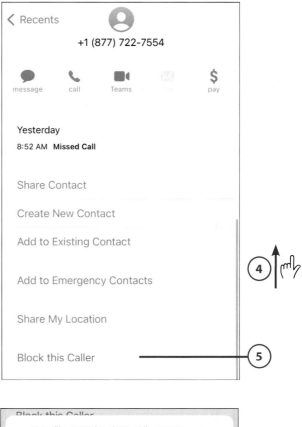

Finding and Securing Your iPhone with Find My iPhone

If your iPhone is lost, stolen, or misplaced, you can use iCloud's Find My iPhone feature to locate your phone and to secure it by locking it or, in the worst case, erasing its data. Hopefully, you'll never need to use this capability, but it's good to know how just in case.

Keep It Secure

For better security, you should configure a passcode along with Face ID or Touch ID, as described in Chapter 2, "Getting Started with Your iPhone." When a passcode is required, it must be entered to unlock and use an iPhone. If you don't have a passcode, anyone who picks up your iPhone can use it. This section assumes you have a passcode set. If not, go back to Chapter 2 and set one before continuing— unless you've already lost control of your iPhone. If you don't have a passcode set and lose your iPhone, you're prompted to create a passcode when you put it in Lost Mode.

When you activate Find My iPhone, you can perform the following actions on your iPhone:

- **Play Sound**—A sound is played on the iPhone and an alert appears on its screen. This provides information to whoever has the device—such as if you've loaned the device to someone and want it back. The sound can help you locate the device if it is in the same general vicinity as you.

- **Lost Mode**—This locks the iPhone so no one can use it without entering the passcode. This can protect your iPhone without changing its data. You can also display contact information on the iPhone's screen to enable someone to get in touch with you.

- **Erase Your iPhone**—This erases the iPhone's memory as the "last chance" to protect your data. You should only do this in the worst-case scenario because when you erase the iPhone, you lose the ability to find it again using Find My iPhone.

Whenever you use one of the Find My iPhone actions, such as playing a sound or using the Lost Mode, you receive emails to your Apple ID (such as iCloud) email address showing the action taken along with the device, date, and time on which it was performed.

To use Find My iPhone, you must enable it on your iPhone using the iCloud Settings as explained in Chapter 4. (It's enabled by default, so unless you have disabled it, it should be ready for you to use when needed.)

Performing the three actions is similar. First, use the Find My Phone app on the iCloud website or another device to locate your iPhone. Then, choose the action you want to perform. You're guided through the rest of the process to complete the action you select. The following section provides step-by-step instructions to locate and lock your phone. Playing a sound or erasing your iPhone is similar.

Using Find My iPhone to Put Your iPhone in Lost Mode

You can put your iPhone into Lost Mode by doing the following:

1. Use a web browser to move to icloud.com.
2. Enter your Apple ID.
3. Enter your Apple ID password.
4. Click the right-facing arrow to sign into your iCloud account.

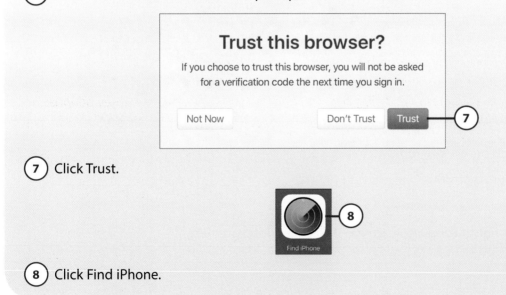

(5) If two-factor authentication is enabled, click Allow.

(6) Enter the verification code at the prompt.

(7) Click Trust.

(8) Click Find iPhone.

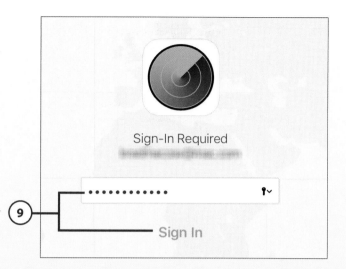

9 Enter your Apple ID password and click Sign In. All the devices associated with your Apple ID and on which Find My iPhone is active are located.

10 Click the All Devices menu. All your devices are shown on the My Devices list.

11 Click your iPhone. If it can be located, you see it on a map. You can use the map's controls to zoom into the iPhone's location so you get the best idea of where it currently is. You also see a dialog box that enables you to perform the three actions described earlier.

Battery level

The iPhone's current location

Actions you can take

12 Click Lost Mode. (If you haven't entered a passcode, you're prompted to create one by entering and verifying it.)

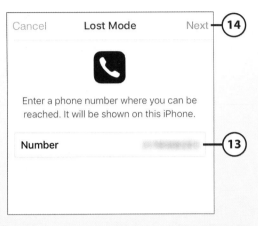

(13) Enter the phone number you want someone to call if your phone is found.

(14) Click Next.

(15) Enter the message you want to appear on the iPhone. This can be your request that whoever has the device call you, instructions for returning the device, and so on.

(16) Tap Done. The iPhone is locked and the message you entered displays on the screen. While the iPhone is in Lost Mode, the Lost Mode icon on the Find My iPhone screen rotates, and you see the Lost Mode indicator.

**This iPhone is locked
and in Lost Mode**

**This iPhone is locked
and in Lost Mode**

The message and
phone number
you entered

The person who
finds the phone
can tap to call
the number you
entered

The phone can still be
used for emergency calls

When the device enters Lost Mode, it's locked and protected with the existing passcode. The phone number and message you entered appear on the iPhone's screen. The person finding your phone can tap Call to call the number you entered.

In addition to being locked, the phone is placed in Low Power Mode so the battery's charge lasts as long as possible. Apple Pay is also disabled.

The iPhone remains in Lost Mode until it's unlocked. You see the current status in the iPhone's window on the iCloud Find iPhone website. When you get your phone back, you can unlock it. You need to re-enter your Apple ID password to restart iCloud services and re-enable Apple Pay.

>>>Go Further

USING FIND MY iPHONE

Here are some more tips on using Find My iPhone:

- When you choose Play Sound, a sound plays on your iPhone until you tap Stop. This is most useful to locate a phone near you, such as in your home. It doesn't protect the phone's content in any way. If the phone is unlocked, it isn't locked when you play the sound so someone can pick it up and use it. (If you have Auto-Lock configured, the phone locks automatically after the set time passes.)

- To update the locations of your devices, refresh the web page and reopen the All Devices list.

- When you send a message to an iPhone that is in Lost Mode or when you erase the phone's data, you might want to include additional contact information in the message, such as your name and where you're currently located. You could also offer a reward for the return of the iPhone. Include enough information so that if someone wants to return the device to you, she has what she needs to be able to do so.

- If you lose your iPhone, use an escalation of steps to try to regain control. Look at the iPhone's location on the Find my iPhone map. If you are very sure no one else has the phone and it's near you, play the sound because it might help you find it again. If the iPhone doesn't appear to be near your current location or it appears to be but you can't find it, put it in Lost Mode. This hopefully prevents someone else from using it while you locate it and allows someone who finds it to contact you. If you lose all hope of finding it again, you can erase the iPhone to delete the data it contains. When you erase your phone, you lose the ability to track your iPhone and have to rely on someone finding it and contacting you. This is a severe action, so you don't want to do it prematurely.

- Erasing an iPhone is a bit of a double-edged sword. It protects your data by erasing your device, but it also means you can't use Find My iPhone to locate it anymore. You should use this only if you're pretty sure someone has your device. After you erase it, there's no way to track the iPhone's location. How fast you move to erase your iPhone also depends on whether you've required a passcode. If you do require a passcode, you know your device's data can't be accessed without that code, so it could take some time for a miscreant to crack it, and you might want to wait a little while before erasing it. If your iPhone doesn't have a passcode, you might want to erase it sooner to prevent someone from accessing your content.

- As you use Find My iPhone, you receive email notifications about various events, such as when a device is locked, when a message you sent is displayed, and so on. These are good ways to know something about what is happening with your iPhone, even though you might not be able to see the iPhone for yourself.

- If Find My iPhone can't currently find your iPhone, you can still initiate the same actions as when the iPhone is found, although they won't actually happen until the iPhone becomes visible to the Find My iPhone service again. To be notified when this happens, check the Email me when this iPhone is found check box. When the iPhone becomes visible to Find My iPhone, you receive an email and then can take appropriate action to locate and secure it.

- There's an app for that. You can download and use the free Find My app on an iOS device to use this feature (it's installed on iPhones by default and is located in the Extras folder). For example, you can use this app on an iPad to locate an iPhone.

Maintaining an iPhone's Power

Keep your eye on the battery icon so you don't run out of power

Obviously, an iPhone with a dead battery isn't good for much. As you use your iPhone, you should keep an eye on its battery status. As long as the battery icon is at least partially filled, you're okay. As the iPhone gets low on power, the battery status icon becomes almost empty. Two separate warnings alert you when the battery lowers to 20% and then again at 10%.

Be Precise

On a iPhone with a Home button, you can choose to display the current percentage of battery charge next to the icon at the top of the screen. To configure this, open the Settings app, tap Battery, and set the Battery Percentage switch to on (green).

This iPhone is in Low Power Mode

When you see the low-power warnings, you can tap Low Power Mode to reduce power and extend the life of your battery. This mode slows down the phone's processor and stops automatic downloads, email push, and any other processes that consume lots of power. Although this reduces the performance of your phone, it can significantly extend the time until your iPhone runs out of power. While in Low Power Mode, the battery status icon is yellow.

Manual Low Power Mode

You can put your iPhone in Low Power Mode at any time by opening the Settings app, tapping Battery, and then setting the Low Power Mode switch to on (green). You also can open the Control Center and tap the Low Power Mode icon (the battery). (If you don't see the battery icon on your Control Center, you can add it as explained in Chapter 6, "Making Your iPhone Work for You.") You might want to put your iPhone in Low Power Mode if you know you won't be able to charge your iPhone for a long time and don't need some of the functions it disables.

If you keep going, whether you choose to run in Low Power Mode or not, the iPhone eventually runs out of power and shuts down. Once it shuts down because of a low battery, you won't be able to unlock or use it until you connect it to a charger, and the battery has recharged enough for the phone to operate, which can take a few minutes. If you try to use it before it has enough power, you see the red, empty battery icon on the screen.

Dim to Save

Your iPhone's screen is one of its major users of battery power. You can operate your iPhone with lower power use by dimming its screen. One way to do this is to open the Control Center and use the Brightness slider to lower the screen's brightness. Of course, at a lower brightness, the screen might be harder to see, so there is a trade-off between visibility and power use.

This iPhone is being charged

Fortunately, it's easy to avoid running out of power by keeping your iPhone charged. Connect the iPhone to the charger or a computer using the USB cable, and it charges automatically. Or, if you have an iPhone 8, 8 Plus, X, or later models, you can place the phone on a Qi wireless charger to charge it. While the phone is charging, you see the charging icon in the upper-right corner of the screen, and if you wake the iPhone, a large battery icon showing the relative state of the battery appears on its screen. When charging is complete, the battery status icon replaces the charging icon in the upper-right corner of the screen, the large battery icon disappears, and you see the iPhone's wallpaper if it's locked, or you see whatever screen you happen to be using if it isn't locked.

If your iPhone moved into Low Power Mode automatically before you started charging it, it also moves out of Low Power Mode automatically when it has been recharged sufficiently. When this happens, you see an onscreen message saying that Low Power Mode has been turned off.

Topping Off

It's a good idea to keep your iPhone's battery topped off; this type of battery actually does better if you keep it charged rather than letting it run down all the way before recharging. Periodically, say every month or two, you might want to let your iPhone run completely out of power and then recharge it to maximize its life and to reset its power management software. The key point is, as Apple puts it, to "keep the electrons moving." So, don't let your iPhone sit with no activity for long periods of time.

Charge It!

If you are going to be unable to charge your phone in the standard ways for long periods of time, consider purchasing a portable charger. These units have a battery you can use to recharge your iPhone wherever you are. These can be really useful, such as when your iPhone is running out of power and you need to arrange a rideshare ride (such as Lyft or Uber). You simply connect your phone to the charger to recharge your battery on the move. (Make sure you keep the charger's battery charged, too, so that it is ready when you need it.)

Solving iPhone Problems

Even a device as reliable as your iPhone can sometimes run into problems. Fortunately, the solutions to most problems you encounter are simple. If a simple solution doesn't work, a great deal of detailed help is available from Apple, and even more is available from the community of iPhone users.

The problems that you can address with the simple steps described in this section vary and range from such issues as the iPhone hanging (won't respond to commands) to apps not working as they should. Use the following tasks to address problems you encounter.

Restarting an App

If an app you are using locks up, displays an error, or isn't working the way it should, try shutting it down and restarting it.

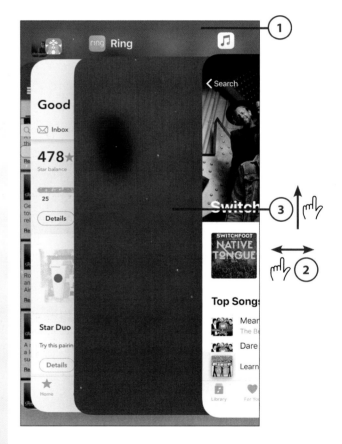

1 Open the App Switcher by swiping up from the bottom of the screen and pausing toward the middle of the screen (iPhones without a Home button) or pressing the Touch ID/Home button twice (iPhones with a Home button).

2 Swipe to the left or right until you see the app that has frozen.

3 Swipe up on the app to shut it down.

4 Tap outside the App Switcher (iPhones without a Home button) or press the Touch ID/Home button (iPhones with a Home button) to close the App switcher and move back to the Home screen (not shown in figure.)

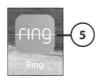

5 Open the app again. If it works normally, you're done. (It's generally a good idea to restart your iPhone when you've had to force an app to quit.)

Restarting Your iPhone

Whenever your iPhone starts acting oddly, restarting it should be the first thing you try. It's easy to do and cures an amazing number of problems. To restart your iPhone, do the following:

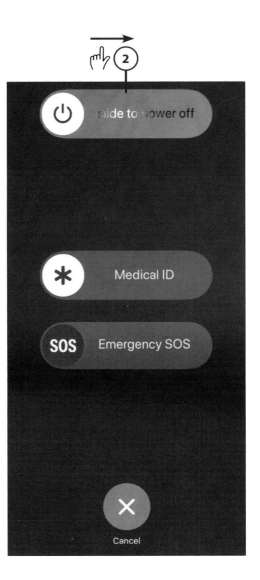

(1) Press and hold the Side button and one of the Volume buttons (iPhones without a Home button) or the Side button only (iPhones with a Home button) (not shown in figure).

(2) Drag the slide to power off slider to the right. The iPhone powers down and the screen goes completely dark.

(3) Press and hold the Side button until you see the Apple logo on the screen (not shown in figure). The iPhone restarts. When the Home screen appears, try using the iPhone again. If the problem is solved, you're done.

Can't Restart?

If your iPhone won't restart normally using the previous steps, press and hold the Side and Lower Volume buttons (iPhones without a Home button) or Side and Touch ID/Home buttons (iPhones with a Home button) at the same time until the iPhone shuts off (about 10 seconds) and the screen goes completely dark. Then, press the Side button to restart it.

Resetting Your iPhone

If restarting your iPhone doesn't help, try resetting your iPhone using the following steps:

(1) On the Home screen, tap Settings.

(2) Tap General.

(3) Swipe up the screen until you see the Reset command.

(4) Tap Reset.

(5) Tap the Reset command for the area in which you're having problems. For example, if you're having trouble with Wi-Fi or other networking areas, tap Reset Network Settings, or if you decide your Home screens are a mess and you want to go back to the default layout, tap Reset Home Screen Layout. If you're having lots of issues in multiple areas, you can restore your iPhone by tapping Reset All Settings or Erase All Content and Settings as explained in the next section (be careful because this erases everything on your iPhone).

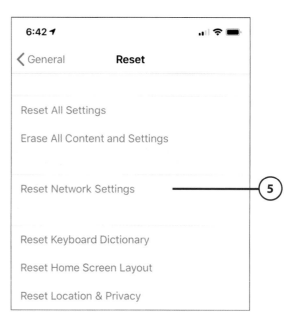

6. If prompted to do so, enter your passcode; if you don't see a passcode prompt, skip to the next step.

7. Tap the confirmation of the reset you are doing. The reset is complete and that area of the iPhone is reset to factory conditions.

8. Reconfigure the area that you reset (not shown in figure). For example, if you reset your network settings, you need to reconnect to the networks you want to use. If you did a Reset All Settings or Erase All Content and Settings, you can restore your iPhone from a backup, or if you don't have a backup, you need to basically start from the beginning as if your iPhone was just taken out of its box.

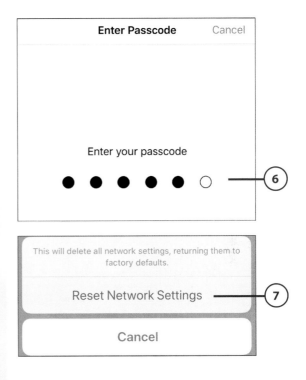

Restoring Your iPhone

The most severe action you can take on your iPhone is to restore it. When you do this, the iPhone is erased, so you lose all its contents and its current iOS software is overwritten with the latest version. If you have added information to your iPhone since it was last backed up, that information is lost when you restore your iPhone—so be sure to back up your iPhone regularly, and if you don't, be careful before doing this. If none of the other tasks in this section corrected the problem, restoring the iPhone should.

To restore your iPhone, perform the following steps:

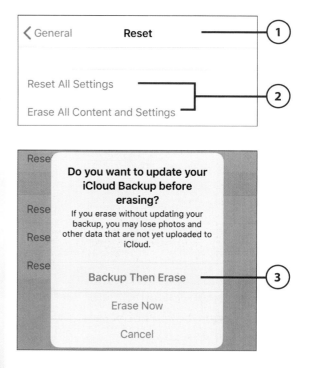

1. Follow steps 1 through 4 in the previous task to move to the Reset screen.

2. Tap Erase All Content and Settings if you want to erase everything (including photos, music, and so on) from the phone or Reset All Settings if you just want to restore the iOS software and leave the content in place.

Patience, Patience

If you see a prompt explaining that a process is underway, such as backing up, choose the option to let that process finish before continuing.

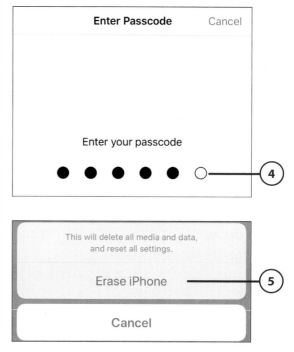

3. Tap Backup Then Erase. This ensures that the information currently on your iPhone gets backed up so you can restore it later. If you want to use an older backup to restore your iPhone for some reason, you can tap Erase Now instead.

4. Enter your passcode.

5. Tap Erase iPhone.

6 Tap Erase iPhone again.

7 Enter your Apple ID password.

8 Tap Erase.

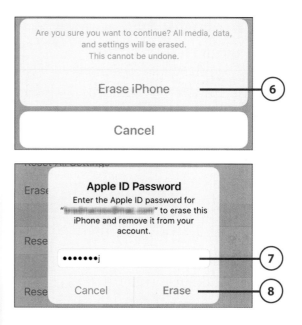

The restore process begins. The iPhone is erased and the software restored. When that process is complete, you see the Hello screen and can walk through setting up the basic configuration of the phone, including connecting to Wi-Fi, setting up Face ID or Touch ID, and so on. You're guided through each step, so just follow the onscreen directions to work through the process.

A very important step is the Apps & Data screen on which you choose how you want to restore your iPhone; in most cases, you want to choose Restore from iCloud Backup so that your iPhone is restored to the condition it was in at the most recent backup.

After you work through the onscreen prompts, the data from the backup is copied onto the iPhone. When that is done, the iPhone should be like it was before the restore without the problems that caused you to restore it.

Have Another iPhone Already Set Up?

At the Quick Start prompt, you can choose to restore based on another iPhone you already have configured. If that is the case, hold the other iPhone close to the one you're restoring and follow the onscreen prompts to complete the process. You'll use the other iPhone's camera to frame a symbol on the screen of the iPhone that you're restoring to start the process. The information is copied from the other iPhone to the one you're restoring, so the one you're restoring will be set up just like the one from which you are restoring it.

Finding a Missing App

I've had more than a few emails from people on whose phones an app icon seems to have disappeared. In almost all of these cases, the app's icon had inadvertently been moved into a folder and so was not visible in its expected location. If this happens to you, perform the following steps to restore the app to where it should be:

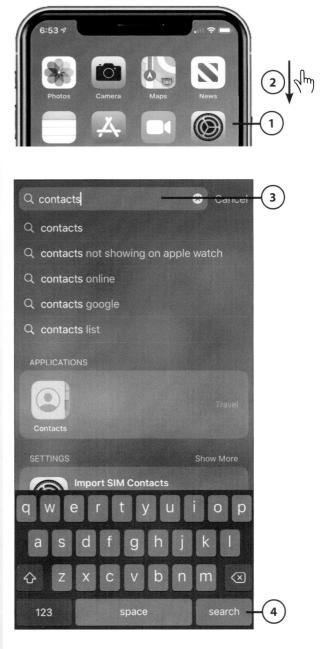

1 Move to a Home screen.

2 Swipe down from the center top area of the screen. The Search tool opens.

3 Tap in the Search bar and enter the name of the app that's missing. As you search, you see a list of items that match your search. If the app is found, you know it's still on the iPhone, and its icon was just accidentally misplaced; you can continue with these steps to find it.

If the app you are looking for isn't found, it's been deleted from the iPhone, and you need to reinstall it. Download and install it again using the App Store app (don't worry; if it has a license fee, you don't need to pay for it again). (Installing apps is covered in Chapter 5, "Customizing Your iPhone with Apps and Widgets.")

4 Tap search.

(5) Note the current location of the app, which is shown on the right side of the screen. If no location is shown, the app is on one of your Home screens, in which case you need to browse through the screens to find it, and you can skip the rest of these steps.

(6) Tap Cancel. You move back to the Home screen.

(7) Open the folder where the app is currently located.

(8) Touch and hold on the app icon until the Action menu appears.

(9) Tap Edit Home Screen.

All Your Apps in One Place

All the apps installed on your iPhone are available in the App Library. To open it, swipe all the way to the left while on a Home screen. When the App Library appears, you can browse or search for the app for which you are looking. You can open from the App Library or move it from the App Library onto a Home screen. Refer to Chapter 5 for the details about the App Library.

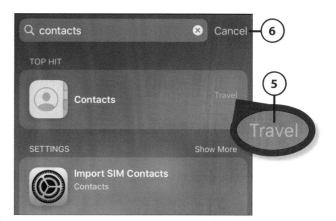

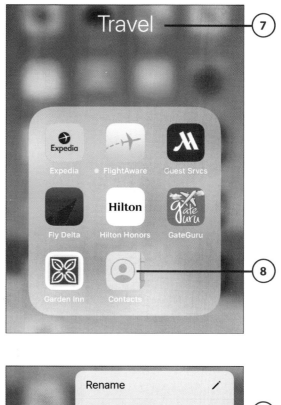

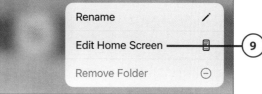

10 Drag the app's icon out of the folder and place the icon where you want it to be (refer to "Customizing Your Home Screens" in Chapter 7, "Customizing How Your iPhone Looks," for details of moving apps on the Home screens).

11 Tap Done (iPhones without a Home button) or press the Touch ID/Home button (iPhones with a Home button) to lock the icons in their current positions.

Other Ways to Find Apps

You can always get to apps directly from the Search tool. When you perform step 3 in the previous task, you can just tap the app's icon to open it; you don't need to know where the icon is.

Making an iPhone's Screen Rotate Again

If your iPhone stops changing to the horizontal from the vertical orientation when you rotate the phone, you probably have inadvertently enabled the Portrait Orientation Lock. Unlock it again by doing the following:

(1) Open the Control Center by swiping down from the upper-right corner of the screen (iPhones without a Home button) or swiping up from the bottom of the Home screen (iPhones with a Home button).

Status Icons on iPhones with a Home button

If you have an iPhone with a Home button, the Portrait Orientation Lock icon appears in the top-right of the Home screen, so you can see it without opening the Control Center.

The orientation of this iPhone is locked

(2) Look at the top of the screen. If you see the Portrait Orientation Lock icon, the iPhone's orientation is locked.

(3) Tap the Portrait Orientation Lock button. Note that when Portrait Orientation Lock is on, the button's graphic is red. When it's off, the button's graphic is white.

(4) Confirm that the Portrait Orientation Lock icon at the top of the screen has disappeared. The iPhone now changes orientation when you rotate it.

Solving the Quiet iPhone Problem

If your iPhone stops ringing or making other sounds you believe it should, perform the following steps:

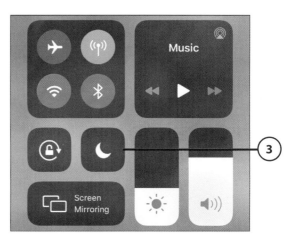

1. Make sure the iPhone isn't muted. The Mute switch on the left side of the iPhone should be in the position toward the front of the phone. If you see color in the switch, the phone is currently muted. Slide the switch toward the front of the iPhone to unmute it. You should hear sounds again.

2. With the iPhone unlocked, press the upper volume switch on the left side of the iPhone to make sure that the volume isn't set to the lowest level. As you press the button, a visual indicator of the current volume level appears on the screen. As long as you see the bar partially filled on this indicator, you should be able to hear sounds the iPhone makes.

3. Make sure that the iPhone is not in Do Not Disturb mode by opening the Control Center and ensuring that the Do Not Disturb icon is off (the moon is white); if it is turned on (the moon is purple), tap it to turn it off. In Do Not Disturb mode, sounds are not made, such as a ring when you receive a phone call.

4 Try a different app or task. If you aren't hearing sound from only one app and everything else sounds normal, you know there's a problem with the app itself. Try restarting it; if that doesn't work, try restarting your iPhone. You might have to try deleting and reinstalling the app. (If you had to pay for the app when you first downloaded it, you don't have to pay again to reinstall it.) If you still don't hear sound from any app, you know the problem is more general.

5 Move to Sounds & Haptics in the Settings app.

6 Make sure the volume slider is set to at least the middle position.

7 Tap Ringtone.

8 Tap one of the default ringtones. You should hear it. If you do, you know the problem is solved. If not, continue.

9 If you aren't hearing sounds when the EarPods aren't plugged in, connect the EarPods and repeat the sound you should be hearing. If you hear the sound only when the EarPods are plugged in, try removing the EarPods and plugging them back in a few times to see if that solves the problem. If not, your iPhone needs to be serviced by an Apple authorized repair center, or you can make an appointment to take it to the Genius bar in an Apple Store.

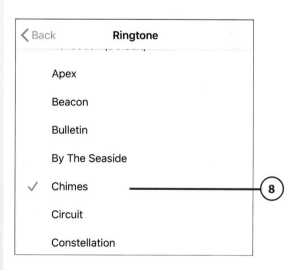

Restart

If none of these steps resolve the quiet iPhone problem, try restarting your iPhone. Or, you might want to try restarting it first. In most cases, you'll find the solution is that the iPhone somehow got muted or placed into Do Not Disturb mode.

Do Disturb!

If Do Not Disturb was on, but you hadn't intended it to be, open the Settings app and tap Do Not Disturb to view its settings. Check to see if an automatic schedule has been set. If the Scheduled button is on (green) then a schedule for starting and stopping Do Not Disturb has been active on the phone. Slide the Scheduled switch off (white) to prevent it from automatically turning on Do Not Disturb.

Getting Help with iPhone Problems

If none of the previous steps solve your problem, including restarting your phone, you can get help in a number of ways:

- **Apple's website**—Go to www.apple.com/support/. On this page, you can access lots of information about iPhones, iTunes, and other Apple products. You can browse for help, and you can search for answers to specific problems. Many of the resulting articles have detailed, step-by-step instructions to help you solve problems and link to more information.

  When you move to the Apple Support page, you can choose the product you need help with, such as iPhone. This takes you to a page with iPhone information. Click Get Support (you might have to browse to the bottom of the screen to see this), and then select one of the problem areas shown or search for the problem. You'll see various problems related to what you selected. Select the type of problem you have. As you continue to get more specific about your problem, you'll find options to use Chat to contact Apple Support or to have Apple Support call you.

- **Web searches**—One of the most useful ways to get help is to do a web search for the specific problem you're having. Just open your favorite search tool, such as Google, and search for the problem. You're likely to find many resources to help you, including websites, forums, and such. If you encounter a problem, it's likely someone else has, too, and has probably put the solution on the Web.

Been Around a Long Time

The iPhone has been around quite a while. It's currently on the fourteenth version of its operating system. There have also been many different models of the iPhone. When searching for help, make sure the information you find is relevant to your phone. Look at the date the information was posted; if it's more than a year or so old, try to find newer information. There is a huge amount of content about the iPhone on the Web, but much of it is outdated and not relevant to current iPhones or iOS 14.

Anyone Can Post

Keep in mind that some of the information you find hasn't been tested and is posted by people who might, or might not, know what they are talking about. Try to stick to more reputable sites, such as those associated with publications.

Pushing a Product or Service

Be wary of a site that promotes a specific app or service as the solution to a problem. Sometimes, help sites are really just marketing tools to get people to download an app or use a service that they have to pay for. In the worst cases, sites offering apps or services to help solve problems are meant to trap people into something that they don't need or that actually causes problems.

- **Genius bar at an Apple Store**—All Apple Stores have a Genius bar staffed with people who will try to help you with problems. If you can't resolve a problem on your own, going to an Apple Store is often a great solution because the Apple geniuses can help you with any problem, whether it's software or hardware. In many cases, the help they provide is free, too (though you might be charged for some hardware problems). You need to make an appointment before going to the Genius bar and can do so by going to www.apple.com/retail/geniusbar/. (If an Apple Store isn't convenient for you, use the information above for Apple's website to reach chat or phone support instead.)

- **Cell Phone Provider**—You can take your phone to a local retail store of the provider from whom you get cellular service, such as AT&T or Verizon, to try to get help. Some of these locations have customer support personnel who will be able to help you with technical issues. Other locations might not have such staff. If you have a store convenient to your location, this can be a good option to try.

- **Me**—You're welcome to send an email to me for problems you're having with your iPhone. My address is bradmiser@icloud.com. I'll do my best to help you as quickly as I can. (I do respond to every email I receive, though at times it might take me a few days to get back to you. I'm not able to respond to FaceTime or Messages requests, so please use email to reach me.)

Glossary

3D Touch   The iPhone 6s and later models have a screen that is pressure sensitive. The 3D Touch feature enables you to press (you don't have to press hard, just apply some pressure) on the screen to perform different actions, such as opening the Quick Action menu when you touch and hold on an app's icon.

3G/4G/LTE/5G   Names of various kinds of cellular data networks that you can use to connect your iPhone to the Internet. Your iPhone automatically uses the fastest cellular network available to it. LTE or 5G are the fastest, 4G is slightly slower, and 3G is slower than that. Because your iPhone automatically chooses a network to use, you don't need to do anything, but it's helpful to understand which network your iPhone is using.

AirDrop   Apple's technology that enables content to be easily shared among nearby iPhones, iPads, and Mac computers. No complicated setup is required, you simply tap the people with whom you want to share content. For example, you can use AirDrop to quickly and easily share photos you've taken on your iPhone with other people near you. You can also use it to send content to other devices you own, such as sending a photo from your iPhone to a Mac computer.

Airplane mode   In this mode, your iPhone's cellular sending and receiving functions are disabled. You're supposed to use this mode when traveling on airplanes to ensure your iPhone doesn't interfere with the operation of the plane. You can also use it at other times, such as if you don't want to receive any phone calls.

AirPrint   Apple's technology that enables iPhones and iPads to print wirelessly to AirPrint-compatible printers without requiring any setup (such as installing printer drivers).

Animoji   An animated overlay on your image as it appears in FaceTime or in Messages. You can choose different Animojis, and the Animoji replaces your head in the image. It moves as your face moves; for example, when you smile, the Animoji smiles. Animojis are only available on iPhones without a Home button.

app   Short for application, these are programs that your iPhone runs to accomplish a wide variety of tasks. Your iPhone comes with a number of apps installed by default, such as Mail and Safari. You can download thousands of other apps from the App Store using the App Store app.

App Store   The store operated by Apple that enables you to find and download apps for your iPhone. The App Store app enables you to access the App Store from your iPhone. Apple has very tight control over the apps that appear in this store, so you don't need to worry about apps you download damaging your iPhone or breaching its security.

App Switcher   The iPhone feature that enables you to quickly change apps. When you activate the App Switcher, you see the apps you're currently using or that you've recently used. You can browse the apps, and then tap one to use it. You can also use the App Switcher to force a running app to quit, such as when the app isn't performing correctly.

Apple ID   When you access iCloud, download items from the App or iTunes Stores, or access other Apple services, you need to have a user account called an Apple ID. The Apple ID consists of the email address associated with your user account and a password. You also use it to connect your iPhone to Apple services, such as iMessage.

Apple Pay   This is Apple's payment service through which you can register credit and debit cards and then use those cards to make payments quickly and easily. Apple Pay is more secure than using a credit or debit card directly because your account number is not involved in Apple Pay transactions. Once configured, you can quickly pay for something by holding your phone near a register and activating Apple Pay. You can also use Apple Pay for online purchases. It also enables you to securely send funds to other individuals using an iPhone or iPad.

badge   A red circle with a number that appears on app or folder icons that indicates the number of new "items," such as emails for the Mail app or messages for the Messages app.

Bluetooth   Technology that enables devices to communicate with each other wirelessly. Your iPhone uses Bluetooth to work with wireless headphones, speakers, keyboards, automobile audio systems, and much more. Using a Bluetooth device is a two-step process. First, pair the device with your phone. Second, connect to the Bluetooth device you want to use.

bookmark   This is a saved location on the Web. When you visit a web page or website, you can save its URL as a bookmark so you can return to it with just a few taps instead of typing its URL. Safari allows you to save and organize your bookmarks on your iPhone.

cellular data network   The technology that enables your iPhone to communicate data similarly to how it can place a phone call. This enables your iPhone to access the Internet from just about any location in the world. Your provider offers various types of cellular data networks that have different speeds. Your iPhone automatically manages its cellular data connection, but you can determine when the connection is used by enabling or disabling it. Cellular data is provided with your cell phone account. Some plans have a limit on the amount of data you can use per month. If you exceed this amount, you have to pay additional fees. Increasingly popular are accounts having unlimited data, which means you can use as much data as you want without additional charges (sometimes, surpassing a threshold results in your iPhone's connection being limited to lower speeds until the next billing cycle).

Control Center   When you swipe down from the upper-right corner of the screen (iPhones without a Home button) or swipe up from the bottom of the screen (iPhones with a Home button), the Control Center appears. You can use its controls to do things such as enabling or disabling features or modes, such as Airplane mode. You can also configure some of the icons you see on the Control Center so that you have quick access to the ones you find most useful.

Do Not Disturb mode   In this mode, your iPhone doesn't ring or make other noises or sounds caused by notifications. This keeps your iPhone from bothering you at times you don't want it to, such as when you're sleeping. You can set exceptions so that important calls come in even when Do Not Disturb is turned on.

emoji   Icons, such as a smiley face, that are used in emails and text messages to indicate emotion or to show an object (for example, using an icon of an airplane instead of writing the word *airplane*). The iPhone includes many emojis you can use.

Face ID   This is the ability for your iPhone to record "your face" on your phone so that it recognizes you. You can then perform a number of actions by "looking" at your phone. For example, you can configure Face ID so that you can unlock your iPhone by looking at its screen.

Facebook   A social media service that people and organizations use to share information. Facebook users have a "page" on which they can present information about themselves along with photos and videos. Other people, who have been tagged as "friends," can view that information and exchange comments. Facebook is available on the iPhone through the Facebook app.

FaceTime   The app and service that enables you to easily have video or audio-only conversations with others. You can use the iPhone's camera to share video content, either in "selfie" mode where your image appears on the screen or you can "show" people other images using the camera on the backside of the iPhone. The FaceTime app makes it very easy to have video conversations; you simply select the person you want to talk to and see; the app handles the details for you. You can receive FaceTime requests and respond to them the same way you answer a phone call. You can include up to 32 people in a FaceTime session.

Fetch   Your iPhone can receive many different kinds of information from the Internet, such as new emails, contacts, and so on. Fetch is when your iPhone retrieves information from various servers (such as email servers) at specific intervals; for example, every 15 minutes. Fetch uses less battery than Push does.

Find My iPhone   The Apple service that can track the location of your iPhone and enable you to protect your iPhone by locking or erasing it. You can track your iPhone using the Find iPhone app on another device, such as an iPad, or using a web browser, such as Safari or Google Chrome, on a computer.

haptic feedback   This is vibratory feedback for certain events. For example, when you make a selection, such as a date, the phone vibrates slightly to confirm you have made a selection.

Home screens   The screens on your iPhone where app icons are stored. The Home screens are the starting point for most of the tasks you learn about in this book.

iCloud   An Apple service that provides storage space on the Internet and a host of features that enable you to share information, such as contacts and calendars, among many devices. iCloud is integrated into the iPhone; you need an account to use it. (An iCloud account is free with 5GB storage and $0.99 per month for 50GB; larger storage options are available.) One of the best uses for iCloud is to store photos and videos because they are backed up so you don't lose them even if something happens to your iPhone, and you can access your photos and videos from many different devices.

iMessage   Apple's messaging service that enables you to send messages to other iMessage users. Unlike traditional text messages, iMessages are sent over the Internet and don't have limits on the content or quantity of messages you can send. You send and receive iMessages using the Messages app. Those messages can contain text, photos, videos, graphics, and just about any other content you want to share.

Instagram   People use this social media service to share photos, video, and messages with others. You can install the Instagram app on your iPhone to make using Instagram easy and convenient.

iOS   The name of the operating system software that controls your iPhone and enables it to do so many wonderful things. The current version is the fourteenth major release of the software, which is why it is called iOS 14.

iTunes Store   The Apple Store that provides content you can use on your iPhone, including music and movies. You can download this content using the iTunes Store app. You pay for most of the content in the iTunes Store, but you can sample most of it for free to decide if you want to purchase it. After it is downloaded, you can view or listen to the content in the associated app. For example, you can listen to music you download in the Music app.

Lightning   Apple's technology for connecting accessories, such as a charger or EarPods, to the iPhone. The iPhone has a Lightning port on the bottom side.

link   A link is a photo or other graphic, text, or object on a web page that has a URL attached to it. When you tap a link, you move to the URL and open the web page associated with it. Most text links are formatted with a color so you can distinguish them from regular text. Links can also be attached to images, such as photos or other kinds of graphics.

Lock screen   To secure the information on your iPhone, you can lock it so that you have to enter a passcode to use most of its functionality. When you wake up a phone, you see the Lock screen that enables you to unlock the phone. You can also perform some actions directly from the Lock screen, such as opening the Camera app to take photos.

Memoji   Memojis are similar to Animojis except the character created is intended to be a representation of you, thus the "me" in the term. You can design a Memoji to look like you, like you wish you look, or something completely different. You can include your Memoji in various communications, such as Messages conversations. You can also use Memoji stickers, which are static versions of a Memoji.

Multi-touch interface   The technology that enables you to control and use an iPhone by touching your fingers to its screen.

Notification Center   The Notification Center displays all the notifications that have been issued recently so you can easily review them and get more detail for items in which you are most interested. You can open the Notification Center by swiping down from the top of the screen when the iPhone is unlocked or swiping up from the center of the screen when it is locked.

notifications   There's a lot of activity happening on your iPhone. Notifications keep you informed about this activity. Notifications can be visual, audible, or vibratory. You can determine the types of notifications you receive for various events, such as when you receive new emails. You can even set up notifications for specific people, such as a sound when you receive a text from someone important.

passcode   A numeric or alphanumeric sequence that is required to unlock an iPhone to make full use of it. It's important to configure a passcode on your iPhone to secure the information stored there. You can use Face ID or Touch ID so that you don't have to enter your passcode each time you need to use it. Instead, you can look at your iPhone's screen (iPhones without a Home button) or touch the Touch ID/Home button (iPhones with a Home button).

personal hotspot   When acting as a personal hotspot, your iPhone can share its cellular data connection with other devices so that those devices can access the Internet.

podcast   An episodic audio or video program that you can listen to or watch using the Podcasts app. You can subscribe to podcasts so that the episodes are downloaded to your phone automatically.

Predictive Text   The iPhone feature that attempts to predict the next text you want to type; you can tap the text on the Predictive Text bar to enter it. Predictive Text learns over time, so it gets better at predicting the text you're going to write. You can use Predictive Text to type faster or to correct mistakes. Predictive Text can also recommend emojis when you type certain words, such as a plane emoji when you type the word "plane."

Push   Your iPhone can receive many different kinds of information from the Internet, such as new emails, contacts, and so on. Push is when information is moved from a server (such as an email server) directly to your iPhone as soon as the new information appears on the server. Push causes your information to be the most current, but also uses more battery than other methods.

Quick Action menu   When you touch and hold on an app's icon, the Quick Action menu appears. You can choose the action you want to perform on this menu.

roaming   The provider from which you get an account to enable your iPhone to be used for phone calls and cellular data covers a specific geographic region, such as the United States. When you take your iPhone outside of that region, a different provider provides services for you to use; this is called roaming. Roaming is important because it often involves additional charges that can be quite expensive.

Safari   This is the default web browser on your iPhone; it is also the default web browser on Apple Mac computers.

search engine or search page   The Web contains information on every topic under the sun. You can use a search engine/page to search for information in which you are interested. There are a number of search engines available, with Google being the most popular. You access a search engine through a web browser. Safari uses Google by default, but you can choose Bing, DuckDuckGo, or Yahoo! as your default search engine if you prefer one of those.

selfie   A photo you take by using the camera on the front side of the iPhone. Selfies can include anything you want them to, but usually the person taking the photo is included in the photo, thus the name *selfie*. Selfies are very popular, especially when someone is someplace or doing something interesting or unusual. It's also common to take a selfie to capture a group of people, such as when they are sitting at a table together.

Settings app   This app enables you to configure and customize your iPhone and the apps you use.

Siri   Apple's voice recognition technology that enables you to speak to your iPhone to perform tasks and dictate text. You can perform just about any task using Siri. For example, you can place phone calls, send text messages, read text messages, or get information, just by speaking to your iPhone.

Sleep   The power-saving mode the iPhone moves into after a period of inactivity or when you press the Side button while it is awake. In Sleep mode, the screen goes dark and some processes stop to conserve battery power. To use the iPhone again, you wake it. If the iPhone is both locked and asleep, you wake it and then unlock it to be able to work with it.

social media   Services and apps that enable people and organizations to share their information, photos, videos, comments, and so on. Social media services can be accessed through apps on your iPhone or via web browsers, such as Google Chrome or Safari, on a computer. Popular social media services include Facebook, Instagram, and Twitter.

Touch ID   The sensor and associated software that enables you to record fingerprints and use them to enter your passcode to unlock your iPhone and passwords in various apps, such as the App Store app when you're downloading apps. All iOS 14-compatible, iPhones with a Home button support Touch ID.

Touch ID/Home button   The circular button on the bottom of the front side of some iPhone models. This button serves several purposes. You press it to wake and unlock your iPhone. When you're using your iPhone, pressing it takes you to the Home screen. Pressing it twice opens the App Switcher. It can also be configured to perform other actions. When you use Touch ID, touching a recorded fingerprint to this button enters the associated passcode or password. For example, you can unlock your iPhone simply by touching a recorded finger to the Touch ID/Home button.

URL   A Uniform Resource Locator (URL) is a web page's or website's "address" on the Web. URLs allow you to direct your web browser to specific locations on the Web. Most URLs you deal with consist of text, such as www.apple.com or www.aarp.org. Some URLs are more complicated because they take you to specific web pages instead of a website. An example of this is www.aarp.org/health, which takes you to the Health web page on the AARP website. You seldom have to type URLs because you usually access web pages by tapping links or using a bookmark, but it's good to know what they are and how to use them.

wake   To save power, your iPhone goes to sleep after a period of inactivity or when you put it to sleep. When sleeping, the iPhone's screen goes dark. To use it again, you wake it by touching the screen (iPhones without a Home button), pressing the Side button, pressing the Touch ID/Home button (iPhones with a Home button), or raising the phone (if the Raise to Wake feature is enabled on your iPhone).

Wallet app   This app stores Apple Pay information along with boarding passes, store discount cards, and other information so that you can quickly use this information, such as to scan a boarding pass when boarding a plane.

wallpaper   The image you see on the Lock screen and Home screens. You can determine the images you want to see on your iPhone using the Wallpaper option in the Settings app. You can use default images that come with your iPhone, or you can use any photo you take using the iPhone's cameras or that you download onto your iPhone, such as from an email.

web browser   This is the software you use to view web pages. There are many different web browsers available. Examples include Safari—which comes preinstalled on your iPhone—as well as Google Chrome, Internet Explorer, and Firefox. They all allow you to view and interact with web pages, and each has its own set of features. Some are available on just about every device there is, such as Google Chrome, whereas some are limited to certain devices, such as Internet Explorer that only runs on Windows computers.

web page   This is a collection of information (text and graphics) that is available on the Web. A web page is what you look at when you use the Web.

website   This is a collection of web pages that go together. For example, most companies and organizations have websites that contain information they use to help their customers or members, provide services, market and sell their products and services, and so on. A website organizes the web pages it contains and provides the structure you use to move among them.

Wi-Fi   Wi-Fi (Wireless Fidelity) technology that enables iPhones, computers, and many other devices to communicate with each other wirelessly. Your iPhone connects to Wi-Fi networks primarily in order to connect to the Internet. Wi-Fi networks are fast and make accessing the Internet easy and convenient.

Widget Center   When you swipe to the right from a Home screen, the Widget Center opens. Here, you can use the widgets that are configured on your Widget Center.

widgets   Widgets are "mini" versions of the apps that are available on the Widget Center; you can also add them to Home screens. Widgets enable you to quickly get information or accomplish a task using its app.

Index

Answers to Your Technology Questions

The My...For Seniors Series is a collection of how-to guide books from AARP and Que that respect your smarts without assuming you are a techie. Each book in the series features:

- Large, full-color photos
- Step-by-step instructions
- Helpful tips and tricks
- Troubleshooting help

For more information about these titles,
and for more specialized titles, visit
informit.com/que

the trusted technology learning source